THE RADIO COLLECTOR'S DIRECTORY AND PRICE GUIDE

1936 **Sparks-Withington 557**

"Sparton" blue mirror
Designed by Walter Darwin Teague
in the Streamlined Moderne look of Art Deco.
Collection of W.S. Lettow

THE Radio Collector's DIRECTORY and Price Guide

Robert E. Grinder
George H. Fathauer

Ironwood Press · Scottsdale Arizona

ISBN 0-932541-01-1
Library of Congress Catalog Card Number 86-80477
Printed in the United States of America

Designed by Carol M. Barth/Mark Heliger
Composed in Garth by Commtype, Los Angeles, CA

Contents

List of Tables and Figures

PREFACE

The Radio Collector's Directory and Price Guide provides an orderly scheme for identifying and pricing vacuum tube radio broadcast receivers that were manufactured in the United States for home entertainment between 1921 and 1941. The volume is designed primarily to serve as a reference text for radio collectors, antique specialists, and historians who frequent the various marketplaces where early broadcast receivers are likely to be bought and sold. It is designed secondarily to provide readers with historical perspective on the significance of cultural and technical forces in the phenomenal rise of radio broadcasting and manufacturing in America.

The Radio Collector's Directory and Price Guide is divided into the following three major parts:

(I) Radio in America, 1921-1941, including Chapter 1, The Birth of Radio Manufacturing; Chapter 2, The Regulation of Radio Broadcasting, 1921-1941; Chapter 3, Developments in Radio Broadcast Receiver Design, 1921-1941; Chapter 4, How to Establish Receiver Year of Manufacture.

(II) The Radio Directory and Price Guide. Part II is comprised of 8,500 radio broadcast receivers identified by manufacturer, model number, year of manufacture, source of power, style, and in many, many instances, current market price. The information presented in the directory is largely self-explanatory; however, as an aid to readers in interpreting the data, comprehensive definitions of the modes of classification introduce the directory listings.

(III) The Trade Name Directory. Part III is divided into two major sections by which readers may quickly cross-reference trade names and manufacturers. Section 1 includes a directory of trade names which references radio manufacturers; Section 2 presents a directory of radio manufacturers which references trade names.

Although our research for *The Radio Collector's Directory and Price Guide* has enabled us to identify and date definitively 8,500 different models of radio broadcast receiver, we are aware of additional models that we do not list. We reviewed dozens of interesting models in radio periodicals, for example, that we could neither identify nor date from the information available to us. We hope, therefore, to produce subsequent editions of *The Radio Collector's Directory and Price Guide,* and we invite our readers to share information with us about either radio broadcast receivers not listed or knowledge of current transactions.

The material for the *The Radio Collector's Directory and Price Guide* was compiled by a longtime radio amateur and a professional radio engineer who share strong interest in the development of radio broadcast receivers, the historical record of the radio manufacturing industry in the United States, and the current market value of the artifacts of that remarkable era. This book is a product of our combined experience, spanning several decades, of collecting, marketing, and servicing vintage radio receivers. It is intended to be both a helpful, informative reference text for identifying the primary characteristics of radio broadcast receivers, manufactured between 1921 and the beginning of World War II, and a guide to the value of these receivers.

Our research efforts led us to countless sources of information. The process of matching model, year, and style to current market value required careful sorting among both historical and contemporary data. We are certainly grateful to those individuals who produced radio periodicals in the 1920s and 1930s. We are particularly indebted to the many radio collectors who responded to our questionnaires and requests for price information. And, we wish especially to express our gratitude to Paul Giganti, William S. Lettow, and William Pugh for their contributions to portions of the directory.

Robert E. Grinder, K7AK
George H. Fathauer

PART 1

RADIO IN AMERICA

1921 - 1941

Radio receivers had been manufactured prior to 1921 mainly for military, weather, and commercial wireless applications, ships at sea, and a small cadre of radio amateurs. But late in 1921 the art of radio sprang—quite unexpectedly—out of the hands of a few entrepreneurs and experimenters, who were engaged in point-to-point communications, into a nationwide public service for news, entertainment, and education. Between 1921 and 1941 billions of radio tubes and millions upon millions of radio broadcast receivers were produced by American manufacturers. Then, with the onset of World War II, December 1941, radio manufacturing for home use ceased for the duration of the war.

Postwar radio manufacturing produced clones of pre-war models until the coming of the solid-state era in the late 1950s. The relatively bulky, power-consuming, tube models were displaced; development accelerated in the production of new models featuring startling advances in high fidelity, style, and size reduction. Manufacturing shifted overseas, where production costs were lower, and lamentably, nearly all of the prewar domestic radio manufacturing companies closed their plants.

The years between 1921 and 1941, therefore, mark an extraordinary era in the history of radio manufacturing.

CHAPTER

1

THE BIRTH OF RADIO MANUFACTURING

RADIO BROADCASTING

Radio broadcasting, in the minds of early radio entrepreneurs, was brought into being as much as an enticement to the public to purchase radio receivers as an agency of public service. David Sarnoff, who was later to become a distinguished pioneer of both radio broadcasting and radio manufacturing—and chairman of the Radio Corporation of America (RCA)—was among the first to recognize its potential. As an assistant traffic manager of the American Marconi Company in 1916, he asserted in a prophetic memorandum to his general supervisor: "I have in mind a plan of development which would make radio a household utility in the same sense as a piano or phonograph. The idea is to bring music into the home by wireless." The young visionary anticipated a "radio music box" that would deliver music, lectures, and accounts of world events. He believed that an enthusiastic public would purchase one million radio music boxes within a period of three years; assuming the selling price of each set to be about $75, he estimated a sales volume of 75 million dollars.

The radio engineering community of the day, however, had it been aware of Sarnoff's memorandum, would have disregarded it for three reasons. First, radio transmitters had advanced in 1916 to the point of utilizing the relatively efficient Poulsen carbon arc design, but like its rotary gap predecessor, it emitted a steady succession of noisy sparks. The "mush" sorely distorted voice transmissions, and radio engineers had no means of overcoming it. Radiotelephony might be technically feasible, they reasoned, but it was commercially impractical.

Second, whereas Sarnoff perceived the potential of radiotelephony to reach many, many people simultaneously, other radio engineers regarded this aspect of radio not as an advantage but as a liability. It was presumed to have a serious drawback; it lacked the privacy of the ordinary wire telephone. What one had to say, everyone else could hear! The engineers saw, nonetheless, that radiotelephony would make the application of radio available to those unskilled in its technical aspects. They believed, for example, that it might be used someday as an efficient means of wireless communication, especially between airplanes and ground forces. Hence, during the early, experimental days of radiotelephony, engineers worked harder at restricting than expanding the size of listening audiences.

Third, at the time the United States entered World War I, members of the radio industry were hopelessly snarled in patent litigations. No single manufacturer was able to build either a transmitter or a receiver adapted to the current technology, because no one could access all the innovations in both circuitry and components. In truth, the radio manufacturing industry was in a state of paralysis, and any notion of producing state-of-the-art "music boxes" seemed farfetched. The United States Navy rescued radio research and development from the predicament by declaring a moratorium in the exercise of radio patents for the duration of the war. It ordered manufacturers to produce for the government radio equipment embodying the latest technical advances regardless of patent ownership. Heretofore, a host of relatively small companies had bargained with one another for patent rights or had fought over them in the courts; now, the government placed wartime contracts in the hands of major electrical companies, where their sizeable engineering staffs were suddenly free to apply the resources of their research facilities to achieving significant improvements in the art of radio transmitting and receiving.

The war years witnessed giant steps forward in the development of vacuum tube wireless equipment. Frank Conrad, a Westinghouse radio engineer, was foremost among the pioneers who facilitated transition from the primitive spark-gap era to that of the vacuum tube. Conrad had become a radio experimenter in 1912 virtually by accident. To settle a five dollar bet regarding the accuracy of a twelve dollar watch he recently had acquired, Conrad built a small receiver in his Pittsburgh home in order to listen to time signals from the National Observatory at Arlington, Virginia. He developed rather swiftly a fascination for radio communication, and he acquired the call letters, 8XK, as a licensed radio amateur.

Although an active operator, Conrad was interested less in chit-chatting in Morse code with other amateurs than in experimenting to improve the effectiveness of his equipment. He thus acquired a great deal of practical expertise in the application of radio. When wartime security precautions led to the suspension of amateur transmitting, the Government commissioned Conrad, on the basis of his scientific work at Westinghouse, to develop vacuum tube apparatus. He was given special privileges to operate radio transmitters, and, as a consequence, in part because of his research, evolutionary new wireless equipment was developed for the United States Signal Corps. In the process, Westinghouse became a major supplier of transmitters and receivers for ship-to-shore and ground-to-airplane communications.

The ban on amateur transmissions was lifted April 15, 1919. Thousands of military personnel had been trained to use wireless equipment during the war, and they had become radio enthusiasts. They swelled the amateur ranks, they built receivers and transmitters, and they filled the air waves with their tests and messages. The majority of them came on the air with spark-gap transmitters. Parts for building the older equipment were relatively inexpensive and readily available; on the other hand, vacuum tubes were very expensive and quite unavailable. A good, commercial, vacuum-tube receiver in 1919 could cost as much as a three bedroom home!

Given his wartime research opportunities, Conrad emerged in the postwar years as a seasoned, sophisticated radio experimenter. He was thus regarded as a leading amateur, and knowing the significance of his work, other operators were eager to report to him on the quality and strength of his signals. These data were important to Conrad because they enabled him to plot the effects of his tinkering, and thus, to study the efficiency of his transmitter. He was accustomed to communicating in Morse code with other amateurs; however, since he had been instrumental in elevating the art of radio to a new plane, he possessed the requisite equipment for transmitting by radiotelephone. So, on October 17, 1919, he placed a phonograph in front of a microphone and played "Old Black Joe" over the air. He did it partly to be different and partly because he had tired of making transmissions in code.

Conrad's listeners were astounded! They were at first a scattered group of experimenters and amateurs. But they increased swiftly in number when they found that Conrad's transmissions could be understood without knowledge of Morse code. People wrote letters to him from miles in every direction surrounding Pittsburgh. They commended him for providing them with so much listening pleasure and they made requests of him for specific records. People even called on the telephone so frequently and at such inconvenient times that Conrad soon announced that he could no longer comply with individual requests; rather, he announced that he would "broadcast" records for two hours every Wednesday and Saturday evenings. Conrad chose to describe his programs as "broadcasting" after the gardening notion of "broadcast," which means to cast or scatter seed over a wide area.

Knowledge of Conrad's concerts passed swiftly

via word of mouth until hundreds, perhaps thousands of people knew about them. When his broadcast schedule exhausted his supply of records, a local music store provided an ongoing supply on the condition that he announce that the records he played could be purchased at the store. Sometimes he interspersed live vocal and instrumental talent between records.

In the summer of 1920, a Pittsburgh department store, the Joseph Horne Company, recognized a commercial possibility to be derived from Conrad's popular broadcasts. The enterprising department store thus began to advertise, in the Pittsburgh *Sun,* inexpensive crystal sets for receiving them. Its first advertisement, September 29, 1920, caught the attention of Harry P. Davis, a Vice President of Westinghouse. Davis saw in the strong public interest in radio entertainment an equally strong marketplace for the manufacturing of radio receivers. He suggested to Westinghouse managers that the company erect a station in East Pittsburgh for the purpose of offering regularly scheduled programs every evening. People, he reasoned, would become as habituated to listening to a radio as to reading a newspaper. Thus radio broadcast entertainment, as Sarnoff had prophesied several years earlier, would be the vehicle for motivating people to purchase radio broadcast receivers.

Two weeks after the first Horne Company advertisement, Westinghouse applied to the Department of Commerce for a license to broadcast from Pittsburgh. Within two more weeks it was authorized to operate as KDKA. The station was readied for its first broadcast, November 2, 1920, the date of the Harding-Cox presidential election returns.

The historic broadcast was made from a small studio on top of one of the Westinghouse buildings in East Pittsburgh. A single room accommodated all of the transmitting equipment, a turntable for records, and a few tables and chairs for the engineers and announcers. Conrad stayed at home that evening. He was prepared to carry on with his amateur transmitter, 8XK, in the event of technical malfunction at East Pittsburgh.

The broadcast began at 6 o'clock election night and continued until noon the following day. It was a nationwide sensation. It was acclaimed by newspapers all over the country. KDKA continued to operate everyday, and within a year 30 stations nationwide, including WBZ, Springfield, Massachusetts; WHA, Madison, Wisconsin; and WWJ, Detroit, had been authorized by the Department of Commerce for broadcast purposes. Actually, WHA, as 9XM, and WWJ, as 8MK, had "broadcast" transmissions even before Conrad's (8XK) historic experiments inspired Westinghouse to inaugurate KDKA programming. These stations may lay claim to having preceded KDKA in giving birth to broadcasting, but KDKA is to be distinguished among all stations as the first to come on the air for the specific intent of promoting sales of radio broadcast receivers.

KDKA stands apart from other radio broadcast stations in another noteworthy respect. At the outset of radio broadcasting nearly all commercial land stations were on the coast, and most radio transmissions consisted of ship-to-shore communications. When the Department of Commerce began to authorize inland broadcast stations, the Mississippi river was made the dividing line for designating K and W call-letters; those to the east were to be assigned calls beginning with the letter W, and those to the west, beginning with K. KDKA, at East Pittsburgh, Pennysylvania, however, is one of the early broadcast stations which was assigned a call beginning with K (as were ships at sea) before the plan went into effect.

THE RISE OF RADIO BROADCAST RECEIVER MANUFACTURING

Early radio stations proliferated faster than those on the air could generate audiences. In the year 1922, the number of authorized stations leaped from 30 to 639. Further, electrical shops, newspapers, colleges—any person or group which could construct a simple radiotelephone transmitter—began broadcasting, legal authorization notwithstanding. People clamored for opportunities to listen, but few receivers were available. Only ten to twenty thousand radio receivers in the United States in 1920 were privately owned. Thus, whenever newspapers announced a new station, people in its vicinity would stand in long lines at retail outlets merely to get on a waiting list to purchase a receiver.

Following the initial KDKA broadcast, RCA, with its extensive control of manufacturing resources, was positioned to dominate the marketplace. It began to distribute under its label broadcast receivers manufactured by both General Electric and Westinghouse. But the production facilities of these great manufacturing companies were no match for the public appetite for broadcast receivers. Consequently, RCA was joined in 1921 and 1922 by hundreds of other companies, many of which switched from manufacturing automobile and electrical accessories to producing radio receivers. Production quarters often were housed in residential garages and basements. Radio amateurs who had built sets for neighbors, and radio engineers, who believed that they could design sets, quit their jobs to go into the radio manufacturing business. As the radio craze swept across the nation, the number of broadcast receivers reaching the marketplace attained staggering proportions.

Radio Retailing has divided radio manufacturing between 1920 and 1941 into three distinct periods: home construction years, 1920-1923; factory-made years, 1924-1929; replacement and extra set years, 1930-1941 (As It Was, 1940). Table 1 presents a historical overview of radio broadcast receiver sales during the three periods. Taking sales as an indication of manufacturing rate, the data shown in Table 1 from *Radio Retailing* reveal for each period, respectively, that 350,000, 14 million, and 71 million—a total of about 85 million—radio receivers were manufactured from the beginning of 1922 to the end of 1941 (Domestic Set Sales, 1938; Radio's Billion-Dollar Business, 1942). The year-by-year records show that during the 1920s manufacturing advanced steadily until 1929, when nearly 4.5 million receivers were manufactured. With the great depression, yearly production fell to a low of 2.4 million in 1932, but in 1933 it again turned upward. By 1936, the 1929 volume had been surpassed; a slight drop in production rate occurred in 1937, but thereafter it rose steadily to its zenith in 1941 when more than 11 million sets were manufactured and sold.

TABLE 1

Domestic Sales of Factory Made Home Radio Receivers

	Year	Number	Total
Home Construction Period	1922	100,000	
	1923	250,000	350,000
Factory Made, Original Purchase Period	1924	1,500,000	
	1925	2,000,000	
	1926	1,750,000	
	1927	1,350,000	
	1928	3,281,000	
	1929	4,435,000	14,316,000
Replacement and Extra Set Period	1930	3,793,800	
	1931	3,312,000	
	1932	2,477,000	
	1933	3,082,000	
	1934	3,304,000	
	1935	4,375,016	
	1936	6,746,290	
	1937	6,278,267	
	1938	7,500,000*	
	1939	9,000,000*	
	1940	10,000,000*	
	1941	11,100,000	70,968,373
			85,634,373

*Extrapolation

Although radio manufacturers produced more than twice as many receivers in 1941 as in 1929, surprisingly, they made less gross profit. Sales in 1929 carried a retail list of 591 million dollars, but in 1941, only 420 million dollars (Domestic Set Sales, 1938; Radio's Billion-Dollar Business, 1942). Changes in the marketplace account for the anomaly. Virtually all of the receivers sold in 1929 were original purchases; furthermore, the average cost of a receiver then was $135, which was the highest of any pre-World War II year. In contrast, only 2.2 million sets were sold in 1941 to homes previously without receivers; on the other hand, 3.3 million were sold as replacements and 5.6 million were sold as second or as extra sets (Radio's Billion-Dollar Business, 1942). The supplemental receivers comprised the inexpensive portable and mantle varieties, which ranged in price roughly from $6.95 to $39.95. At these prices the 5 million extra sets generated little dollar volume.

Given the probability that about 85 million radio receivers were manufactured between 1921 and 1941, can we estimate how many different models were manufactured? Yes, but the task is complicated by two important considerations: first, many early manufacturers left no production records for us to inspect, and second, several retailers and distributors marketed the same set (same manufacturer) using their own particular model numbers, thus inflating the historical record. On the one hand, during the early years of radio manufacturing, hundreds of companies emerged for a few months, perhaps for a year or two, to produce a few models before competition drove them out of business. When David Grimes developed the reflex circuit, for example, where three tubes might do the work of five, in part to reduce both battery drain and manufacturing costs, tubes were relatively expensive. Manufacturers who employed the then economical reflex circuit thus captured a sizeable share of the market for a few years in the mid 1920s. When tube prices dropped dramatically, however, their models were no longer as competitive as those of manufacturers with simpler circuitries, and many of them, having limited resources, went out of business. On the the other hand, many of the larger, successful manufacturers, in the 1930s especially, often invested a great deal of their research resources in developing a particular chassis configuration, which they might then insert in a dozen or more table and console models. The same chassis design also might appear in the marketplace under two or more manufacturer's labels. As the *RCA-Victor Service Data Manual* (RCA, 1944) indicates, many of the RCA Victor, General Electric, Westinghouse, and Graybar models shared exactly the same chassis.

We can, nonetheless, offer a reasonable estimate of the number of different models manufactured. The master index for the Rider "Perpetual Trouble Shooter's" manuals, volumes 1-15 (Rider, 1947), for example, lists about 30,000 prewar models. We cannot conclude from these data, however, that only 30,000 different models of radio broadcast receivers were produced between 1921 and 1941. As anyone who has used the Rider manuals knows, Rider omitted the products of several manufacturers. Let us, then, sharpen our perspective with corroborative data. The fifth edition of *The Mallory Radio Service Encyclopedia* (Mallory, 1946) carries this helpful proclamation on its front cover: "replacement recommendations for all prewar receivers." The Encyclopedia includes roughly 25,000 models, but contrary to its claim of inclusiveness, it covers relatively few models manufactured prior to 1930. It also omits several thousand radio broadcast receivers of the 1930s and 1940s for which, apparently, Mallory did not produce replacement parts. However, when we sum the pre-1930 listings included in McMahon's (1973) directory with those of Rider and Mallory—taking into account the omissions in these two "comprehensive" volumes—we estimate that approximately 32,500 models of radio broadcast receivers were manufactured between 1921 and 1941.

References

As it was in the beginning. (1940, November). *Radio Retailing.*

Domestic set sales . . . 1922-1937. (1938, January). *Radio Retailing,* p. 26.

Mallory radio service encylopedia. (5th ed.). (1946). Indianapolis: P. R. Mallory & Co., Inc.

McMahon, M. E. (Ed.). (1973). *Radio collector's guide.* Palos Verdes Penninsula, CA: Vintage Radio.

Radio's billion-dollar business. (1942, January). *Radio Retailing,* pp. 16-17.

RCA-Victor Service Notes Volume I. (1944). Camden, NJ: RCA Victor Division of Radio Corporation of America.

Rider, J. F. (1947). *Master index for Rider manuals, volumes 1-15.* New York: John F. Rider Publishing Co., Inc.

CHAPTER

2

THE REGULATION OF RADIO BROADCASTING

Early in the twentieth century, commercial, military, maritime, and amateur radio telegraphy stations shared the known radio spectrum on a functional basis. Since it was believed then that the effectiveness of radio transmissions increased in direct proportion to their wavelength, commercial and military stations, which conducted point-to-point long distance communications, used the longest practical frequencies, somewhere in the range of 20,000 (15 kHz.) to 700 (428.57 kHz.) meters. Because space for antennas on ships was limited, maritime communications ranged from about 450 (666.67 kHz.) to 600 (500 kHz.) meters. And the radio amateurs, to ensure that they did not interfere with the other services, were assigned the shorter, presumably useless, wavelength of 200 meters. Accurate frequency measurement was impossible then, and stations often wobbled far from their assigned wavelengths. Ship-to-ship communication, for example, might range as low as 300 meters and amateur operators often transgressed into the 375 meter range.

Radio communications were regulated in the United States from 1912 to 1927 by the Radio Act of 1912. The Bureau of Navigation, Department of Commerce, administered the Act. No one in the Bureau had given much thought to broadcast stations when Westinghouse applied in October, 1920, for authorization to operate a station [KDKA] for the purpose of providing home entertainment to the general public. The Bureau of Navigation thus treated the Westinghouse application as if it were from a merchant ship, and accordingly, it assigned it an operating frequency of 600 meters. Indeed, KDKA operators were required to interrupt their initial transmissions from time to time and to monitor 600 meters to ensure that their signal was not interfering with maritime distress signals. On the whole, however, regulation of broadcasters during 1921 was relatively haphazard. The din and cacophony between 300 and 600 meters reached crescendo proportions, as amateurs and anyone else who could build a transmitter began broadcasting programs for friends, much as Conrad had done. Consequently, in January, 1922, the Department of Commerce amended the 1912 radio regulations to require that all applicants for broadcast stations secure a new form of commercial license for broadcasting and that amateur radiotelephone stations cease broadcasting entirely. And importantly, all broadcast stations were moved to 360 meters (833 kHz.), a relatively uninhabited spot in the radio spectrum.

By early Spring, 1922, a multitude of stations had been licensed to operate on 360 meters. Even though most of them were using low power, they were interfering with one another. Complaints poured into the Department of Commerce from manufacturers, broadcasters, and the general public. The latter now was spending millions to purchase broadcast receivers, and it wanted pleasurable, uninterrupted listening experiences. Early radio sets were comparatively expensive, too, and since the initial purchasers often were both affluent and leading citizens in their communities, they had the clout to raise complaints with legislators in Washington whom they believed could do something about the abominable interference. Hence, in the winter of 1922, Herbert Hoover, Secretary of Commerce, convened the first of a series of four National Radio Conferences, a conference at which he said in his opening remarks:

> We have witnessed in the last four or five months one of the most astounding things that has come under my observation of American life. This Department estimates that today more than 600,000 (one estimate being 1,000,000) persons possess wireless telephone receiving sets, whereas there were less than fifty thousand such sets a year ago. We are indeed today upon the threshold of a new means of widespread communication of intelligence that has the most profound importance from the point of view of public education and public welfare. (quoted in Leinwoll, 1979)

To cope with the growing proliferation of broadcast stations, the recommendations of conference participants led to the assignment to radio broadcasting of a second wavelength at 400 meters (750 kHz.). Broadcast stations with "high power," 500 to 1,000 watts, were to operate on 400 meters whereas those with less than 500 watts were to remain on 360 meters. The inadequacies inherent in the regulations generated by the 1922 Conference swiftly became apparent as hundreds of new stations came on the air. Several of the new stations operated high power transmitters and offered excellent programming; others, however, were low power stations licensed by local businesses, churches, and schools for specific purposes. Although many of these stations passed into oblivion rather quickly, interference was especially tumultuous in urban areas, because most of the early receivers could not distinguish between stations on 360 and 400 meters when proximity made them comparably strong in signal strength.

Secretary Hoover, therefore, called a second conference in May, 1923, to resolve the next phase of problems facing radio broadcasting. This conference recommended that the nomenclature for station assignment be changed from "wavelength" to "frequency" and that frequencies be distinguished from one another by "kilocycles" (see Chapter 3). Rules were established to regulate station separation according to region and frequency; the United States was divided into five zones and stations were to be distributed ten kilocycles apart. To minimize interference further, broadcast stations also were classified on the basis of three levels of transmitter power: Class A, about 500 watts; Class B, 500 to 1,000 watts; and Class C, relatively low power. Consequently, the relatively high power stations could be systematically separated within given regions by several broadcast channels. The recommendations of the 1923 conference led to enormous improvements. However, as technological advances ensued during

1923 and 1924, broadcasters were challenged to increase power and, thereby, reach larger potential audiences. They were beginning to sell airtime to advertisers who insisted upon larger and larger guaranteed audiences before they would invest sizeable funds in the new media for selling their products. The largest radio audiences, of course, were in urban areas where concentrations of people were greatest. Urban areas thus drew a disproportionate share of the powerful stations, who competed intensely for listeners, often by simply boosting power output to override or swamp the signals of competitors. Broadcasters and audiences alike in coastal areas were particularly susceptible to congestion and interference because ships were still entitled to use frequencies in the broadcast band.

By late 1924, all the channels in the broadcast band were used up, and Secretary Hoover called a third conference to make further recommendations for improvement. Participants at the third conference recommended, first, that maritime communications be excluded from the broadcast band of frequencies; that channels for broadcast stations be extended to include all frequencies between 550 and 1,500 kilocyles; and third, that increases in power over 1,000 watts be encouraged on the grounds that not only would range increase but that listeners nearby the more powerful stations would receive them with less static and background noise.

Bedlam persevered, nonetheless. Low-power and high-power stations alike increased power. Several broadcasters were beginning to break the 5,000 watt power barrier. So for the fourth time, in November, 1925, Secretary Hoover held another conference. Court rulings, however, were now seriously undermining his efforts. For example, the Court of Appeals in the District of Columbia had ruled in 1923 that although the Secretary of Commerce had the right to assign an operating frequency to a station, the Department had no right to deny a license to any applicant. The ruling voided the grand plan of the government for allocating stations across the broadcast band on the basis of power. Yet a semblance of cooperation prevailed in the broadcast industry for a few more years, which enabled the Department of Commerce to exercise "gentle persuasion." Then, in 1926, the Courts ruled that the Secretary of Commerce had no authority in the first place to establish any regulations regarding radio transmissions. The right of the Secretary of Commerce to assign frequencies to radio broadcast stations and to regulate their power had been set aside.

As the frequency allocations that had been so carefully worked out previously went by the board in 1926, stations jumped from frequency to frequency on the basis of parochial interests. The practice of increasing power with the intent of drowning out competing stations became commonplace. Results were predictable. Approximately five million radio receivers were in the hands of the public in 1926, and in spite of significant improvements in both receiver selectivity and sensitivity, most listeners were unable to enjoy a single interference-free program.

Clearly, the times called for Congress to enact new regulations. President Coolidge rose to the occasion. He asked the lawmakers to remedy the problems wreaking havoc with the new communications media. The Dill-White Radio Act of 1927 emerged from their deliberations. A five member Federal Radio Commission (FRC) was created. The Commission was empowered to issue broadcast licenses, to assign specific frequencies to individual stations, to specify transmitter power, and to issue regulations for the purpose of minimizing interference. One of the first acts of the FRC was a sweeping reallocation of stations across the broadcast channels. And importantly, to relieve congestion, approximately 150 of the 732 stations operating in 1927 were required to give up their licenses during the first year of the Act.

Although the Radio Act of 1927 introduced a great deal of order to radio broadcasting, tensions mounted yearly between the FRC and the Department of Commerce. Congress had assigned the former responsibility for broadcasting and left to the latter responsibility for regulating all other radio services. Lines of authority thus were still unclear. Therefore, within seven years, Congress passed the Communications Act of 1934, which established the Federal Communications Commission. The Commission took over the radio regulatory functions of both the Department of Commerce and the FRC. The Commission was given broad authority and charged with responsibility for regulating in the United States all aspects of interstate and foreign radio communications.

Subsequent growth in radio broadcasting has been orderly. During the late 1930s, the FCC authorized experimental frequency modulation and television broadcasting. These services, however, were not commercially available on a large scale until after World War II. Further, under the FCC, the broadcast band was widened in the 1930s to encompass all channels between 540 and 1,600 kHz. Careful attention to problems of interference enabled the FCC to increase by the end of 1941 the number of broadcast stations in the nation to 915.

Reference

Leinwoll, S. (1979). *From spark to satellite: a history of radio communication.* New York: Charles Scribners' Sons.

CHAPTER

3

DEVELOPMENTS IN RADIO BROADCAST RECEIVER DESIGN

The pioneer manufacturers of radio receivers, in a series of remarkable accomplishments, elevated the quality of broadcast reception immensely between 1921 and 1941. The successful manufacturers were those who attracted the attention of buyers by introducing design innovations. Hundreds of companies were out of the starting blocks in the 1920s, but by 1941, radio broadcast receiver manufacturing was concentrated in the hands of only 82 companies (Radio's Billion-Dollar Business, 1942).

The survivors among the radio manufacturers offered receivers to the public which were superior to those of their competitors in circuitry, control functions, listening quality, and physical construction. The leading radio engineers worked fervently, that is, to increase standards of performance, simplify operating features, enhance the quality of broadcast receptions, and anticipate trends in cabinet styles. The historical record of their successes in each of the four areas, as indicated below, chronicles the progress that they made not only in meeting but in creating consumer demand for radio broadcast receivers.

CIRCUITRY (TECHNICAL) CONSIDERATIONS

Early radio engineers experimented with circuitry to improve *sensitivity,* the capability of a receiver to pick up desired stations, and *selectivity,* the capability of a receiver to accept the broadcast of a desired station and to reject the unwanted signals of adjacent stations.

The designers of the first radio broadcast receivers concentrated primarily on problems of sensitivity. Broadcast stations were few and far between in 1921 and 1922, and the listening public wanted receivers that could hear as many of them as possible. Furthermore, when all of the broadcast stations were clustered on 360 meters, radio manufacturers could afford to ignore the question of selectivity. Eventually, however, as broadcast stations began to spread across the dial, radio manufacturers were obliged to develop receivers that would detect a great many stations while simultaneously preventing the relatively strong ones from obliterating the weaker signals.

Crystal Detector Circuit

Radio manufacturers turned initially to the crystal detector, because tubes and parts for the vacuum tube circuitry developed during World War I were as yet too expensive. A station was tuned in on a crystal set by adjusting either a capacitance or an inductance to the station frequency and detected by probing the crystal surface with a fine wire (cat whisker) to find its most sensitive spot. The operation of the receiver required endless retuning, reprobing, retuning, and so forth.

Crystal sets were insensitive, unstable, and their lack of signal amplification restricted them to head phone use. A station fifty miles away might be picked up with a good antenna, and when that happened, it was regarded as a real accomplishment. Nonetheless, crystal sets were relatively inexpensive, and they were easy to produce. The Joseph Horne Company, for example, advertised crystal sets as a means of listening to Frank Conrad's broadcasts.

Regenerative Circuit

The prospect of copying distant stations entranced many radio listeners, and they gave up sleep willingly to tune in faraway stations during the wee hours when local stations were off the air. But all the diligence in the world would not bring in an exotic station on a crystal set. Radio manufacturers thus turned to the regenerative circuit, which utilizes a vacuum tube, to obtain greater sensitivity. Although the regenerative receiver required a cumbersome battery supply, it was superior to the crystal set in sensitivity, selectivity, and audio output power. Its basic circuitry is simple, and as tubes and parts began to be produced in huge quantities, it proved to be both easy and inexpensive to manufacture. Consequently, virtually every radio broadcast receiver manufacturer soon ceased producing crystal sets for the general public.

Three companies captured a major share of the early broadcast receiver market with quality regenerative receivers: RCA (Westinghouse) introduced the RA/DA and RC series; Clapp-Eastham entered the market with the RADAK series, and Grebe produced the initial CR series. The receivers of these manufacturers included typically a detector and two stages of audio amplification; in most instances the detector unit and that of the audio amplifier could be purchased separately. Although only two wavelengths, 360 and 400 meters, were allocated to broadcasting in 1922, listeners also wanted to receive signals of ships at sea on 600 meters (500 kHz.) and of amateurs on (200-250) meters (1200-1500 kHz.). The manufacturers, therefore, designed early regenerative receivers to cover a tuning range from about 600 meters to 250 meters, a range that corresponds closely to that of present-day AM radio broadcast receivers.

In the regenerative receiver, radio-frequency (RF) energy from a broadcast station is fed back from the output of the detector tube to its input. The ensuing "feedback" greatly increases sensitivity, but it renders the receiver difficult to adjust and susceptible to functioning as a low-power transmitter. For example, changes in atmospheric, battery, and tube conditions from moment-to-moment and day-to-day make it impossible to adjust a regenerative receiver so that it receives a given station consistently at the same dial settings.

Early broadcast enthusiasts probably would have tolerated difficulties with station logging, but they were unable to cope with the propensity of the regenerative receiver to oscillate. The problem originates from a coincidence—the point to which feedback must be advanced for maximum sensitivity is virtually the same point at which the regenerative receiver pops into oscillation. When a regenerative receiver breaks into oscillation, it becomes a low-powered transmitter. This produces an uncomfortable howl in the head phones or loud speaker, and whistles appear in any nearby receiver that is tuned to the same station. The regenerative receiver was especially disruptive of listening pleasure in congested urban neighborhoods, where dozens of them in close proximity would swish their jarring transmissions back and forth across the dial (Lescarboura, 1924). The regenerative receiver thus became such a public nuisance in the early 1920s that it threatened, along with the interference that broadcast stations created among themselves, to short-circuit early expectations and hopes for both the radio broadcasting and the radio manufacturing industries.

Tuned-Radio-Frequency Circuit

The regenerative receiver had vanished as a force in the marketplace by 1926. Except for the very low priced offerings of a few manufacturers, the inherent nuisance factors of the regenerative receiver led to its demise. Radio engineers had begun to replace it about 1923 with circuits using tuned "radio frequency" (RF) amplification stages in front of a non-regenerative detector. The Tuned-Radio-Frequency (TRF) circuit quickly evolved into a basic design involving five tubes (two RF amplifiers, one detector, and two audio frequency (AF) amplifiers. In the process, the TRF receiver contributed immensely to the credibility of the radio manufacturing industry. It overcame the horrendous disadvantages of the regenerative receiver, and it satisfied consumer expectations regarding selectivity, especially as broadcast stations were now beginning to be separated in 10 kHz. segments.

The TRF receiver of the day employed conventionally three tuned circuits, and therefore, three main dials, one for tuning each of the RF stages and one for tuning the detector stage. Although the detector in the TRF receiver, or "three-dialer," as it came to be known, was non-regenerative, the tuned RF stages that preceded it were prone to oscillate much like a regenerative receiver, causing howls in the receiver loudspeaker and annoying radiation. Radio engineers eliminated the problem by greatly reducing the potential sensitivity of the RF stages. Radio listeners now could purchase a receiver that would be inoffensive to their neighbors, but it was compromised relative to the sensitivity that they so highly prized.

Neutrodyne Circuit

Radio engineers were diverted from their problems with the regenerative and TRF receivers after Professor L. A. Hazeltine, Stevens Institute of Technology, invented the "neutrodyne" circuit. Hazeltine recognized that a small amount of "negative" feedback could be used to cancel or balance out the positive feedback that produced oscillation. In brief, Hazeltine developed a circuit to neutralize the RF stages of a basic TRF receiver, and thereby, to achieve stable tube operation and prevent oscillation. Radio broadcast listeners were thrilled. The "squeal-less" neutrodyne receiver provided both RF amplification and station selectivity. The circuit was even so stable that the numbers at which stations appeared on the three dials could be logged for future reference.

The first receivers to carry the neutrodyne label were the F. A. D. Andrea FADA 160 and the Freed-Eisemann NR-5. From 1924 through 1927, any "neutrodyne" radio broadcast receiver was certain to sell. Several radio manufacturers, like F. A. D. Andrea and Freed-Eisemann, who elected to pay the royalty fee for use of the Hazeltine patent, based their entire productions on the neutrodyne circuit.

Second-Generation TRF Circuit

Suddenly, however, about 1927, neutrodyne receivers dropped precipitously into obsolescence. A new generation of "screen-grid" vacuum tubes reached the laboratories of radio engineers. The new tubes resolved internally problems of oscillation, and the necessity for neutralization evaporated. Radio manufacturers then set their sights on developing second-generation TRF receivers using the new screen-grid tubes.

These new TRF receivers, including the Atwater Kent model 55, the Colonial Radio series 32, and the Philco series 95, dominated the marketplace until 1932, when RCA, under threat of antitrust suits, released its superheterodyne patents to licensees. For the next few years the superheterodyne circuit wholly dominated radio manufacturing. Virtually every other receiver circuit dropped out of production until, a few years later, manufacturers revived the TRF circuit for very low priced mantle and table receivers.

Superheterodyne Circuit

RCA obtained sole rights to the superheterodyne circuit shortly after Edward A. Armstrong, a distinguished young inventor patented the circuit in 1920. RCA engineers recognized its potential and refused steadfastly to allow anyone to market superheterodyne receivers during the first ten years of the radio manufacturing industry. Consequently, everyone in the industry except RCA was forced to deal with the inadequacies of the regenerative and TRF receivers. RCA, in the meantime, proudly presented vastly superior models featuring the advantages of the superheterodyne. The history of radio manufacturing surely would have been different had early radio engineering laboratories been free to experiment with the superheterodyne circuit.

A first-generation superheterodyne was offered by RCA to the general public in 1924. It was a table model named the Radiola Superheterodyne type AR-812. This model was followed with a console version, the Radiola Super VIII (AR-810), a relatively deluxe receiver which came complete with a built-in, rotatable loop antenna and a horn speaker. RCA also subsequently manufactured in the 1920s several different models of the superhet circuit, including Radiolas 24, 25, 26, and 28.

In order, perhaps, to prevent piracy of its superheterodyne secrets, RCA sealed the chassis of each of its early superheterodyne models in a metal "catacomb," the top of which was protected with lead seals stamped with the RCA logo. RCA service literature indicated how to examine a catacomb for defects, but not how to service it. Dealers were required both to return the old one and to obtain a new one from RCA (Wolkonowicz, 1977).

Since RCA refused to license other manufacturers to produce superheterodynes, the only way a radio enthusiast could avoid purchasing the superior receiver from RCA, at a relatively steep price, was to build it. A number of radio parts manufacturers thus generated a lively business producing special transformers and coils primarily for home experimenters. For a time, then, during the 1920s, it was actually possible for an individual to build a better receiver than radio manufacturers other than RCA could market.

Before RCA released its superheterodyne patents, it established a substantial technical lead on its competition by bringing out in 1930 the Radiola 80, using the new type 24 screen-grid tube (Carlson, Holmes, and Wunderlich, 1930). The Radiola 80 set new standards for performance in sensitivity, selectivity, fidelity, and audio power output. More significantly for the radio industry, it revealed to the world that radio manufacturing had attained a new plateau in the evolution of broadcast receivers.

The chief distinction of the superheterodyne circuit is that it changes the broadcast frequency to a lower, fixed frequency where amplification and selectivity can occur much more efficiently. The superheterodyne circuit makes the RF amplification process largely independent of the frequency of the original signal. The frequency of the broadcast signal is "mixed" with another signal from an internally or locally generated oscillator to produce a third signal at a much lower frequency. The superheterodyne receiver is tuned by changing the frequency of the oscillator in proportion with incoming signals to ensure that the difference or "mixed" frequency remains constant. The new signal is then fed to a fixed intermediate-frequency (IF) amplifier, where, after amplification, it is sent to a "second detector," at which the audio-frequency component is removed and fed to the audio amplifier.

One of the major advantages of the superheterodyne circuit over that of the TRF receiver is the uniformly high amplification and selectivity produced by the IF amplifier. In the TRF receiver, for example, the factors that affect both sensitivity and selectivity operate at the frequency of the incoming broadcast signal, and they must be adjusted to that frequency. This means that the RF amplifiers in a TRF receiver must be sufficiently modifiable to fit every frequency in the broadcast band, which makes consistent performance unattainable. In contrast, the operating characteristics of an IF amplifier in a superheterodyne are independent of the broadcast frequency to which the receiver is tuned. Since a significant proportion of the RF amplification takes place at a fixed, intermediate frequency, maximum amplification is relatively uniform over the entire broadcast band.

The widespread adoption of the superheterodyne circuit in the 1930s enabled radio manufacturers to meet growing public demand for shortwave capability in broadcast receivers. Shortwave broadcast stations all over the world were coming on the air in the early 1930s, and prospects of hearing faraway foreign broadcasts provided listeners heady stimulants for acquiring new receivers. The superheterodyne circuit was admirably suited for adaption to multiple band reception, and manufacturers capitalized on it to produce all-wave receivers, which, in turn, generated an enormous new market for radio receivers.

CONTROL FUNCTIONS: LIGHT-SOCKET POWER, SINGLE DIAL TUNING, AND AUTOMATIC VOLUME CONTROL (AVC)

The radio manufacturing industry recognized at the outset that it had to produce broadcast receivers that would be increasingly easy to operate. As radio enthusiasts multiplied across the nation, one particular message reverberated from retailer showroom to showroom: make radio sets foolproof operationally—eliminate tedious adjustments (Langley, 1927). Successes came slowly, but one triumph followed another. Who could argue with the radio engineer who proclaimed during a moment of self-congratulatory exuberance:

> The man, woman or child of 1938 turns the dials or pushes the buttons of a radio set with as much matter-of-fact expectancy of program reception, as they turn the kitchen faucet and confidently know that the water will flow. This despite the fact that the world has had waterworks ever since the dawn of the Christian era, while radio is only 18 years old. (Rost, 1938, p. 30)

Let us, then, consider how radio manufacturers simplified control functions in three major aspects of receiver operation: light-socket power (from battery to alternating current power), single dial and calibrated tuning; and automatic volume control (AVC).

Light-Socket Power

Early broadcast receivers, which employed vacuum tubes, obtained their power from batteries. At least two types of battery were required. A rechargeable "A" storage battery was used to supply low voltage and high current for tube filaments. A non-rechargeable "B" battery, comprised of small dry cells wired together in blocks of fifteen to thirty, was used to supply the higher voltage and low current requirements of the receiver detector and amplifier circuits. A family who listened to broadcasts several hours per week usually recharged its "A" battery once a week. About every third time the "A" storage battery was recharged, the "B" battery had to be discarded. The interminable recharging and replacement of batteries constituted an enormous nuisance, not to mention expense, for radio listeners.

Radio engineers wasted precious little time in finding ways of adapting their receivers to light-socket power. The Philadelphia Storage Battery Company (later to become PHILCO) marketed in 1924 the first "B" battery eliminators, and the public eagerly snapped them up. Several competitors joined the Philadelphia company in selling eliminators for both the "A" and "B" batteries. Prospective broadcast receiver buyers thus could purchase in 1926 battery eliminators as accessories for any receiver. Then, also in 1926, Zenith introduced deluxe console receivers in which both the "A" and "B" battery eliminators were included in the cabinet as integral components of the receiver.

At about the same time, RCA released for general radio manufacturing use a series of newly developed tubes whose filaments could be operated directly from ac current. A new generation of radio broadcast receivers, designed specifically for batteryless, ac operation was at hand. RCA spearheaded the movement, in 1927, with the Radiola 17, one of the first receivers designed specifically for ac operation. And in the same year the Atwater Kent Company brought out its Model 38, an ac receiver in which the power supply was fully self-contained.

Various refinements in light-socket power appeared in the 1930s. International Radio developed the ac *or* dc circuit in 1931 in order that the same receiver could be powered by ac in urban communities and by direct current in areas of the country where only dc power was available. This circuit eliminated the power transformer, which greatly reduced cost and weight. It appealed to manufacturers, and from 1935 to 1941 about 20 percent of the receivers in the marketplace were of the ac/dc variety. Manufacturers of radio tubes also introduced in the mid 1930s tubes with different filament voltages, which enabled set manufacturers to wire various tubes in series so that the sum of their filament requirements would approach the voltage at the light bulb socket.

The new tubes led to the development of midget and mantle receivers. Manufacturers flooded the marketplace in the late 1930s with these little sets, hoping to stimulate greater demand for second and third receivers in the home. Many of them were simple TRF units mounted in colorful molded plastic cases. The last pre-World War II power arrangement emerged when three-way ac/dc/battery portable receivers appeared on the market in 1939 using a new series of low-power battery tubes.

Single Dial and Calibrated Tuning

Single dial tuning

Early broadcast listeners often lost their patience during the period when the several tuned circuits in their receivers had to be adjusted individually. After spending a great deal of time retuning the dials of their receivers for optimal reception, stations would slip into the static as a result of atmospheric changes or a weakened battery. Although the instabilities inherent in the regenerative and TRF receivers made it difficult to attain precise dial settings, Magnavox in 1924 and Atwater Kent, Grebe, and other manufacturers in

1925 made valiant attempts to provide reliable single dial tuning. Their efforts proved less than satisfactory until the development of screen-grid tubes, which enabled radio engineers to stabilize the tuned circuits. Once they were empowered to track stages simultaneously, it became feasible to "gang" variable condensers together so as to reduce the number of dials required to tune a receiver.

Radio manufacturers initially reduced the number of dials from three to two. In the conventional TRF receiver, one dial tuned the first RF stage and the other, the second RF stage and the detector. As tracking techniques improved, the number of dials dropped to one. By 1927 single dial tuning had become so popular among the public that radio manufacturers who ignored the trend faced a hostile marketplace.

Calibrated tuning

Radio evolved in the late nineteenth century out of technological advances in electrical engineering. When very early radio engineers caused an electromagnetic wave to be "radiated" in air space, they discovered two noteworthy facts. First, they found that such a wave travels at the velocity or speed of light. Second, they discovered that by devising tuning circuits they could vary the intensity or frequency of its magnetic field alternations or cycles. What, they asked, should be our unit of measurement for these alternations? Since the speed of light travels about 300 million meters per second, the radio engineers chose to divide this rate of speed by the number of cycles being generated per second to determine the distance one radio wave travels during the time it takes to complete one alternation. Thus, a radio signal that alternates 300,000 times per second travels 1,000 meters in completing one cycle. It has a *wavelength* of 1,000 meters.

Early radio engineers were accustomed to thinking in the tangible dimensions of physics, so they focused on units of *wavelength* to reference the location of their signals in the radio spectrum. Furthermore, since a high proportion of pre-broadcast ship-to-shore radio communications took place around 1,000 meters, in the interest of brevity, members of the radio community preferred to speak with each other in terms of meters rather than cycles—it was simply easier to write or say 1,000 meters than 300,000 cycles (Morecroft, 1927). As a consequence, when KDKA and other stations initiated radio broadcasting, newspapers dutifully published their transmitting frequencies solely in units of meters.

Although "wavelength" became the standard measure by which to refer to a station in the radio spectrum, none of the early radio broadcast receivers were calibrated. The instabilities inherent in the receivers precluded this convenience. Instead, the manufacturers of the TRF receivers of the early to mid 1920s elevated the external, three-inch circular dial of molded, polished, black bakelite to the height of its popularity. Each dial was marked with an arbitrary scale of 100 divisions, or channel numbers as they were known then, which covered 180 degrees. A dial usually read from left to right, from 0 to 100, which meant, as an artifact of the way early receivers were designed, that as the numbers increased, the wavelength decreased. For example, listeners would turn a broadcast receiver dial counterclockwise to go from 400 to 360 meters. It is noteworthy that many Atwater Kent breadboard models employed dials where the numbers read from right to left, or counterclockwise. It was a noble experiment to help unsophisticated listeners match the order of newspaper listings to that of their dials, but Atwater Kent eventually abandoned the practice, perhaps, because the "backward" dial perplexed more listeners than it enlightened.

The cycle superseded wavelength as the designation for specifying station operating frequencies when efforts to distribute stations across the broadcast band intensified. (To facilitate printing and speaking, the frequency of a station was divided by 1,000 and given in units of kilocycles; whereas people resisted 300,000 cycles as a frame of reference, they were comfortable with 300 kilocycles.) As stations were increasingly spread across the broadcast band in 10 kc. intervals, it was relatively easy to indicate a station assignment in kilocycles, since all that was necessary was three or four numbers (no decimals) and, since the stations were separated in multiples of 10 kc., the last number was always a cipher. In contrast, wavelength listings had to be expressed in three figures plus a decimal point. Newspapers and radio stations thus began conveying radio station frequencies to the public in kilocycles rather than wavelengths. The corresponding wavelength assignment first appeared in parentheses alongside the kilocycle designation; then, it was dropped entirely. How much easier it was for an announcer to say "we are broadcasting on a frequency of 1,200 kc." than to state "we are on a wavelength of two hundred and ninety one and one-tenth meters."

As single dial tuning became the vogue, a major goal of radio manufacturers became that of making the numbers on the dial correspond to the frequencies of broadcast stations. The Ferguson 8 (Ferguson Radio Corp.) was one of the earliest receivers to be marketed with a calibrated dial; however, the calibration was in meters, and by this time, 1925, listeners were becoming accustomed to reckoning their dial settings in kilocycles. To meet consumer demand for dials calibrated in kilocycles, and to make it really easy to find stations on the dial, American Bosch introduced in 1927 the Cruiser series. Each model in the series featured an illuminated dial calibrated both in arbitrary numbers, from 0 to 100, and in kilocycles.

Surveys and retailer feedback revealed to radio manufacturers that listeners on the average did not take time to tune their receivers properly. Often they blamed receivers for poor performance when in fact the problem was their inaccuracy. To overcome this dilemma, radio manufacturers turned to various forms of tuning facilitators. On the one hand, they began experimenting in the late 1920s with techniques of automatic tuning. Montgomery Ward, especially, pioneered a mechanical form of automatic tuning, which was popular in the 1930s. The tuning dial was rotated to a predetermined setting with a lever or disc similar to that in a telephone dial. On the other hand, manufacturers also developed in the 1930s a host of different means of visually indicating proper tuning, including "magic eye," "target," "shadow," and "colorama" tuning. Later, about 1938, various radio manufacturers used electric motors to facilitate automatic tuning. The motor was activated by pressing pushbuttons to obtain rotation of the dial to desired stations. Although electric motors were featured in a few table models, most of the receivers with electric tuning were deluxe consoles, of which the RCA 811K is prototypical. And along with the development of automatic tuning, radio manufacturers were able to offer more affluent buyers, as an option, a remote tuning attachment.

Automatic Volume Control (AVC)

When single dial tuning became commonplace, its significance as a factor in simplifying the operation of radio broadcast receivers was limited because listeners still had to use two hands as they adjusted their receivers to different stations. Receivers designed to be sensitive to weak signals invariably had too much audio gain for powerful local stations. For this reason, broadcast listeners had to place one hand on the tuning dial and the other on the volume control in order to prevent "blasting" when tuning from a weak to a strong station. After radio engineers solved the sensitivity problems of early receivers, they began to recognize the need for some sort of automatic volume control.

A solution escaped them, however, until Harold A. Wheeler, a graduate student at Johns Hopkins

University, who was in his first year of employment at the Hazeltine Corporation, invented an automatic control circuit early in 1926. Wheeler discovered a way of using the detector stage of a receiver to control "automatically" the RF amplification of the stages preceding it.

Radio engineers, at the time, nonetheless, were preoccupied in making the transition from battery to ac receivers, and they underestimated the intensity of consumer desire for automatic volume control. But in 1928 the Philco Company, eager to strengthen its foothold in the marketplace, contracted with the Hazeltine Corporation for design of a state-of-the-art broadcast receiver. Wheeler's AVC circuit was incorporated in the new design. The receiver, introduced to the public in 1929 as the Philco 95 series, was an instant success (Holland and MacDonald, 1929). The AVC action in the Philco 95 provided automatic reduction in audio gain for strong signals and noticeable increase in gain for weak signals. It was unnecessary to touch the volume control once it was adjusted. The development of AVC thereby brought true single dial operation to radio listeners. It is easy to understand why, by the time a patent was issued in 1932 for Wheeler's invention, virtually every receiver on the market employed his AVC circuit.

LISTENING CONVENIENCE

In the early days of radio manufacturing, loud speakers were made and sold independently of receivers, even after the marketing of AC receivers. When the first radio broadcast receivers reached retailer showrooms, the only method of hearing station programs was by use of a "head phone," which hung from the end of a long cord. After loud speakers were placed on the market, receivers were designed so that head phones and loud speakers could be used interchangeably. Several manufacturers provided separate jacks on the front panel for each audio stage of the receiver. Listeners were encouraged to plug head phones into the lower-power stages, thus saving battery energy by cutting off the subsequent audio stages. Eventually, in the late 1920s, head phone jacks disappeared from radio broadcast receivers. Nonetheless, loud speakers still dangled at the end of a cord, since, even after the introduction of ac operated models, loud speakers were made and sold independently of receivers.

The earliest forms of loud speaker employed a single head phone as a driving unit for several varieties of long, exponential (curved) horns. Magnavox was among the first of the manufacturers to improve upon the horn speaker, marketing in 1924 a balanced armature magnetic-type. Next to emerge in the marketplace were cone type loud speakers, whereby a paper cone was driven at its apex or center by a balanced magnetic unit. The earliest cone or "hornless" loud speaker was made up of two paper cones, but by the mid 1920s they had been replaced by single cone loud speakers, which were constructed in a variety of metal, wooden, and plastic housings.

General Electric engineers in 1925 developed a "dynamic" loud speaker which used a moving coil to drive a paper cone, the edges of which were clamped to a fixed rim. The edges of the cone were made of soft, flexible material, which permitted the entire cone to vibrate. RCA was first to market the new, dynamic loud speaker, as Model 104, for use in its ac versions of the Radiola 25 and Radiola 28. As a consequence of patent restrictions, however, which persisted through the late 1920s, radio manufacturers other than RCA were unable to utilize in their models the vastly superior dynamic loud speakers.

TRENDS IN CABINET DESIGN

Manufacturers of radio broadcast receivers in the 1920s conceptualized what their products should look like as they went along. An early practice with both crystal sets and vacuum tube receivers consisted simply of fastening the parts down to a board to hold the connectors in place. Furthermore, at the beginning of radio manufacturing, tubes, batteries, and sound reproducers were regarded as separate components, which were to be sold individually. For example, in 1922 an inexpensive regenerative broadcast receiver might have cost only $39, but tubes and batteries might have been another $25, and prices for a sound reproducer might have ranged from $3 for head phones to $35 for a horn speaker. Add $10 to cover antenna materials, and an average listener might have expended from $97 to $129 for a fairly basic, no frills, broadcast receiver. But that might not be the end of it! A suitable table on which to place the receiver also had to to be obtained; its functional value and cost were increased when it provided space beside the receiver for a horn speaker and a shelf underneath for batteries.

The breadboard sets, however attractive their exposed parts might have been to radio enthusiasts, were annoying to those in the household responsible for dusting them. The majority of radio manufacturers, therefore, turned to a bewildering variety of square or rectangular boxes and cabinets in which to house their wares. These early designs were more important for their functional than for their stylistic considerations. The first consoles, for example, were relatively large. Batteries in the earliest models required considerable space. A compartment in the bottom of the cabinet, usually reached from the rear, was provided for the batteries, and it was made high enough so that numerous connections could be made easily when batteries had to be changed.

The introduction of wooden cabinets led to a series of design changes in the front panel of early receivers. The initial models placed only the control devices on a bakelite or wooden panel—other components like coils, transformers, and tube sockets were placed on a base, which might also be bakelite or wooden. Next came sets where the panel was metal and the base was omitted, which meant that all components were mounted on the metal panel. This stage of receiver design simplified servicing and construction, and hundreds of thousands of sets of this type were built and sold. Then, desire to conceal unsightly wiring led radio manufacturers to return to the bakelite or wooden base method of mounting parts.

Eventually, the economics of manufacturing in the late 1920s resulted in a reduction in hand assembly procedures, which in turn, led to the development of the punched steel chassis, where all the wiring could be contained inside the base. By this time, the metal front panel, with its standard three-inch dials, rheostat knobs, and sound reproducer jacks, was giving way to radio receivers with decorative wood panels. Radio manufacturers endeavored thereby to provide their receivers with controls that so blended with the decorative scheme of the panel that they were relatively inconspicuous. Indeed, the standard three-inch tuning dial and its associated knobs disappeared in many receivers, and in their place a variety of ingenious controls emerged in the form of drums or rims which protruded through decorative slots, as in the Grebe MU-1. Some manufacturers placed the tuning dial(s) behind an engraved escutcheon. By tuning a small knob on the panel, listeners could see the frequency of the station to which they were tuned through a window in the escutcheon.

Contemporary radio collectors generally marvel at the extraordinary quality that many early radio manufacturers built into their receivers. Radio engineers in the 1920s believed that they could overcome limitations in sensitivity, selectivity, and fidelity by rigidly mounting parts, matching variable condensers, anchoring wiring, and firmly shielding compartments. For example, the chassis of many of the 1927 FADA neutrodyne receivers were made of 1/16 pressed automobile body steel. Each chassis was supported on a three-point suspension to enhance rigidity, and thereby, to prevent the parts mounted on it from slipping out of alignment. The FADA tuning condensers were matched to an accuracy of 1/8 of one percent, and the shaft on which the tuning drums and variable condensers were mounted were gauged to a tolerance of .0005 of an inch. An accuracy of one percent in capacity throughout the tuning range was considered satisfactory (Rider, 1927).

An emphasis on sturdiness in 1928-29 led many prominent radio manufacturers, including Atwater Kent and Crosley Radio, to house many of their receivers in all metal cabinets. The necessity of providing access to unsightly batteries was gone, and now, "small" table receivers could be fitted into a living room in such a way as to add to the decorative scheme. Radio manufacturers seldom made a pretense at simulating furniture with their metal cabinets, but they did try to ensure that they were in harmony with floor lamps and other metal fixtures.

Wooden cabinets emerged as an important aspect of model design in the late 1920s when radio retailers began to purchase receiver chassis and speakers from different manufacturers, and then, to combine them in cabinets for resale to buyers. Many of the sets sold in 1928 were mounted in floor-type, console cabinets, but neither sets nor loud speakers were designed for the cabinet. Radio manufacturers, however, swiftly picked up on the trend toward housing receivers and speakers in one unit, and by 1929, the practice had become commonplace.

The average broadcast receiver of the 1920s was relatively high priced and was more or less a luxury item for the household. Standards of performance were largely nonexistent at first, and the fact that a receiver could copy distant as well as local stations with a certain degree of reliability rendered it satisfactory and acceptable to most consumers. Progress in the technical dimensions of radio manufacturing, however, resulted in a steady increase in standards of performance. By the late 1920s and early 1930s acceptable standards had been attained for sensitivity, selectivity, and fidelity, and manufacturers could not deviate very far from these standards and still expect to have a marketable broadcast receiver. A basic configuration had emerged in the radio industry. It was the alternating-current, tuned-radio-frequency set, it employed no batteries, it was self-contained with the necessary power supply, and it used a single control tuning system (Felix, 1928). People thus began to choose among receivers on the basis of cabinet design. A representative of the industry put his view of the significance of cabinets in this way:

> Each new cabinet, in a sense, is an experiment in art, and it will succeed or fail, partly on the excellence of the equipment which it houses, but to a much greater degree because of the discriminating taste of the buyer in choosing, from an artistic standpoint, the furnishings for his home. (Langley, 1929, p. 71)

The emphasis on cabinets led radio manufacturers to market cabinets that were styled to follow the general trend of furniture design. Sometimes they camouflaged their models as pianos, ice boxes, reading lamps, tables, treasure chests, etc. These efforts were inspired initially by desire to hide the receiver, and later, by interest in making the receiver stand out in a novel or dramatic fashion. Affluent households demanded deluxe models, encompassing ornate "modern" and "period" styles. Middle-class consumers, too, began to expect cabinet styles that would complement the furniture in their homes. Various shapes and sizes of cabinets thus emerged from design studios to harmonize with any household setting regardless of use, whether in the bedroom, kitchen, or living room. Furthermore, in the 1930s, the styling of dials became integral to the styling of cabinets. Some dials were made very large and very readable, and station names were placed on them for added ease in locating desired stations.

Since the main purpose of a receiver is to reproduce the original signal of a broadcast station as accurately as possible in the listener's home, the issue of sound reproduction or fidelity also became an important feature of cabinet design in the 1930s (Carson et al., 1937). Radio engineers recognized proper control of the acoustic properties of the cabinet, in which the loud speaker and chassis were placed, to be an important factor in sound reproduction. They saw, too, that proper acoustics could not be obtained with a conventional open back cabinet. The acoustic properties in such cabinets varied in accordance with their placement near walls and surrounding objects. Radio engineers, therefore, developed ways of closing the back of the cabinet, and by 1938, many models appeared with single or multiple tone chambers, acoustic labyrinths, and high frequency horns. As the pre-World War II era of radio manufacturing drew to a close, advances in cabinet design enabled listeners, who could afford the deluxe receivers, an exciting new dimension in high fidelity sound reproduction.

References

Carlson, W. L., Holmes, R. S., & Wunderlich, N. E. (1930). The engineering behind the R. C. A. Radiola superheterodyne. *Radio News, 12,* 398-400, 464.

Carson, B. R., Chittick, K. A., Cole, D. D., & Perry, S. V. (1937). New features in broadcast receiver design. *RCA Review, 2,* 45-59.

Felix, E. H. (1928). An interview with J. V. L. Hogan. *Radio Broadcast, 12,* 353-354.

Holland, W. E., & MacDonald, W. A. (1929). The Philco 95 screen-grid plus. *Radio Broadcast, 16,* 111-112.

Langley, R. H. (1927). Meters, kilocycles, or "channel numbers"? *Radio Broadcast, 12,* 18.

Langley, R. H. (1929). The trend in cabinet design. *Radio Broadcast, 15,* 69-71.

Lescarboura, A. C. (1924, May). "This is station———": Putting radio programs on the air and taking them out of the air. *Scientific American,* pp. 302-303, 338, 354-358.

Morecroft, J. H. (1927). *Principles of radio communication.* New York: John Wiley & Sons.

Rider, J. F. (1927). Facts about the FADA "special" set. *Radio Broadcast, 12,* 128-130.

Rost, O. F. (1938, October). The radio industry's job. *Radio Retailing,* pp. 30-31.

Wolkonowicz, J. (1977, December). The Radiola superheterodyne story. *Radio Age,* pp. 1-3.

CHAPTER

4

HOW TO ESTABLISH RECEIVER YEAR OF MANUFACTURE

How old is it? What year was it produced? When was it was manufactured? Queries about the age of a radio broadcast receiver predominate whenever collectors of radio broadcast receivers come across an unfamiliar model.

When the model is not among the thousands identified in *The Radio Directory and Price Guide,* we need not despair—a reasonably good estimate of the year of manufacture can be attained. The information presented in Chapter Four enables anyone—novice, seasoned collector, or antique specialist—to establish when a receiver, which was manufactured between 1921 and 1941, first appeared in the market place.

Three sets of facts provide corroborative evidence for identifying when a radio broadcast receiver was manufactured. The year(s) in which: (1) its tube complement was first introduced; (2) its major, readily observable control functions and features were widely adopted; and (3) its cabinet style first achieved popularity. As the detailed commentary of Chapter Four reveals, the periods during which particular characteristics of radio broadcast receivers became prominent may be matched to establish a good estimate of year of manufacture.

BY TUBE COMPLEMENT

The broadcast receiver will have been manufactured no earlier than the year in which its "newest" tube was introduced to the radio manufacturing industry. Since tube types were generally superseded a scant few years after manufacturers adopted them, reference to Table 2 and Figure 1 permits specifying, within a tolerable margin of error, the date any given broadcast receiver was manufactured. (Consider, for example, how tube complement serves to date the RCA model 120 radio broadcast receiver. The set employs tube types 2A7, 2B7, 2A5, 58, and 80. The type 58 was introduced in 1932 and the type 80, a few years earlier. The types 2A5, 2A7, and 2B7 appeared in 1933. The latter suggest that the RCA model 120 was manufactured in 1933. Our directory shows that the model 120 was in fact manufactured in 1933.)

The historical development of the vacuum tubes shown in Table 2 divides chronologically into two periods. The first extends up to 1927, the second emerges when tubes designed for battery power gave way to a new generation of tubes produced for ac operation.

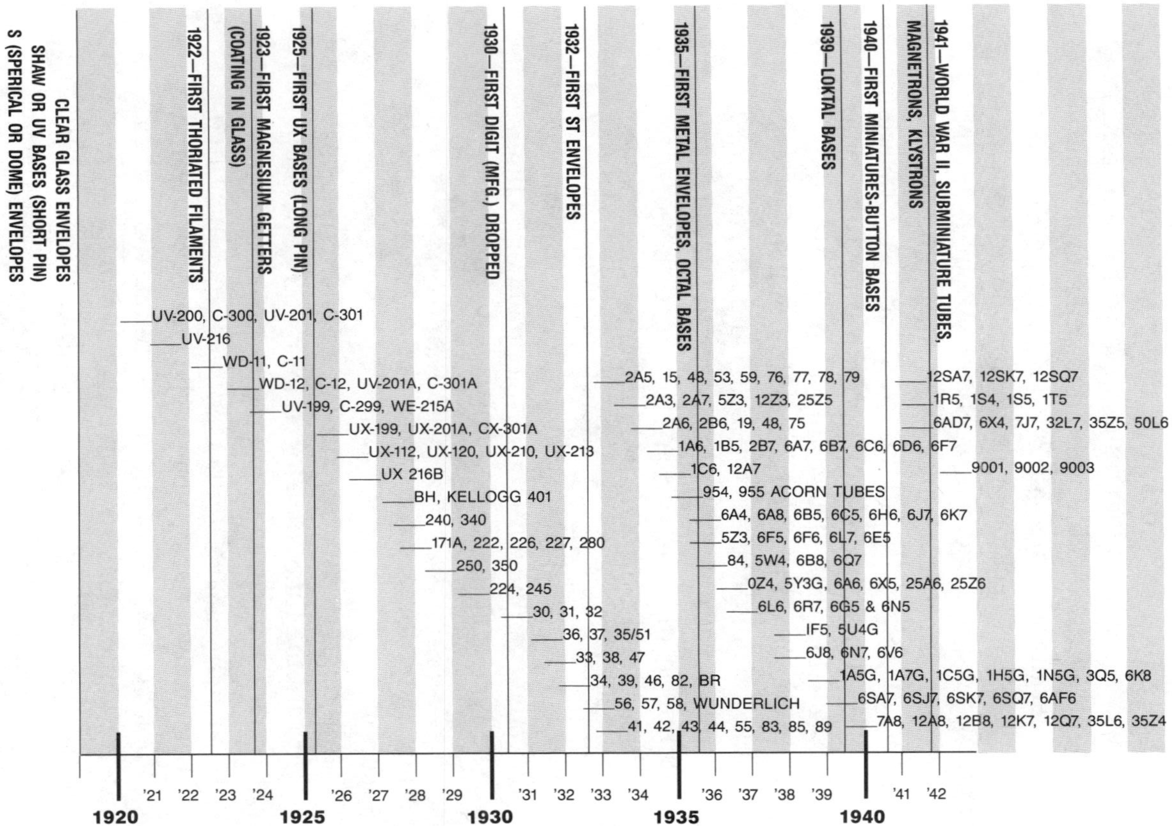

Table 2: Vacuum Tube Development, 1921 - 1941

Early vacuum tubes were costly and unreliable; however, with the stimulus of World War I, tube manufacturing techniques improved rapidly. Shortly after the war, the type 200 and the type 201 tubes (Figure 1.A) became available for use in manufacturing radio broadcast receivers. About 1922, the General Electric Company began producing the type 199 (Fig. 1.B) and the 201A (Fig. 1.A, C) while the Westinghouse Company introduced the type WD-11 (Fig. 1.B). These tubes, among the first to have thoriated filaments, required much less filament current than the types 200 and 201, and they enabled manufacturers to reduce the battery consumption of their sets by immense proportions. Subsequently, the demand for higher loud speaker output led tube manufacturers to produce types 112 and 171 (Fig. 1.C). Then, elimination of "B" batteries became possible with the development of rectifiers. The type 280 (later type 80) rectifier tube, which was released in 1927, was destined to be the mainstay for several years of receiver power supplies.

Although alternating current operation was feasible with the tubes developed for battery operation, further advances in vacuum tube development led to a new generation of tubes manufactured specifically for ac receivers. The types 226 (26) and 227 (27), the first of the mass-produced ac tubes, were announced in 1927. Radio manufacturers rushed accordingly to redesign their receivers.

The first screen-grid tube, type 222, was also introduced in 1927, but it was designed for dc filament operation. Manufacturers, now interested

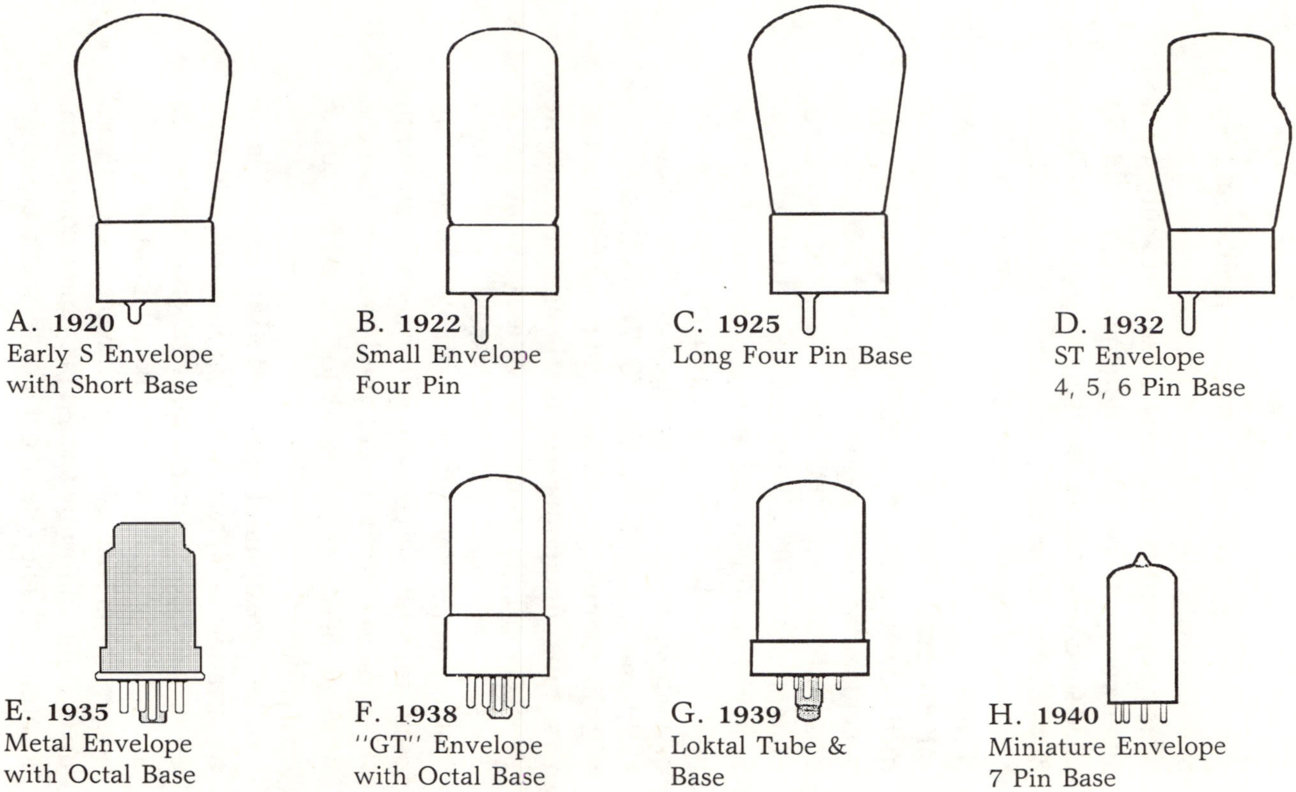

Figure 1: Vacuum Tube Configuration

(Not to Scale)

primarily in receiving tubes which could be operated with alternating current on the filaments, were unmoved by the new screen-grid tube until an ac version, type 224 (24), appeared in 1929. By 1930 superheterodyne receivers employing the type 224 or 24 in RF and IF amplifier stages were clearly out-performing the best of the conventional TRF receivers of the 1920s.

Between 1931 and 1933 manufacturers began to turn out automobile receivers using such newly developed ac or dc 6-volt tubes as the 36, 37, 38, and 39. The year 1932 also witnessed the introduction of the ST shaped bulb (Fig. 1.D). The new bulb or envelope design evolved to strengthen the internal structure of receiving tubes. Development of a whole series of highly efficient tubes followed, including types 56, 57, 58, 76, 77, and 78.

Radio manufacturers began to produce low cost, five tube (including rectifier) superheterodyne receivers in 1933 with the new types 2A7 and 6A7. These tube types, identical except for filament voltage, combined within a single envelope an oscillator and a screen-grid mixer section, and with them, radio manufacturers could greatly reduce the number of tubes required for superheterodyne operation.

Tube development appeared to be approaching a plateau in 1934, but in fact, the base and envelope of virtually every forthcoming receiving tube was about to change. Most radio manufacturers in 1935 adopted octal base tubes, which were designed with up to eight uniformly spaced pins and a convenient locating lug. The line of octal base, metal envelope tubes (Fig. 1.E), which replaced those with the ST shaped glass envelope, included the 6J5, 6J7, 6A8, and 6F6. Within a year or two after the widespread adoption of the metal types, a host of glass envelope equivalents emerged. This new line was designated on tubes by either the suffix "G," that is, 6J5G, 6K5G, and so forth, which corresponded to the ST shape, or the suffix "GT" (Fig. 1.F), that is, 6A8GT, 6F6GT, which were about the same size as their metal counterparts. The metal tubes and the G series gradually disappeared from use in the early 1940s, but the glass envelope, octal base GT types survived into the postwar years.

The 6L6, a relatively high output, high efficiency, and low distortion audio output tube proved to be one of the major accomplishments of the metal tube era. It was the product in 1936 of extensive laboratory work, and it was followed in 1937 by the 6V6, a replica of smaller physical size and lower output.

By 1939, as competition from the GT clones diminished use of metal tubes, a new line of low-loss, efficient glass tubes with a "loktal" base (Fig. 1.G), including the 7A4, 7A5, 7B4,and 7C5, was made available to the radio manufacturing industry. Almost simultaneously, in 1939, the development of miniature glass tubes for battery operation inspired a new generation of truly portable receivers. The new types 1R4, 1R5, 1S4, 1T4, and so forth, were enclosed in small glass envelopes with a new, 7-pin glass base (Fig. 1.H). The miniature vacuum tubes were produced during World War II in a wide variety of filament voltages, and, except for the larger, higher power tubes, served as the standard for postwar receivers until the advent of the solid-state era. Their structure was small, and the efficiency provided by their short leads made them ideally suited for forthcoming developments in high-frequency FM broadcasting.

BY CONTROL FUNCTIONS AND FEATURES

Many readers will find it helpful, in establishing receiver year of manufacture, to review the discussion of radio broadcast receiver development in Chapter 3. Only the readily observable, easily identified control functions and features, which were common among receivers that were manufactured between 1921 and 1941, are described here. This information, along with that pertaining to cabinet styles below, is especially useful in establishing relatively quickly the period during which a given receiver became prominent in the marketplace.

CONTROL FUNCTIONS

Dials

Composition (1922-1927):

Round, 2-4 inches, large flange, fluted knob. Molded bakelite, polished black finish with scale markings in contrasting white.

Window:

Small vision (1927-1929)

Translucent drum or dial strip inside cabinet Often illuminated. Includes embossed escutcheon plate.

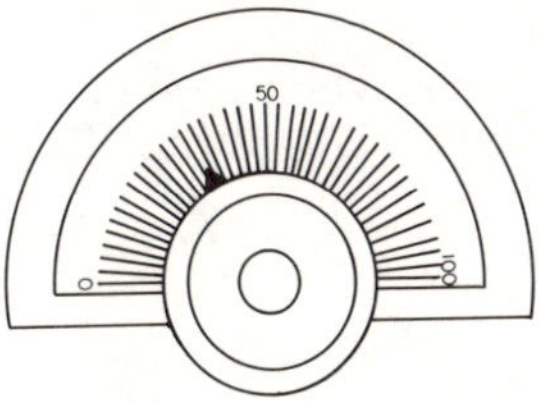

Full vision (1930-1932)

Airplane:

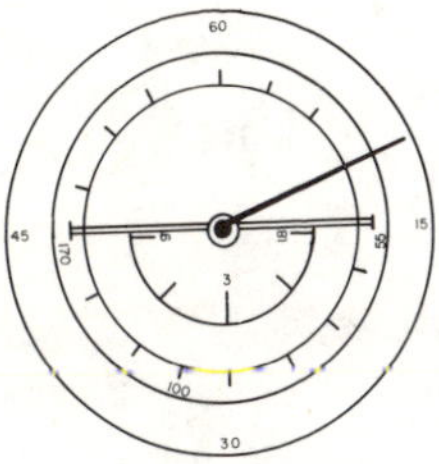

Small diameter (1933-1941)

Large diameter (1935-1941)

Round dial over which a pointer rotates. Term comes from similarity to an aircraft instrument dial.

Slide rule (1938-1941):

Pointer moves in a straight line across a rectangular scale resembling that of a slide rule.

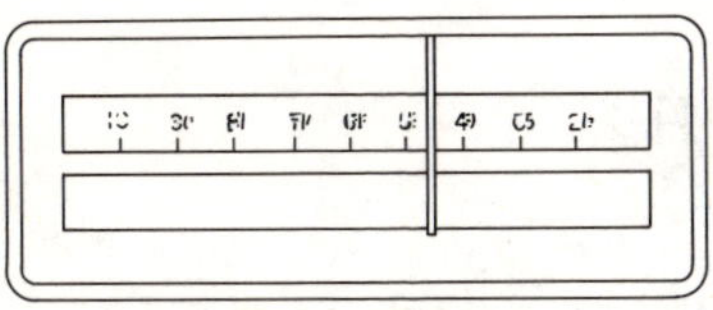

Tone controls (1930-1941):

An additional knob (or two) on the front panel that is similar to the receiver tuning and volume control knobs.

Pushbutton tuning (1938-1941):

"Buttons" on the front panel which, when pushed, automatically tune a receiver to given stations.

FEATURES

AC operation (1927-1941):

Linecord and ac plug will be prominently visible.

Calibration in kilocycles (1928-1941):

Dial will be calibrated from lower to upper limit of broadcast band in kilocycles rather than in arbitrary (0-100) numbers.

Built-in speakers (1930-1941):

Cabinet will be of sufficient size to house loud speaker; "grill" area for sound output will be prominent.

Shortwave bands (1933-1941):

Dial will be calibrated in both broadcast and short-wave frequencies; switch on panel for bands.

Tuning indicator (1935-1941):

A device on the front panel that indicates when a receiver is tuned accurately to a station. (Most popular device is a "tuning eye.")

Loop antenna (1923-1927, external):

An antenna consisting of one or more complete turns of wire, which will be wound on a rectangular, wooden frame. Generally will be separately mounted.

Loop Antenna: (1939-1941, internal):

An antenna consisting of one or more complete turns of wire, which will be wound either on a wafer frame mounted inside the cabinet or along the inside walls of the cabinet.

BY CABINET STYLE

Manufacturers of radio broadcast receivers devised three basic housing configurations for their models: table, console, and portable. The table set was constructed to be set on a table, mantle, or shelf. The console set was built to be freestanding. The portable was designed to be carried about, supported by anything that would bear its weight, and operated anywhere, in or outside of buildings.

Although the table, console, and portable models represent basic style categories, within each of these broad classifications, a great many different style variations arose. Descriptions and illustrations of the more significant of these style variations, including notations of the years during which they were prominent, are presented in Figure 2 and detailed below.

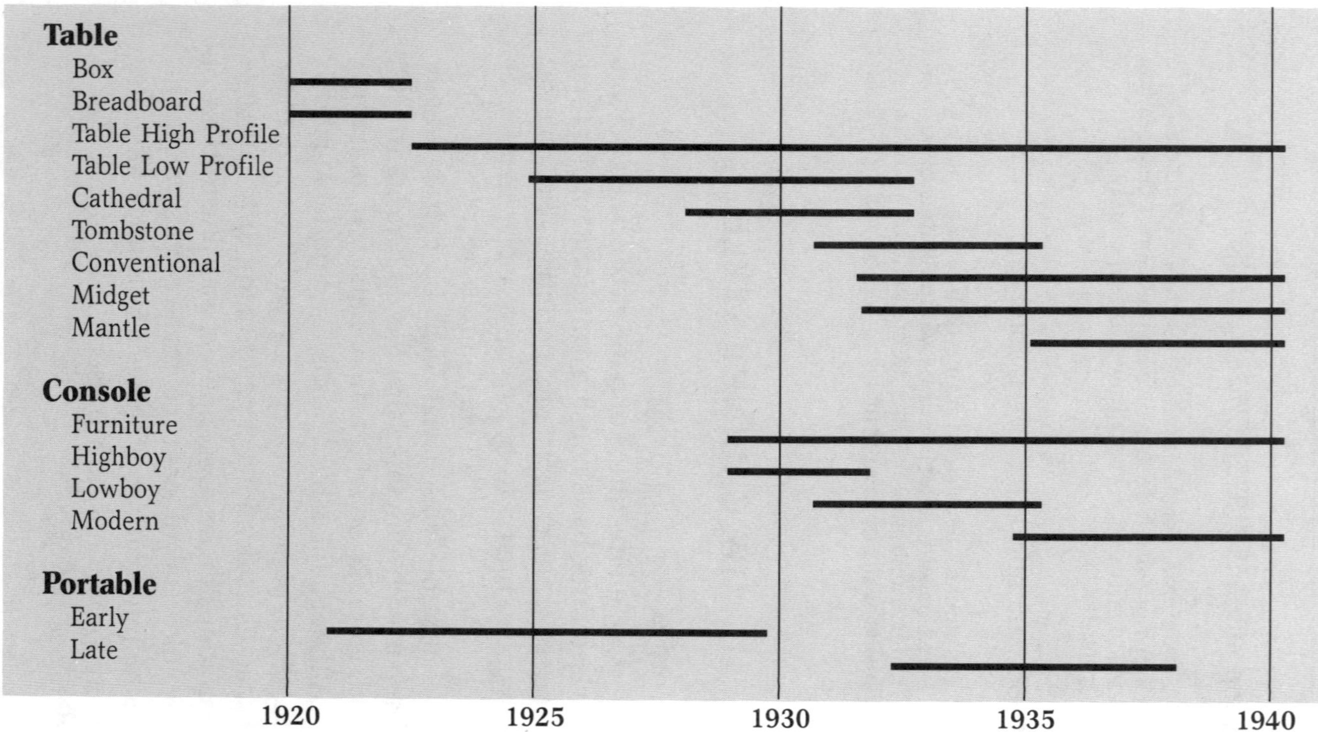

Figure 2: Cabinet Styles 1921-1941

Table Styles

Box (early 1920s)
The earliest radio broadcast receivers of the 1920s tended to be built in rectangular wooden boxes with bakelite panels. Tubes often protruded either vertically or horizontally to allow visual inspection of filament glow.

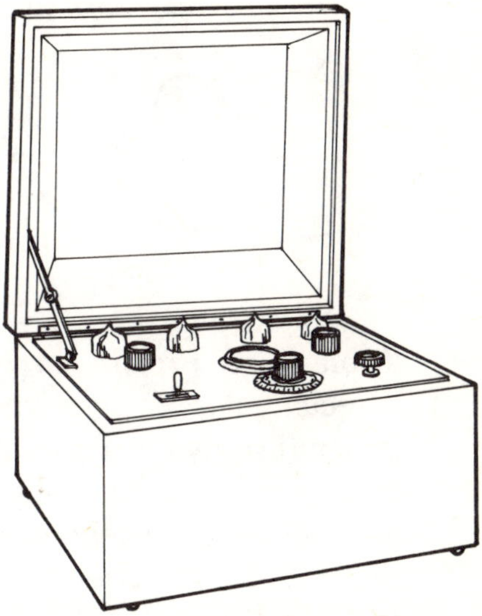

Breadboard (early 1920s)
The "breadboard" designation stems from the practice of early manufacturers, notably Atwater Kent, of assembling receiver components on a flat, wooden surface resembling a breadboard. The manufacturer thus conveniently avoided entirely the relatively complex considerations of cabinet design.

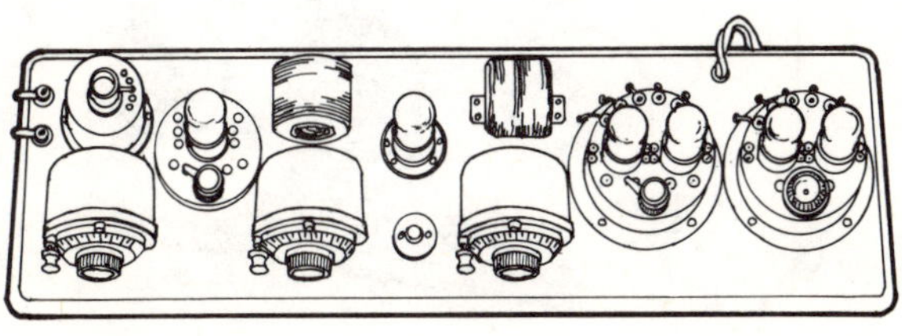

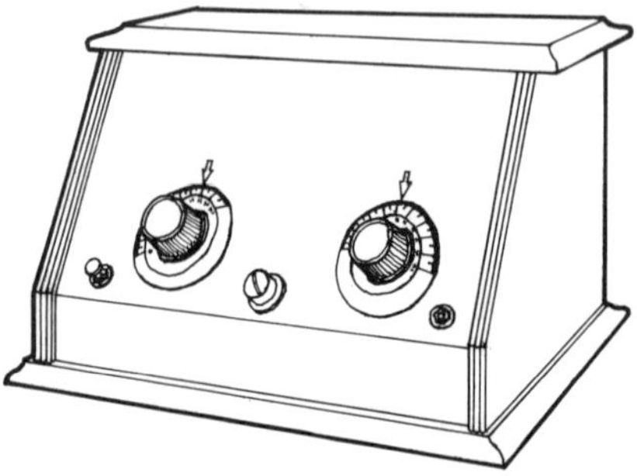

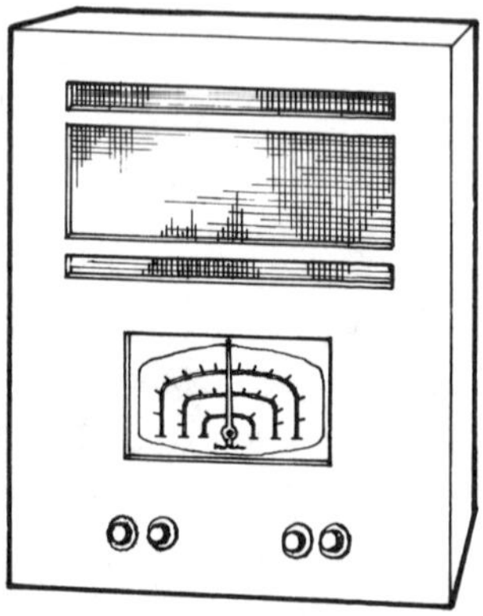

Table High Profile (early 1920s to 1940s)
A rectangular receiver where the height dimension is significantly greater than that of length. Although high profile cabinets appeared in both the 1920s and the 1930s, receivers of the two eras are easily distinguished by their tube complements and control functions.

Table Low Profile (mid 1920s to early 1930s)
A rectangular receiver where the length dimension is significantly greater than that of height. The table low profile category encompasses the popular "three dialer" of the 1920s. This receiver unit was

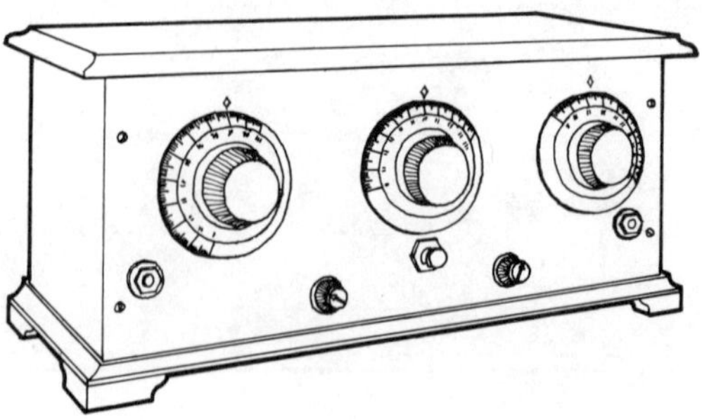

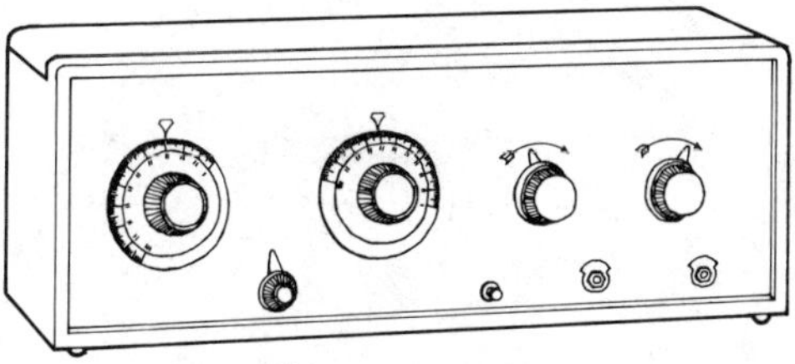

generally housed in a handrubbed waxed or varnished, "artistic" oak or mahogany cabinet with a hinged top. Tubes were usually hidden from view by either a vertical, or a sloping vertical, bakelite, engraved panel. As the illustrations show, the familiar cabinet style of the the "three dialer" held its popularity among radio designers even after technical advances in tuning progressed from external dials to the small window dial.

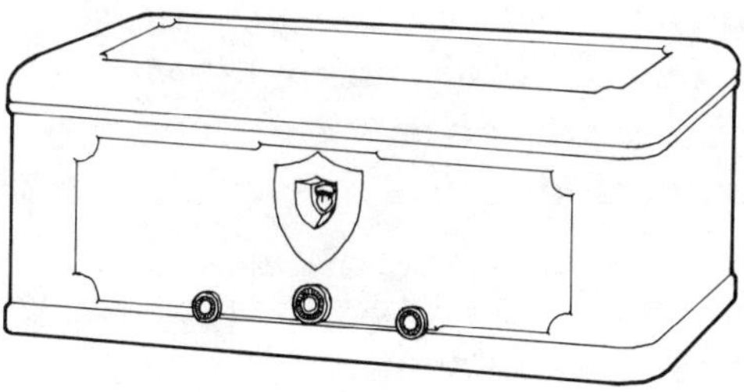

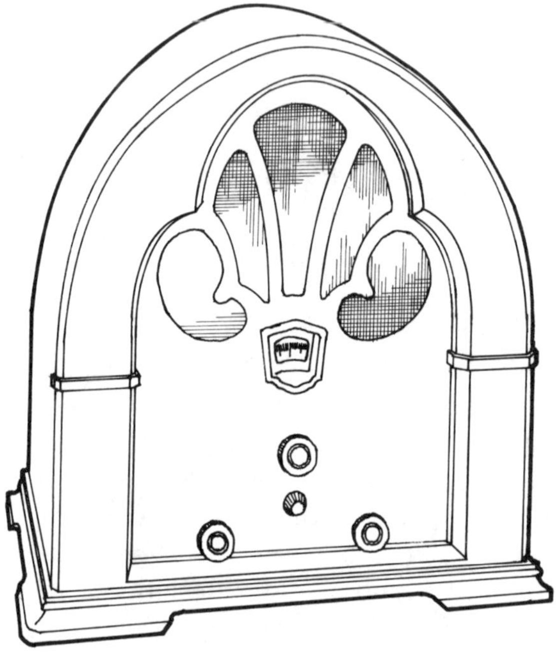

Cathedral (late 1920s to early 1930s)
Also known as *Gothic* or *Upright*. This style of table model cabinet is distinguished by its rounded or curved top. It was always a high profile receiver; its prototypical dimensions were about 20″ x 15″ x 15″.

The cathedral model and its many variants arose when radio manufacturers began to combine speaker and receiver units. They were forced to cast about for a suitable cabinet style, and their inspired designers created radio broadcast receivers in the image of significant architectural forms.

Tombstone (early to mid 1930s)

Also known as *Upright*. This style of cabinet is similar to that of the cathedral except that its top was relatively more flat than curved. It was also a high profile receiver with the same general dimensions as the cathedral.

Radio manufacturers adopted "upright" as their designation for this high-profile, squared-off, cabinet style; however, product reviewers for the radio retail industry renamed it "tombstone" in the 1930s when they noted its physical resemblance to the markers commonly used in graveyards.

Conventional (early 1930s to 1940s)

Table models that do not fit into one of the distinct cabinet styles noted above may be referred to as "conventional." Such cabinets ranged roughly in height from 7 to 15 inches, in width from 7 to 18 inches, and in depth from 6 to 11 inches. Conventional cabinets are designated in the directory (*see* Part II) simply as "table" styles.

Midget (early 1930s to 1940s)

Also known as *Compact*. A designation reserved for receivers of small proportions. The parameters of the designation changed as manufacturers acquired the technical proficiency to produce smaller and smaller table styles.

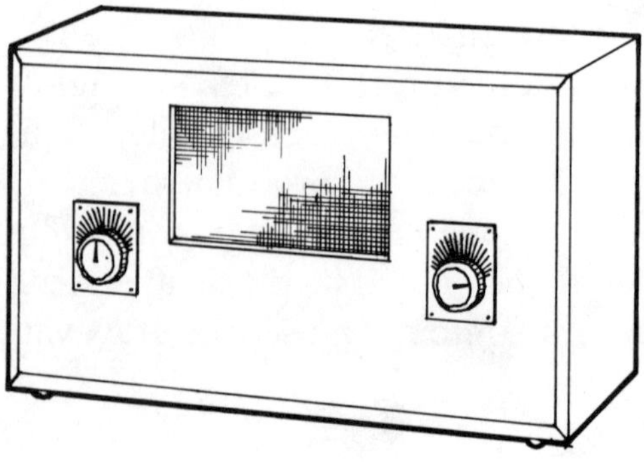

Radio manufacturers began to use the term "midget" about 1930 to describe an assortment of different styles, including small cathedrals and tombstones. A broadcast receiver 15 inches in height, 15 inches in length, and 10 inches in depth, for example, would have been regarded between 1930 and 1934 as a midget model. By 1939, a receiver probably would not have been designated as midget if its largest dimension exceeded about six or seven inches. At this point in time, a receiver had to be half the size of a mantle model (see below) to be considered a midget.

We endeavored to use the midget or compact designations in the directory (*see* Part II) as they would have been used by radio designers in the early 1940s (each dimension being about 6 inches or less). In a few instances, however, where manufacturers during the early 1930s had advertised their receivers as either "midget" or "compact," we elected to adopt their designations.

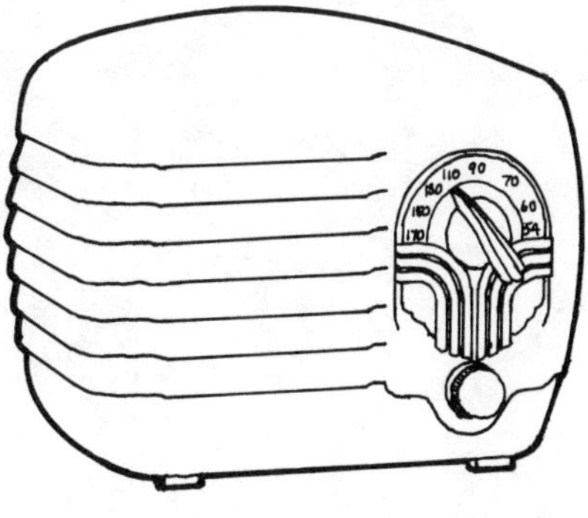

Mantle (mid 1930s to 1940s)

This style of cabinet may be differentiated from other table styles (except midget) by its small size. The radio industry never adopted standardized H x W x L criteria for its mantle designation; however, to facilitate receiver identification, the designation is used in the directory when each dimension of a model (height, width, and length) is less than 10 or 11 inches but greater than 6 inches.

The "mantle" designation arose to describe receivers that could be moved easily from shelf to table or from room to room. The mantle model met the depression period need for low priced, economy receivers. It was mass produced with minimal features and technical refinements. Manufacturers thus turned to distinguishing their mantle models cosmetically in order to stimulate buyer appeal. As a consequence, mantle sets constitute some of the most distinctive receivers of the 1930s. Many of these novelty sets are today highly prized collector items.

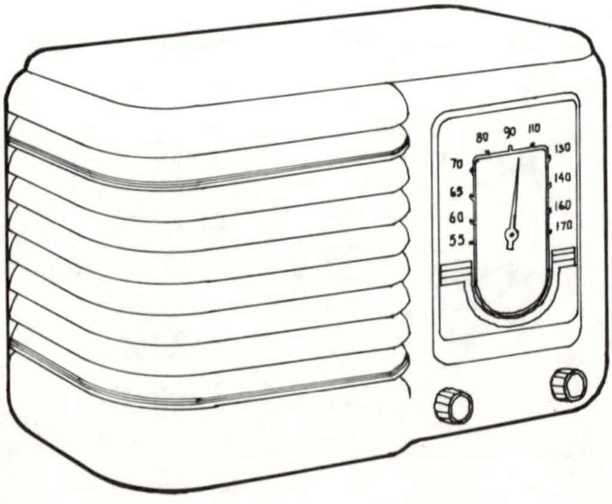

Console Styles

The freestanding console style first appeared in the mid 1920s. It ranged in size from that of a table model set simply on four slender legs to that resembling a large piece of furniture such as a desk, secretary, or piano.

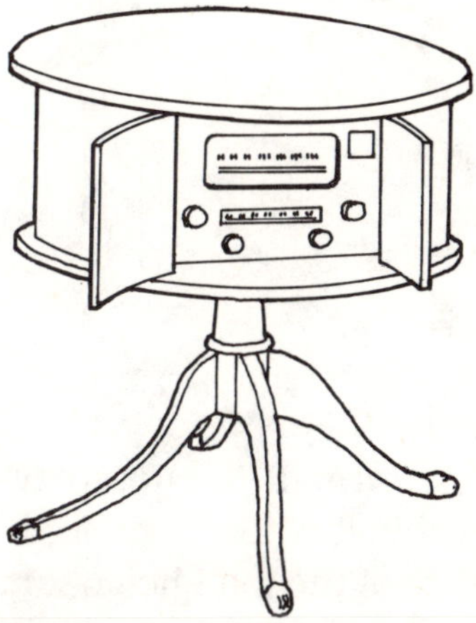

Console furniture (late 1920s to 1940s)
A console style designed to double as a piece of household furniture—a grandfather clock, desk, piano, chest, bureau, or table. The year of manufacture of such a receiver is distinguishable primarily by its tube complement and control functions.

Console furniture styles appeared relatively frequently in the guise of myriad forms of table—end, coffee, cocktail, chairside, conventional, etc. A receiver housed in a console table served a dual function. Radio listeners could tune in a station, close the cabinet, and thereby, use the radio cabinet as a table even while the receiver was operating.

Console Highboy (late 1920s to early 1930s)
A console style cabinet whose leg length comprised 50 percent or more of the total height of the cabinet would have been designated highboy in the late 1920s and early 1930s. Radio manufacturers adopted a variety of leg styles—turned, Queen Anne, Hepplewhite, etc.—by which to distinguish their console highboys.

Console Lowboy (early 1930s to mid 1930s)
A console style cabinet whose leg length comprised less than fifty percent of the total height of the cabinet would have been designated lowboy from the early 1930s to the mid 1930s.

Console Modern (mid 1930s to 1940s)

A freestanding console style in which the body of the cabinet extends to the floor. The design of the "modern" console displaced the distinctive leg styles of the earlier console styles. Manufacturers designated occasionally the console modern style as either highboy or lowboy; however, contemporary collectors tend to reserve the "highboy" or "lowboy" terminology primarily for the earlier consoles.

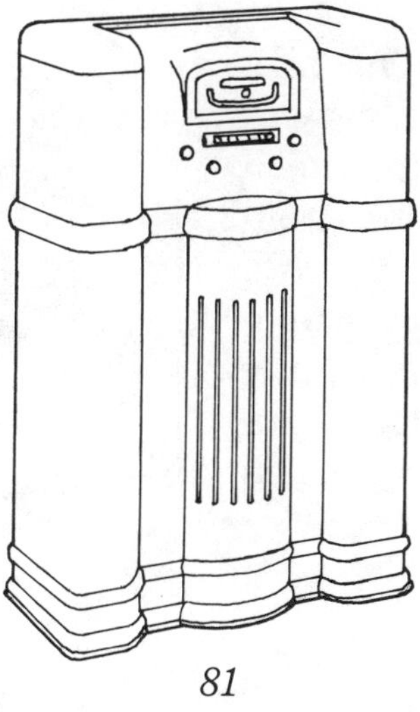

Portable Styles

A style of radio broadcast receiver that is designed to operate from internal sources of power, to be easily and safely transported from one site to another, and to function consistently under different environmental conditions. Portable receivers captured the interest of radio listeners at the outset of the manufacturing era.

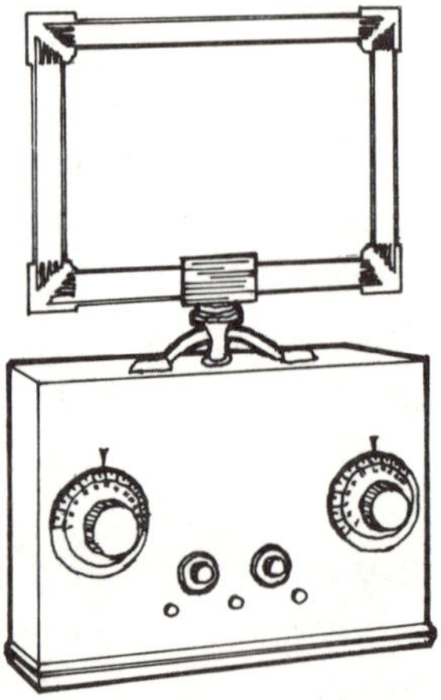

Early Portables (1920s)
The portables of the 1920s were usually enclosed in cabinets the size of small suitcases. An antenna usually swung out from the cabinet or assembled above it during operation.

Late Portables (mid 1930s)
When low current, high gain tubes were devised in the mid 1930s, which enabled operation from small batteries and with built-in antennas, manufacturers were finally able to produce lightweight, easily transported portables.

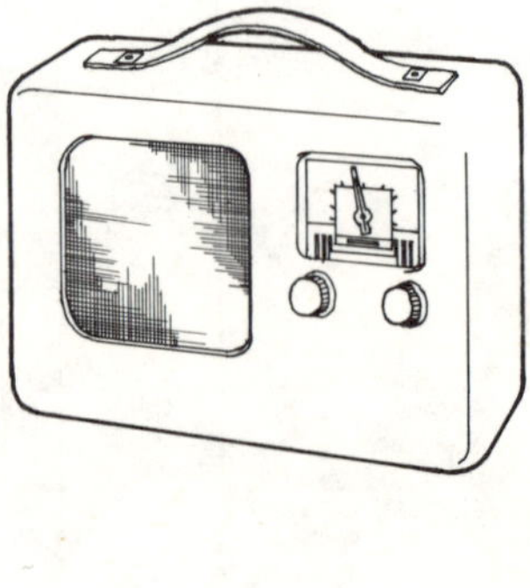

1922 **Crosley Ace Type V**
Collection of R.E. Grinder

1923 **Atwater Kent 10**
"Radiodyne" with Truetone horn speaker
Collection of G.H. Fathauer

1923 **Freed-Eisemann NR-5**
Collection of R.E. Grinder

1928 **RCA Radiola 60**
Collection of R.E. Grinder

1929 **Atwater Kent 46**
Collection of R.E. Grinder

1931 **Atwater Kent 90**
Collection of R.E. Grinder

1932 **International Kadette**
Collection of R.E. Grinder

1936 **Philco 37-610-T**
Collection of W.S. Lettow

1937 **Zenith 5-J-217**
Collection of W.S. Lettow

1938 **Grigsby-Grunow 511 "Majestic"**
Collection of W.S. Lettow

1938 **Zenith 5-R-312**
Collection of W.S. Lettow

1939 **RCA 94-BP-1**
Collection of R.E. Grinder

PART II

The RADIO DIRECTORY AND PRICE GUIDE

The six columns of *The Radio Directory and Price Guide* describe receivers by manufacturer, model number, year of manufacture, source of power, cabinet style and features, and market price. The data are listed, first, alphabetically by *manufacturer,* and second, by *model number* within manufacturer. Readers are cautioned, when searching for a model number by manufacturer, to be certain that the name affixed to the cabinet of the receiver is indeed that of the manufacturer. Several hundred manufacturers of radio broadcast receivers are listed in the directory, and these manufacturers generated well over 1,000 trade names for their products. Highly successful advertising campaigns in the 1930s made household names, for example, of Admiral, Arvin, Majestic, Silvertone, and Sparton. These familiar lines of broadcast receiver were manufactured or distributed, respectively, by Continental Radio & Television Corp., Noblitt-Sparks Industries, Inc., Grigsby-Grunow Corp., Sears, Roebuck & Co., and Sparks-Withington Co. Since a high proportion of radio broadcast receivers were in fact marketed by their trade names, readers who fail to locate an "apparent" manufacturer in the Radio Directory should consult Part III of this volume, *The Trade Directory*. Section 1 of Part III includes a directory of more than 1,200 trade names, which are listed alphabetically with reference to manufacturer. In Section 2 of Part III, manufacturers are listed alphabetically with reference to their trade names. The two comprehensive sections are exceptionally useful sources for identifying either the manufacturers of models whose nameplates carry only trade names or the different trade names of manufacturers.

A definition is given below for each of the six classifications used in the directory to describe the characteristics of a radio broadcast receiver.

Manufacturer

The "manufacturer" designation in the directory denotes the name of the company or corporation that presented itself to the general public as having produced the radio broadcast receiver. The names of manufacturers shown in the directory correspond to those in the Rider and Gernsback service manuals.

Radio manufacturers often adopted distinct, recognizable trade names to enhance the marketing and advertising of their products. Although most of the entrepreneurs named their companies with either their own surnames or the names of the cities in which they were located, many of them found these identities to be too prosaic for use in nationwide sales campaigns. These entrepreneurs thus adopted "trade names" designed to stimulate enticing images which the general public would associate with their products. Many of them added the suffix "dyne" to words in the radio vernacular to create such catchy trade names as radiodyne, audiodyne, claradyne, hyperdyne, and so forth. (The suffix is derived from the Greek word "dynamos," which signifies "power.") A few mercantile giants, like Montgomery Ward (Airline) and Sears, Roebuck (Silvertone) not only marketed receivers via trade names but also contracted with outside firms for the production of their models. These huge retail outlets were primarily distributorships; nonetheless, the radio industry and general public held them accountable, as they would any manufacturer, for standards of performance and quality of construction.

Model

The number or name a manufacturer (or distributor) assigned to each of its radio broadcast receivers. Some manufacturers used chassis as well as model numbers to identify their products. Whenever it was available, however, we used the set model number in the directory.

Year of Manufacture

The earliest year in which a model was advertised, included in a product review, or announced as being in production.

A debate often ensues whenever two or more radio collectors gather to discuss the date a broadcast receiver was manufactured. Although opinions may differ by a year or more, each protagonist is likely to assert stubbornly that he or she is right. And surprisingly, each of them may be, for depending on one's perspective, there are usually several dates upon which to draw. First, there is the date at which the experimental prototype of the receiver was announced to have emerged from the design laboratory; second, there is the date at which production was started; third, the date at which the model was advertised to the public; and fourth, the date at which the model eventually reached retail showrooms. Radio manufacturers in the late 1920s and the 1930s frequently advertised new models for the forthcoming year as much as six months before releasing the receivers to retailer showrooms. How, then, does one choose among the several dating options to determine when a radio broadcast receiver was manufactured? Take your choice! We have elected to date a model by the earliest year in which we could find evidence that it was definitely in existence. For example, in the instance of Philco models, the model 42-345 was advertised as a 1942 release, yet we know, as the directory shows, that it was first manufactured in 1941. Consequently, collectors and historians who are attuned to the year assigned by manufacturers to a given model may find the "year of manufacture" that we specify in the directory to be approximately a year earlier than they anticipate.

Power

The source of power required by the radio broadcast receiver. AC="light-socket" power or alternating current; DC=direct current, which may have been "light-socket" supplied or obtained from wet and/or dry batteries; BAT=direct current power obtained solely from wet and/or dry batteries.

Style

Cabinet styles are identified in the directory by their housing configurations. As noted in Chapter 4, manufacturers of radio broadcast receivers devised three basic housing configurations for their models: table, console, and portable. A great many different style variations arose within these broad classifications; consequently, such cabinet styles as the breadboard, cathedral, tombstone, mantle, and midget are identified specifically in the directory (see Chapter 4 for generic illustrations of these and many other receiver cabinet styles which are noted in the directory).

Price

The prices shown in Part II, *The Radio Directory and Price Guide* present our estimate, as derived from comprehensive analyses of recent sales information, of the market values of radio broadcast receivers in relatively good condition (see below).

The prices shown for the battery-operated receivers of the 1920s are generally for sets *without* vacuum tubes. Early broadcast receivers used tubes that today are somewhat rare and sometimes exotic. The current value of such tubes fluctuates markedly among collectors. The widely used types range from $5 to $20, but tube prices may extend into the hundreds of dollars. A set of tubes for an early receiver, therefore, may command higher prices that the receiver itself. The prices shown for the ac, ac/dc, and battery-powered portables of the late 1920s and 1930s, in contrast, are for receivers *with* tubes. Vacuum tubes for these later model receivers were manufactured by the millions, and even today, they are relatively easy to obtain and inexpensive to buy.

Sales of vintage radio broadcast receivers occur, of course, in antique stores throughout the world. Many more sales, however, are transacted during neighborhood garage sales, community swap meets, regional and national vintage radio swap meets, estate auctions, and antique radio club auctions. A great many sales result, too, from the distribution of mailing lists by collectors who buy and sell actively out of their homes or who are closing out their collections (for reasons of poor health, death in the family, change of residence, etc.), and from the "buy and sell" advertisements carried in antique and vintage radio periodicals.

The source associated with procurement of a radio broadcast receiver is an important correlate of market value. In general, a receiver will command a higher price in the Western than in the

Eastern United States. And a receiver obtained at a swap meet or an auction often will cost less than one purchased from an individual, a dealer-hobbyist, who operates from his or her home, or a professional dealer. A professional dealer, for example, must expend valuable time finding the set, displaying and advertising it, and perhaps, restoring it. A receiver may thus cost twice as much in a dealer's showroom as at a swap meet, but after all, it is there and available when a buyer desires it.

A vintage receiver is, in truth, really worth only what a collector is willing to pay for it! On the one hand, a 1928 Atwater Kent model 40 with loudspeaker (a table model with power supply in a metal cabinet) might bring only $50 in a regional swap meet held for serious radio collectors, but as much as $250 in an antique store. On the other hand, a rare, pre-broadcast era Grebe CR-7 detector would sell for about $400 to a sophisticated collector at the swap meet, but might bring only $50 in the antique store. The unsophisticated public will pay a great deal more for the former because it probably "plays" whereas the classic Grebe may lack the requisite sources of battery power, headphones, and tubes to make it operable.

Given the range in market values created by circumstances associated with the sale of vintage radio broadcast receivers, we list in the directory our best estimate of the "fair" market value of a receiver. In other words, each figure is based on the "average price" a knowledgeable buyer would expect to pay today for that receiver. The estimated price of a receiver represents a relatively precise figure when data were available to us for a half-dozen or more sales from several different markets. Receivers for which no "fair" market price is given are those receivers for which we lacked sufficient information, about either model characteristics or sales transactions, for developing a reasonable estimate of their value.

In instances where we lacked sales information, but possessed sufficient technical, stylistic, and manufacturing details for making close comparisons among models, we made "educated estimates." Our educated estimates are based upon close, collective consideration of the following criteria. Each criterion taken alone is a relatively crude index of the market value of a radio broadcast receiver, but weighed together qualitatively, they enable us to provide a close approximation of its actual value.

Age. Age or year of manufacture is often related directly to current market value, but age as factor is extremely variable across individual cases and its importance is often discounted by other considerations.

Original Price. Original price is an insignificant factor in establishing current market value. Its influence is felt only in models that were originally very expensive production units. Such top of the line models often embodied advanced technical and stylistic refinements.

Scarcity. Scarcity is poorly correlated with current market value. On the one hand, more than half of the radio manufacturers were tiny "cottage" industries. A great many of them copied popular designs, produced a few hundred receivers, and then, went under. Their nondescript models may be scarce today, but there is nothing about these sets to make them attractive to collectors. On the other hand, several manufacturers—Clapp-Eastham, Crosley, DeForest, Grebe, Magnavox, and RCA, among others—either pioneered new advances or mounted creditable advertising campaigns. As a consequence of factors based partly on earned reputation and partly on advertising hype, certain of their models, whatever their level of scarcity, command very high prices.

Novelty or innovation. Novel or innovative factors are the most significant of all considerations underlying market value. This criterion encompasses circuitry, mechanical design, and cabinet style. Sometimes such considerations are difficult to appraise tangibly; receivers in the class of an Atwater Kent breadboard model or a Sparks-Withington mirror-glass model have attained near reverential standing among collectors. Their current market appeal greatly transcends the indifference with which they were regarded when they were first manufactured.

Operating features. Operating features are generally highly correlated with market value. This fact holds particularly for broadcast receivers produced during the 1930s when manufacturers competed fiercely to lead the way in simplifying tuning procedures and in augmenting new advances in high fidelity.

Size. The physical size of a radio broadcast receiver correlates inversely with market value. Few collectors or admirers of early radio receivers possess sufficient space to store or display in their homes more than one monolithic console. Restoration of the big consoles is usually costly, since highly unique knobs, dials, escutcheons, etc. are difficult to repair or replace. Moreover, transporting a console from one site to another often presents problems. Many of the magnificent home entertainment centers of the 1930s are thus dinosaurs in the current marketplace.

Condition. Condition, as it is with all antiques, is a very important criterion associated with market value. The condition of an early radio broadcast receiver may range from "mint" to "junker." In general, cosmetic condition is a more important determinant of market value than operating condition. For example, die cast parts, such as variable condensers, are often costly to repair, cabinets are difficult to restore, and panels and dials are hard to etch. It takes time to restore an abused receiver to a semblance of its original condition, and service charges may range up to several hundred dollars.

The market value of a receiver listed in the directory is based on the assumption that with minimal effort it can be placed in operating condition and elevated to a pleasing level of esthetic appeal. In other words, to pay our price or more for a radio receiver in poor condition is tantamount to a "bad buy," to pay our price for a receiver in excellent condition is to obtain a "fair bargain," and to pay less than the market value we specify for a receiver in premier, mint condition is to get a very good "deal."

The RADIO DIRECTORY AND PRICE GUIDE

To convey the maximum amount of information regarding receiver features in very limited space, the abbreviations below are used after the basic style designation.

ABBREVIATION	DEFINITION
AUTO	AUTOMATIC TUNING
ANT	ANTENNA
BC/SW or B/S	BROADCAST/SHORTWAVE BANDS
BC/SW/LW	BROADCAST/SHORTWAVE/LONGWAVE BANDS
CAB	CABINET
DL or D	EXTERNAL TUNING DIAL
ET	ELECTRICAL TUNING
HB	HIGHBOY (CONSOLE)
HP	HIGH PROFILE (TABLE)
LB	LOWBOY (CONSOLE)
LP	LOW PROFILE (TABLE)
MTL	METAL
PAN	PANEL
PB	PUSHBUTTON
RC	REMOTE CONTROL TUNING
REC	RECORDER
R/P or RP	RADIO/PHONOGRAPH
SP or SPKER	SPEAKER
T	TUBE
TUN EYE or TE	TUNING EYE

MANUFACTURER	MODEL	YEAR	POWER	STYLE	PRICE
A					
A-C ELECTRICAL (DAYTON)	AC-63	1928	AC	TABLE 7 TUBE 1 DIAL	
A-C ELECTRICAL (DAYTON)	AC-65	1928	AC	TABLE 7 TUBE 1 DIAL	
A-C ELECTRICAL (DAYTON)	AC-66	1928	AC	TABLE LP 7 TUBE 1 DIAL	85
A-C ELECTRICAL (DAYTON)	AC-98 NAVIG.	1929	BAT	TABLE LP 9 TUBE 1 DIAL	90
A-C ELECTRICAL (DAYTON)	AC-9960	1929	BAT	CONSOLE LB	145
A-C ELECTRICAL (DAYTON)	AC-9970	1929	BAT	CONSOLE LB	185
A-C ELECTRICAL (DAYTON)	AC-9980	1929	BAT	CONSOLE LB	150
A-C ELECTRICAL (DAYTON)	AC-9990	1929	BAT	CONSOLE HB	180
A-C ELECTRICAL (DAYTON)	XL-5	1925	BAT	TABLE LP GLASS ENC. 5T 3DL	150
A-C ELECTRICAL (DAYTON)	XL-5	1925	BAT	CONSOLE 5 TUBE 3 DIAL	145
A-C ELECTRICAL (DAYTON)	XL-5 POLYDYNE	1925	BAT	TABLE LP 5 TUBE 3 DIAL	65
A-C ELECTRICAL (DAYTON)	XL-10	1925	BAT	TABLE LP 5 TUBE 3 DIAL	65
A-C ELECTRICAL (DAYTON)	XL-15	1923	BAT	CONSOLE 5 TUBE 3 DIAL	110
A-C ELECTRICAL (DAYTON)	XL-20	1925	BAT	TABLE 5 TUBE 3 DIAL	75
A-C ELECTRICAL (DAYTON)	XL-25	1926	BAT	TABLE 5 TUBE 2 DIAL	85
A-C ELECTRICAL (DAYTON)	XL-25	1926	BAT	CONSOLE 5 TUBE 2 DIAL	100
A-C ELECTRICAL (DAYTON)	XL-30	1926	BAT	TABLE 6 TUBE 2 DIAL	80
A-C ELECTRICAL (DAYTON)	XL-30	1926	BAT	CONSOLE 6 TUBE 2 DIAL	
A-C ELECTRICAL (DAYTON)	XL-50	1927	AC	TABLE 6 TUBE 1 DIAL	
A-C ELECTRICAL (DAYTON)	XL-60	1927	BAT	CONSOLE 6 TUBE 1 DIAL	
A-C ELECTRICAL (DAYTON)	XL-61	1928	BAT	TABLE 6 TUBE 2 DIAL	
A-C ELECTRICAL (DAYTON)	XL-70	1927	BAT	TABLE 7 TUBE 2 DIAL	
A-C ELECTRICAL (DAYTON)	XL-70	1927	BAT	CONSOLE 7 TUBE 2 DIAL	
A-C ELECTRICAL (DAYTON)	XL-71	1929	BAT	TABLE	135
A-C ELECTRICAL (DAYTON)	XL-72 NAVIGATOR	1929	BAT	TABLE LP 7 TUBE 1 DIAL	
ACME APPARATUS	ACMEFONE	1922	BAT	TABLE W/SPEAKER 3 TUBE 1 DIAL	175
ACME APPARATUS	S	1925	BAT	TABLE LP 5 TUBE 1 DIAL	75
ACME APPARATUS	S-2	1925	BAT	TABLE LP 5 TUBE 1 DIAL	75
ACME ELECTRIC	77	1929	AC	CONSOLE LB 7 TUBE 1 DIAL	
ACME ELECTRIC	78	1929	AC	CONSOLE LB 8 TUBE 1 DIAL	
ACME ELECTRIC	88	1929	AC	CONSOLE LB 8 TUBE 1 DIAL	
ACME ELECTRIC	88	1929	AC	CONSOLE HB 8 TUBE 1 DIAL	
ACME ELECTRIC	AC-7	1928	AC	TABLE 7 TUBE 2 DIAL	
ACME ELECTRIC	AC-7	1928	AC	CONSOLE LB 7 TUBE 1 DIAL	
ACME ELECTRIC	AC-7	1928	AC	TABLE 7 TUBE 1 DIAL	
ADAMS-MORGAN	2	1924	BAT	TABLE LP 2 TUBE 1 DIAL	175
ADAMS-MORGAN	3	1924	BAT	TABLE LP 3 TUBE 1 DIAL	190
ADAMS-MORGAN	4	1924	BAT	TABLE LP 4 TUBE 1 DIAL	210
ADAMS-MORGAN	III PARAGON	1924	BAT	TABLE HP ENCLOSED 1 DIAL	250
ADAMS-MORGAN	III-A	1924	BAT	TABLE HP ENCLOSED 1 DIAL	300
ADAMS-MORGAN	R-A-6	1916	BAT	TABLE LP 3 DIAL	1300
ADAMS-MORGAN	R-A-10	1921	BAT	TABLE LP 3 DIAL	800
ADAMS-MORGAN	R-A-10/D-A-2	1921	BAT	TABLE LP 2 BOXES 3 DIAL	850
ADAMS-MORGAN	R-B-2 PARAGON	1923	BAT	TABLE LP 3 DIAL	250
ADAMS-MORGAN	R-B-2A PARAGON	1923	BAT	TABLE LP ENCLOSED 3 DIAL	300
ADAMS-MORGAN	R-D-5 PARAGON	1922	BAT	TABLE LP 3 DIAL	350
ADAMS-MORGAN	R-D-5/A-2	1922	BAT	TABLE LP 2 BOXES 3 DIAL	750
ADLER	10	1925	BAT	CONSOLE R/P 5 TUBE 3 DIAL	

MANUFACTURER	MODEL	YEAR	POWER	STYLE	PRICE
ADLER	10-SR	1925	BAT	TABLE LP 5 TUBE 3 DIAL	65
ADLER	20-AR	1925	BAT	TABLE LP 5 TUBE 3 DIAL	85
ADLER	50	1925	BAT	CONSOLE R/P 5 TUBE 3 DIAL	
ADLER	199	1925	BAT	TABLE LP 5 TUBE 3 DIAL	85
ADLER	201	1925	BAT	TABLE LP 5 TUBE 3 DIAL	65
ADLER	E	1925	BAT	CONSOLE HB R/P 5 TUBE 3	
ADLER	F	1925	BAT	CONSOLE R/P 5 TUBE 3 DIAL	
ADROIT TOOL	R-5-C	1925	BAT	CONSOLE 5 TUBE 1 DIAL	
ADROIT TOOL	R-5-P	1925	BAT	TABLE LP 5 TUBE 3 DIAL	75
ADROIT TOOL	R-5-S	1925	BAT	TABLE SPEAKER 5 TUBE 3 DL	
ADVANCE AUTO. ACCESSORIES	1926	1925	BAT	TABLE METAL NEUTRO-WOUND 4 DIAL	200
ADVANCE AUTO. ACCESSORIES	1927	1926	BAT	TABLE METAL NEUTRO-WOUND 3 DIAL	175
ADVANCE AUTO. ACCESSORIES	1928	1927	BAT	TABLE METAL NEUTRO-WOUND 3 DIAL	
ADVANCE AUTO. ACCESSORIES	MASTER ALLELEC.	1927	AC	CONSOLE 2 DIAL	
ADVANCE AUTO. ACCESSORIES	SUPER ALLELEC.	1927	AC	TABLE LP 2 DIAL	
ADVANCE ELECTRIC	11	1929	AC	CONSOLE	
ADVANCE ELECTRIC	23	1929	AC	CONSOLE	
ADVANCE ELECTRIC	53	1929	AC	CONSOLE	
ADVANCE ELECTRIC	77	1930	AC	CATHEDRAL MODERN W/CLOCK	150
ADVANCE ELECTRIC	88	1930	AC	CATHEDRAL MODERN	150
ADVANCE ELECTRIC	89	1930	AC	CATHEDRAL	
AERODYNE	2	1925	BAT	TABLE LP 5 TUBE 3 DIAL	75
AINSWORTH RADIO	TORODYNE	1925	BAT	TABLE LP 5 TUBE 3 DIAL	95
AIR-KING PRODUCTS	6E	1935	AC	TABLE	
AIR-KING PRODUCTS	9E	1935	AC	TABLE	
AIR-KING PRODUCTS	9EC	1935	AC	CONSOLE	
AIR-KING PRODUCTS	37	1933	AC	TABLE	
AIR-KING PRODUCTS	52	1933	AC	TABLE	
AIR-KING PRODUCTS	60	1933	AC	TABLE	
AIR-KING PRODUCTS	60A	1933	AC	TABLE BC/LW	
AIR-KING PRODUCTS	155	1935	AC	TABLE BC/SW	
AIR-KING PRODUCTS	222	1938	AC	MANTLE	20
AIR-KING PRODUCTS	251	1938	AC	TABLE	
AIR-KING PRODUCTS	255	1938	AC	TABLE	
AIR-KING PRODUCTS	424	1935	AC	MANTLE	
AIR-KING PRODUCTS	616	1935	AC	TABLE BC/SW	
AIR-KING PRODUCTS	666	1935	AC	TABLE BC/SW	
AIR-KING PRODUCTS	777	1938	AC	TABLE	
AIR-KING PRODUCTS	810	1938	AC	TABLE BC/SW	
AIR-KING PRODUCTS	815	1938	AC	TABLE BC/SW	
AIR-KING PRODUCTS	820	1938	AC	TABLE BC/SW	
AIR-KING PRODUCTS	911	1938	AC	TOMBSTONE BC/SW	130
AIR-KING PRODUCTS	915	1938	AC	TABLE BC/SW	30
AIR-KING PRODUCTS	2207	1938	AC	MANTLE	
AIR-KING PRODUCTS	3905	1939	BAT	PORTABLE	
AIR-KING PRODUCTS	3906	1939	BAT	PORTABLE	

MANUFACTURER	MODEL	YEAR	POWER	STYLE	PRICE
AIR-KING PRODUCTS	3907	1939	AC	TABLE RADIO/PHONO	
AIR-KING PRODUCTS	3909	1939	AC	TABLE RADIO/PHONO	
AIR-KING PRODUCTS	3912	1939	AC	PORTABLE	
AIR-KING PRODUCTS	3914	1939	AC	PORTABLE	
AIR-KING PRODUCTS	3922	1939	AC	MANTLE PLASTIC	
AIR-KING PRODUCTS	3923	1939	AC	MANTLE	
AIR-KING PRODUCTS	3950	1940	AC	PORTABLE LEATHERETTE BC/SW	25
AIR-KING PRODUCTS	3952	1939	AC	TABLE	
AIR-KING PRODUCTS	3953	1939	AC	TABLE	
AIR-KING PRODUCTS	3980	1939	AC	CONSOLE RADIO/PHONO	
AIR-KING PRODUCTS	4129	1941	AC	TABLE RADIO/PHONO	15
AIR-KING PRODUCTS	4130	1940	AC	TABLE RADIO/PHONO	
AIR-WAY ELECTRIC APPLIANCE	41	1924	BAT	TABLE LP 4 TUBE 2 DIAL	75
AIR-WAY ELECTRIC APPLIANCE	51	1924	BAT	TABLE LP 4 TUBE 3 DIAL	85
AIR-WAY ELECTRIC APPLIANCE	52	1924	BAT	CONSOLE 4 TUBE 3 DIAL	
AIR-WAY ELECTRIC APPLIANCE	61	1925	BAT	TABLE LP 6 TUBE 2 DIAL	120
AIR-WAY ELECTRIC APPLIANCE	62	1925	BAT	TABLE HP SPEAKER 6 T 2 DL	
AIR-WAY ELECTRIC APPLIANCE	63	1925	BAT	CONSOLE 6 TUBE 2 DIAL	
AIR-WAY ELECTRIC APPLIANCE	F	1923	BAT	TABLE HP 4 TUBE 2 DIAL	200
AIRO-MASTER	60	1924	BAT	TABLE LP CHEST 5 TUBE 3 DL	95
AIRO-MASTER	75	1924	BAT	TABLE HP SPEAKER 5 T 1 DL	135
AIRO-MASTER	130	1924	BAT	CONSOLE LB 5 TUBE 1 DIAL	175
AIRO-MASTER	150	1924	BAT	CONSOLE HB 5 TUBE 3 DIAL	200
ALADDIN	510	1924	BAT	TABLE LP 4 TUBE 2 DIAL	65
ALGONQUIN ELECTRIC	TF-5	1925	BAT	TABLE LP 1 DIAL	
ALGONQUIN ELECTRIC	TF-6	1926	BAT	TABLE LP 1 DIAL	
ALGONQUIN ELECTRIC	TF-7	1926	BAT	TABLE LP 1 DIAL	
ALL AMERICAN MOHAWK				See also MOHAWK	
ALL AMERICAN MOHAWK	44	1927	BAT	TABLE LP 6 TUBE 1 DIAL	
ALL AMERICAN MOHAWK	55	1927	BAT	CONSOLE HB 6 TUBE 1 DIAL	
ALL AMERICAN MOHAWK	60	1928	AC	CONSOLE 6 TUBE 1 DIAL	
ALL AMERICAN MOHAWK	61	1928	AC	CONSOLE 6 TUBE 1 DIAL	
ALL AMERICAN MOHAWK	62	1928	AC	CONSOLE 6 TUBE 1 DIAL	
ALL AMERICAN MOHAWK	65	1928	AC	CONSOLE 6 TUBE 1 DIAL	
ALL AMERICAN MOHAWK	66	1927	AC	CONSOLE 6 TUBE 1 DIAL	
ALL AMERICAN MOHAWK	70	1928	AC	CONSOLE 8 TUBE 1 DIAL	
ALL AMERICAN MOHAWK	75	1928	AC	CONSOLE 8 TUBE 1 DIAL	
ALL AMERICAN MOHAWK	77	1927	AC	TABLE LP 10 TUBE 1 DIAL	
ALL AMERICAN MOHAWK	80	1927	AC	TABLE LP 9 TUBE 1 DIAL	
ALL AMERICAN MOHAWK	83	1928	AC	TABLE 9 TUBE 1 DIAL	
ALL AMERICAN MOHAWK	84	1928	AC	TABLE 9 TUBE 1 DIAL	
ALL AMERICAN MOHAWK	85	1928	AC	TABLE 9 TUBE 1 DIAL	
ALL AMERICAN MOHAWK	88	1928	AC	CONSOLE RADIO/PHONO 9 TUBE 1 DIAL	
ALL AMERICAN MOHAWK	90	1927	AC	TABLE LP 10 TUBE 1 DIAL	
ALL AMERICAN MOHAWK	93	1929	AC	CONSOLE 10 TUBE 1 DIAL	95
ALL AMERICAN MOHAWK	94	1929	AC	CONSOLE LB 7 TUBE 1 DIAL	
ALL AMERICAN MOHAWK	95	1929	AC	CONSOLE HB 10 TUBE 1 DIAL	
ALL AMERICAN MOHAWK	97	1929	AC	CONSOLE	

MANUFACTURER	MODEL	YEAR	POWER	STYLE	PRICE
ALL AMERICAN MOHAWK	98	1929	AC	CONSOLE	
ALL AMERICAN MOHAWK	99	1927	AC	CONSOLE LB 7 TUBE 1 DIAL	
ALL AMERICAN MOHAWK	115	1927	AC	CONSOLE HB 2 DIAL	
ALL AMERICAN MOHAWK	B-7	1931	BAT	TABLE	
ALL AMERICAN MOHAWK	B-34	1931	BAT	CONSOLE LB	
ALL AMERICAN MOHAWK	B-80	1931	BAT	CONSOLE LB	
ALL AMERICAN MOHAWK	C-6	1930	AC	CATHEDRAL	95
ALL AMERICAN MOHAWK	D-11	1930	AC	CONSOLE LB	
ALL AMERICAN MOHAWK	D-19	1930	AC	CONSOLE LB	
ALL AMERICAN MOHAWK	D-29	1930	AC	CONSOLE LB	
ALL AMERICAN MOHAWK	D-39	1930	AC	CONSOLE HB RADIO/PHONO	
ALL AMERICAN MOHAWK	D-69	1930	AC	CONSOLE HB RADIO/PHONO	
ALL AMERICAN MOHAWK	DUET	1926	BAT	TABLE LP 6 TUBE 2 DIAL	110
ALL AMERICAN MOHAWK	DUET	1927	AC	TABLE 2 DIAL	
ALL AMERICAN MOHAWK	FORTE	1926	BAT	TABLE LP DOORS 7 TUBE 1 DIAL	125
ALL AMERICAN MOHAWK	FORTE	1927	AC	CONSOLE 9 TUBE 1 DIAL	
ALL AMERICAN MOHAWK	H-19	1930	AC	CONSOLE LB	
ALL AMERICAN MOHAWK	H-29	1930	AC	CONSOLE HB	
ALL AMERICAN MOHAWK	H-69	1930	AC	CONSOLE HB	
ALL AMERICAN MOHAWK	LORAINE	1926	BAT	CONSOLE HB 7 TUBE 1 DIAL	275
ALL AMERICAN MOHAWK	LORAINE	1927	AC	CONSOLE HB 9 TUBE 1 DIAL	
ALL AMERICAN MOHAWK	PORTABLE	1925	BAT	PORTABLE LOOP ANT. 2 DIAL	135
ALL AMERICAN MOHAWK	R	1925	BAT	TABLE LP 5 TUBE 2 DIAL	95
ALL AMERICAN MOHAWK	R	1926	BAT	CONSOLE HB 5 TUBE 2 DIAL	150
ALL AMERICAN MOHAWK	S-6	1931	AC	TABLE	
ALL AMERICAN MOHAWK	S-7	1931	AC	TABLE	
ALL AMERICAN MOHAWK	S-8	1931	AC	CONSOLE	
ALL AMERICAN MOHAWK	S-61	1931	AC	CONSOLE	
ALL AMERICAN MOHAWK	SA-65	1932	AC	TABLE	
ALL AMERICAN MOHAWK	SA-66	1932	AC	CONSOLE LB	
ALL AMERICAN MOHAWK	SA-90	1932	AC	CONSOLE LB	
ALL AMERICAN MOHAWK	SA-130	1932	AC	CONSOLE	
ALL AMERICAN MOHAWK	SEXTET	1926	BAT	CONSOLE HB DESK 6 TUBE 2 DIAL	250
ALL AMERICAN MOHAWK	SEXTET	1927	AC	CONSOLE HB 6 TUBE 2 DIAL	
ALL AMERICAN MOHAWK	SG-1	1929	AC	CONSOLE 8 TUBE 1 DIAL	
ALL AMERICAN MOHAWK	SOVEREIGN	1926	BAT	CONSOLE HB 7 TUBE 1 DIAL	275
ALL AMERICAN MOHAWK	SOVEREIGN	1927	AC	CONSOLE 9 TUBE 1 DIAL	
ALL AMERICAN MOHAWK	SW-80	1932	AC	CONSOLE BC/SW	
ALL AMERICAN MOHAWK	Va	1925	BAT	TABLE HP 1 DIAL	140
ALL AMERICAN MOHAWK	X	1925	BAT	TABLE HP ENCLOSED 1 DIAL	185
ALL AMERICAN MOHAWK	XII	1925	BAT	CONSOLE HB ENCLOSED 1 DIAL	225
ALLIED RADIO	KNIGHT 8	1930	DC	TABLE METAL BRONZE	
ALLIED RADIO	ROAMER	1927	BAT	PORTABLE BOX 3 DIAL	95
AMBER	D	1924	BAT	TABLE LP METER 6 T 2 DIAL	
AMBER	DC	1924	BAT	CONSOLE METER 6 T 2 DIAL	
AMBER	T	1924	BAT	TABLE LP METER 5 T 2 DIAL	
AMBER	512-C	1924	BAT	TABLE LP 5 TUBE 3 DIAL	175

MANUFACTURER	MODEL	YEAR	POWER	STYLE	PRICE
AMERICAN BOSCH	4	1935	AC	TABLE BC/SW	
AMERICAN BOSCH	5	1935	AC	TABLE BC/SW	
AMERICAN BOSCH	5-A	1931	AC	TABLE	35
AMERICAN BOSCH	5-B	1931	AC	CONSOLE END TABLE	
AMERICAN BOSCH	16 AMBOROLA	1925	BAT	TABLE LP 6 TUBE 2 DIAL	220
AMERICAN BOSCH	20-J	1931	AC	CONSOLE	
AMERICAN BOSCH	20-K	1931	AC	CONSOLE	
AMERICAN BOSCH	20-L	1931	AC	CONSOLE	
AMERICAN BOSCH	27 AMBORADA	1926	BAT	CONSOLE 7 TUBE 2 DIAL	290
AMERICAN BOSCH	28	1928	AC	TABLE LP 8 TUBE 1 DIAL	115
AMERICAN BOSCH	28-A	1928	AC	CONSOLE 8 TUBE 1 DIAL	
AMERICAN BOSCH	29	1928	AC	CONSOLE	
AMERICAN BOSCH	30	1929	AC	CONSOLE HB R/P DOORS	275
AMERICAN BOSCH	31-H	1931	AC	CONSOLE	
AMERICAN BOSCH	35 CRUISER	1926	BAT	TABLE LP 5 TUBE 2 DIAL	
AMERICAN BOSCH	35 ROYAL	1926	BAT	CONSOLE 5 TUBE 2 DIAL	
AMERICAN BOSCH	38	1928	AC	TABLE	
AMERICAN BOSCH	38-A	1928	AC	CONSOLE	
AMERICAN BOSCH	46	1927	BAT	TABLE LP 1 DIAL	
AMERICAN BOSCH	48	1929	AC	TABLE LP 7 TUBE 1 DIAL	
AMERICAN BOSCH	48-17	1929	AC	CONSOLE 7 TUBE 1 DIAL	
AMERICAN BOSCH	48-18	1929	AC	CONSOLE 7 TUBE 1 DIAL	
AMERICAN BOSCH	48-19	1929	AC	CONSOLE 7 TUBE 1 DIAL	
AMERICAN BOSCH	48-A	1929	AC	CONSOLE LB 7 TUBE 1 DIAL	
AMERICAN BOSCH	48-J	1929	AC	CONSOLE HB 7 TUBE 1 DIAL	
AMERICAN BOSCH	48-L	1929	AC	CONSOLE 7 TUBE 1 DIAL	
AMERICAN BOSCH	48-R	1929	AC	CONSOLE 7 TUBE 1 DIAL	
AMERICAN BOSCH	49	1929	AC	TABLE LP 7 TUBE 1 DIAL	
AMERICAN BOSCH	56	1929	BAT	TABLE	
AMERICAN BOSCH	56-A	1929	BAT	CONSOLE	
AMERICAN BOSCH	57	1927	BAT	CONSOLE LB 7 TUBE 1 DIAL	140
AMERICAN BOSCH	58-A	1930	AC	CONSOLE	
AMERICAN BOSCH	58-B	1930	AC	CONSOLE	
AMERICAN BOSCH	60-D	1930	AC	CONSOLE	
AMERICAN BOSCH	60-E	1930	AC	CONSOLE	
AMERICAN BOSCH	61	1930	AC	CONSOLE HB	160
AMERICAN BOSCH	62-C	1930	AC	CONSOLE	
AMERICAN BOSCH	66 CRUISER	1927	BAT	CONSOLE LB 6 TUBE 1 DIAL	130
AMERICAN BOSCH	66 CRUISER	1927	BAT	TABLE LP 6 TUBE 1 DIAL	130
AMERICAN BOSCH	76 CRUISER	1927	BAT	CONSOLE LB 6 TUBE 1 DIAL	125
AMERICAN BOSCH	76 CRUISER	1927	BAT	TABLE LP 6 TUBE 1 DIAL	
AMERICAN BOSCH	78 CRUISER	1927	BAT	CONSOLE LB 6 TUBE 1 DIAL	
AMERICAN BOSCH	87 CRUISER	1927	BAT	TABLE LP 7 TUBE 1 DIAL	95
AMERICAN BOSCH	87 CRUISER	1927	BAT	CONSOLE LB 7 TUBE 1 DIAL	95
AMERICAN BOSCH	96	1927	AC	CONSOLE	
AMERICAN BOSCH	107	1927	AC	CONSOLE	
AMERICAN BOSCH	116	1927	AC	TABLE	
AMERICAN BOSCH	126	1927	AC	TABLE	
AMERICAN BOSCH	136	1927	AC	CONSOLE	
AMERICAN BOSCH	146	1927	AC	CONSOLE	

MANUFACTURER	MODEL	YEAR	POWER	STYLE	PRICE
AMERICAN BOSCH	156 CRUISER	1927	AC	CONSOLE	
AMERICAN BOSCH	166	1927	AC	TABLE	
AMERICAN BOSCH	176	1927	AC	CONSOLE	
AMERICAN BOSCH	200-A	1932	AC	TABLE CHEST	
AMERICAN BOSCH	200-B	1932	AC	TABLE CHEST DELUXE	
AMERICAN BOSCH	205-A	1932	AC	TABLE	
AMERICAN BOSCH	224-D	1932	DC	CONSOLE	
AMERICAN BOSCH	226-F	1932	BAT	CONSOLE	
AMERICAN BOSCH	236-A	1932	AC	TABLE	
AMERICAN BOSCH	242-E	1932	AC	CONSOLE	
AMERICAN BOSCH	250-M	1932	AC	CONSOLE	
AMERICAN BOSCH	260-C	1932	AC	CONSOLE BC/SW	
AMERICAN BOSCH	260-R	1932	AC	CONSOLE BC/SW	
AMERICAN BOSCH	312-C	1933	AC	CONSOLE BC/SW	
AMERICAN BOSCH	312-G	1933	AC	CONSOLE BC/SW	
AMERICAN BOSCH	350	1933	AC	TABLE	40
AMERICAN BOSCH	352	1934	AC	COMPACT	
AMERICAN BOSCH	355	1933	AC	TABLE	
AMERICAN BOSCH	356 IMPERIAL	1926	BAT	TABLE LP 5 TUBE 2 DIAL	160
AMERICAN BOSCH	357	1934	AC	TABLE	
AMERICAN BOSCH	360-E	1933	AC	CONSOLE BC/SW	
AMERICAN BOSCH	360-M	1933	AC	CONSOLE BC/SW	
AMERICAN BOSCH	360-S	1933	AC	CONSOLE BC/SW	190
AMERICAN BOSCH	360-T	1933	AC	TOMBSTONE BC/SW	140
AMERICAN BOSCH	360-W	1934	AC	TABLE	
AMERICAN BOSCH	360-X	1934	AC	CONSOLE	
AMERICAN BOSCH	360-Y	1934	AC	CONSOLE	
AMERICAN BOSCH	370-E	1933	AC	CONSOLE BC/SW	
AMERICAN BOSCH	370-M	1933	AC	CONSOLE BC/SW	
AMERICAN BOSCH	370-S	1933	AC	CONSOLE BC/SW	
AMERICAN BOSCH	370-T	1933	AC	TOMBSTONE BC/SW	
AMERICAN BOSCH	370-W	1934	AC	MIDGET	
AMERICAN BOSCH	370-X	1934	AC	CONSOLE	
AMERICAN BOSCH	376-BT	1934	BAT	TABLE	
AMERICAN BOSCH	376-N	1935	2/DC	TABLE	
AMERICAN BOSCH	376-S	1934	BAT	CONSOLE	
AMERICAN BOSCH	385	1935	2/DC	TABLE	
AMERICAN BOSCH	386	1935	2/DC	TABLE	
AMERICAN BOSCH	402	1934	AC	TABLE	45
AMERICAN BOSCH	420	1934	AC	TABLE BC/SW	
AMERICAN BOSCH	430-J	1935	AC	CONSOLE BC/SW	
AMERICAN BOSCH	430-T	1935	AC	TABLE BC/SW	
AMERICAN BOSCH	440-C	1934	AC	CONSOLE BC/SW	
AMERICAN BOSCH	440-T	1934	AC	TABLE BC/SW	80
AMERICAN BOSCH	450-H	1935	AC	TABLE	
AMERICAN BOSCH	450-L	1935	AC	CONSOLE	
AMERICAN BOSCH	460-A	1934	AC	TABLE BC/SW	60
AMERICAN BOSCH	460-B	1934	AC	TABLE BC/SW	40
AMERICAN BOSCH	460-R	1934	AC	CONSOLE BC/SW	110
AMERICAN BOSCH	462-A	1934	DC	TABLE BC/SW	

MANUFACTURER	MODEL	YEAR	POWER	STYLE	PRICE
AMERICAN BOSCH	462-B	1934	DC	TABLE BC/SW	
AMERICAN BOSCH	462-Y	1934	DC	CONSOLE BC/SW	
AMERICAN BOSCH	470-G	1935	AC	CONSOLE	
AMERICAN BOSCH	470-U	1935	AC	TABLE	
AMERICAN BOSCH	480-D	1934	AC	CONSOLE BC/SW	
AMERICAN BOSCH	480-R	1934	AC	CONSOLE	
AMERICAN BOSCH	501	1933	AC	MANTLE	
AMERICAN BOSCH	502	1934	AC	COMPACT	
AMERICAN BOSCH	505	1935	AC	TABLE BC/SW	30
AMERICAN BOSCH	510	1935	AC	TABLE BC/SW	
AMERICAN BOSCH	510-E	1935	AC	CONSOLE BC/SW	
AMERICAN BOSCH	515	1936	AC	TABLE BC/SW	
AMERICAN BOSCH	565-K	1935	AC	CONSOLE	
AMERICAN BOSCH	565-W	1935	AC	CONSOLETTE	
AMERICAN BOSCH	575-F	1935	AC	TABLE BC/SW	
AMERICAN BOSCH	575-Q	1935	AC	CONSOLE BC/SW	
AMERICAN BOSCH	585-F	1935	AC	TABLE	
AMERICAN BOSCH	585-Y	1935	AC	TABLE BC/SW	
AMERICAN BOSCH	585-Z	1935	AC	CONSOLE BC/SW	
AMERICAN BOSCH	595-M	1935	AC	CONSOLE BC/SW	
AMERICAN BOSCH	595-P	1935	AC	CONSOLE BC/SW	
AMERICAN BOSCH	600	1936	6/DC	TABLE BC/SW	
AMERICAN BOSCH	601	1936	2/DC	TABLE BC/SW	
AMERICAN BOSCH	602-C	1936	2/DC	TABLE BC/SW	
AMERICAN BOSCH	602-T	1936	2/DC	TABLE BC/SW	
AMERICAN BOSCH	604	1936	AC	TABLE BC/SW	
AMERICAN BOSCH	605	1936	AC	TABLE BC/SW	
AMERICAN BOSCH	605-C	1936	AC	CONSOLE BC/SW	
AMERICAN BOSCH	610	1936	AC	TABLE BC/SW	
AMERICAN BOSCH	620	1936	AC	TABLE BC/SW	
AMERICAN BOSCH	625	1936	AC	CONSOLE BC/SW	
AMERICAN BOSCH	640	1936	AC	TABLE BC/SW	
AMERICAN BOSCH	650	1936	AC	CONSOLE BC/SW	
AMERICAN BOSCH	660-C	1936	AC	CONSOLE BC/SW	
AMERICAN BOSCH	660-T	1936	AC	TABLE BC/SW	40
AMERICAN BOSCH	670-C	1936	AC	CONSOLE BC/SW	170
AMERICAN BOSCH	670-S	1936	AC	CONSOLE BC/SW	
AMERICAN BOSCH	680	1936	AC	CONSOLE BC/SW	
AMERICAN ETHERPHONE	RX-1	1923	BAT	TABLE BOX 1 TUBE 3 DIAL	
AMERICAN ETHERPHONE	RX-3	1924	BAT	TABLE HP 2 TUBE 2 DIAL	
AMERICAN EUROPEAN	PORTO-0-RADIO	1924	BAT	PORTABLE 6 TUBE 1 DIAL	
AMERICAN RADIO	3-A	1923	BAT	TABLE LP 2 TUBE 2 DIAL	
AMERICAN RADIO	4	1923	BAT	TABLE LP 4 TUBE 2 DIAL	
AMERICAN RADIO	5 SUPER	1923	BAT	TABLE LP 4 TUBE 2 DIAL	75
AMERICAN RADIO	AMERICAN BEAUTY	1925	BAT	TABLE LP 5 TUBE 3 DIAL	85
AMERICAN RADIO & RESEARCH	70 CONCERTO	1928	AC	CONSOLE	95
AMERICAN RADIO & RESEARCH	70 NOCTURN	1928	AC	CONSOLE	
AMERICAN RADIO & RESEARCH	70 OPERA	1928	AC	CONSOLE RADIO/PHONO	
AMERICAN RADIO & RESEARCH	70 SONATA	1928	AC	CONSOLE	
AMERICAN RADIO & RESEARCH	81 ARIA	1929	AC	CONSOLE RADIO/PHONO	

MANUFACTURER	MODEL	YEAR	POWER	STYLE	PRICE
AMERICAN RADIO & RESEARCH	81 BEL CANTO	1929	AC	CONSOLE RADIO/PHONO	
AMERICAN RADIO & RESEARCH	81 DUET	1929	AC	CONSOLE RADIO/PHONO	
AMERICAN RADIO & RESEARCH	81 SERENATA	1929	AC	CONSOLE RADIO/PHONO	240
AMERICAN RADIO & RESEARCH	81 SYMPHONY	1929	AC	CONSOLE RADIO/PHONO	
AMERICAN RADIO & RESEARCH	2596/2634	1921	BAT	TABLE 3 TUBE 4 DIAL	
AMERICAN RADIO & RESEARCH	3366	1923	BAT	TABLE 1 TUBE 2 DIAL	400
AMERICAN RADIO & RESEARCH	3500-3	1923	BAT	TABLE HP 4 TUBE 2 DIAL	
AMERICAN RADIO & RESEARCH	3500-4	1923	BAT	TABLE HP 4 TUBE 2 DIAL	
AMERICAN RADIO & RESEARCH	3500-6	1923	BAT	TABLE HP 4 TUBE 2 DIAL	
AMERICAN RADIO & RESEARCH	AC-5	1926	AC	TABLE LP 3 DIAL	135
AMERICAN RADIO & RESEARCH	AC-5-C	1926	AC	CONSOLE LB 3 DIAL	150
AMERICAN RADIO & RESEARCH	AC-6 WARICK	1927	AC	TABLE LP 7 TUBE 1 DIAL	125
AMERICAN RADIO & RESEARCH	AC-6-C BERWICK	1927	AC	CONSOLE LB	190
AMERICAN RADIO & RESEARCH	AC-7 WINDSOR	1927	AC	TABLE LP 7 TUBE 1 DIAL	70
AMERICAN RADIO & RESEARCH	AC-7-C HASTINGS	1927	AC	CONSOLE LP DOORS	225
AMERICAN RADIO & RESEARCH	AC-9	1926	AC	TABLE LP 2 DIAL	145
AMERICAN RADIO & RESEARCH	AC-9-C	1926	AC	CONSOLE LB 2 DIAL	160
AMERICAN RADIO & RESEARCH	DC-6 WARICK	1927	BAT	TABLE LP 6 TUBE 1 DIAL	95
AMERICAN RADIO & RESEARCH	DC-6-C BERWICK	1927	BAT	CONSOLE LB 6 TUBE 1 DIAL	200
AMERICAN RADIO & RESEARCH	DC-7 WINDSOR	1927	BAT	CONSOLE	190
AMERICAN RADIO & RESEARCH	DC-7-C HASTINGS	1927	BAT	CONSOLE LB DOORS	200
AMERICAN RADIO & RESEARCH	NEUTRODYNE	1924	BAT	TABLE LP 3 DIAL	95
AMERICAN RADIO & RESEARCH	S-522	1926	BAT	TABLE LP 5 TUBE 3 DIAL	70
AMERICAN RADIO & RESEARCH	S-522-C	1926	BAT	CONSOLE LB 5 TUBE 3 DIAL	100
AMERICAN RADIO & RESEARCH	S-733	1926	BAT	TABLE LP 7 TUBE 2 DIAL	80
AMERICAN RADIO & RESEARCH	S-733-C	1926	BAT	CONSOLE LB 7 TUBE 2 DIAL	110
AMERICAN SPECIALITY	18	1925	BAT	TABLE LP 5 TUBE 3 DIAL	
AMERICAN SPECIALITY	24	1925	BAT	TABLE LP 5 TUBE 3 DIAL	65
AMERICAN SPECIALITY	39 REPUBLIC	1926	BAT	TABLE LP 5 TUBE 3 DIAL	170
AMERICAN SPECIALITY	GRAND	1923	BAT	TABLE HP SPEAKER 5 T 3 DL	
AMERICAN SPECIALITY	REGAL DECT/AMP	1922	BAT	TABLE 2 BOXES 3 TUBE 2 DL	195
AMERICAN SPECIALITY	STANDARD	1923	BAT	TABLE HP SPEAKER 5 T 3 DL	125
AMSCO PRODUCTS	MELCO SUPREME	1923	BAT	TABLE LP 5 TUBE 3 DIAL	85
ANDREA F. A. D.				See also FADA RADIO	
ANDREA F. A. D.	7 FADA	1927	BAT	TABLE LP LOOP ANT 7 T 2 DL	140
ANDREA F. A. D.	7 FADA	1927	BAT	CONSOLE HB 7 TUBE 2 DIAL	135
ANDREA F. A. D.	8 FADA	1927	BAT	CONSOLE HB LOOP ANT 8T 2DL	150
ANDREA F. A. D.	8 FADA	1926	BAT	TABLE LP LOOP ANT 8 T 2 DL	125
ANDREA F. A. D.	10	1928	AC	TABLE LP 7 TUBE 1 DIAL	125
ANDREA F. A. D.	11	1928	AC	TABLE LP 7 TUBE 1 DIAL	
ANDREA F. A. D.	16	1928	AC	TABLE LP 8 TUBE 1 DIAL	
ANDREA F. A. D.	17	1928	AC	TABLE LP 8 TUBE 1 DIAL	
ANDREA F. A. D.	18	1929	DC	TABLE LP METAL 7 TUBE 1 DIAL	95
ANDREA F. A. D.	20	1929	AC	TABLE METAL	
ANDREA F. A. D.	25	1929	AC	CONSOLE	
ANDREA F. A. D.	30	1928	AC	CONSOLE HB 7 TUBE 1 DIAL	135
ANDREA F. A. D.	31	1928	AC	CONSOLE LB DOORS 7 T 1 DL	150
ANDREA F. A. D.	32	1929	AC	CONSOLE	

MANUFACTURER	MODEL	YEAR	POWER	STYLE	PRICE
ANDREA F. A. D.	35	1929	AC	CONSOLE	
ANDREA F. A. D.	40	1929	AC	CONSOLE HB DOORS	195
ANDREA F. A. D.	41	1930	AC	CONSOLE HB	
ANDREA F. A. D.	42	1930	AC	CONSOLE LB	
ANDREA F. A. D.	44	1930	AC	CONSOLE LB	
ANDREA F. A. D.	45	1930	AC	CONSOLE LB	
ANDREA F. A. D.	45	1931	AC	CONSOLE LB	
ANDREA F. A. D.	46	1930	AC	CONSOLE HB	
ANDREA F. A. D.	47	1930	AC	CONSOLE RADIO/PHONO	
ANDREA F. A. D.	48	1930	AC	CONSOLE LB	
ANDREA F. A. D.	48	1931	AC	CONSOLE LB	
ANDREA F. A. D.	49	1930	AC	CONSOLE HB	
ANDREA F. A. D.	49	1931	AC	CONSOLE HB	
ANDREA F. A. D.	50	1928	AC	TABLE LP 9 TUBE 1 DIAL	120
ANDREA F. A. D.	51	1930	AC	TABLE	
ANDREA F. A. D.	51	1931	AC	TABLE	
ANDREA F. A. D.	53	1930	AC	CONSOLE LB	
ANDREA F. A. D.	53	1931	AC	CONSOLE LB	
ANDREA F. A. D.	57	1930	AC	CONSOLE LB	
ANDREA F. A. D.	57	1931	AC	CONSOLE LB	
ANDREA F. A. D.	65	1930	AC	CONSOLE LB	
ANDREA F. A. D.	65	1931	AC	CONSOLE LB	
ANDREA F. A. D.	70	1928	AC	CONSOLE LB DOORS 9 T 1 DL	165
ANDREA F. A. D.	71	1928	AC	CONSOLE LB DOORS 9 T 1 DL	165
ANDREA F. A. D.	72	1928	AC	CONSOLE LB DOORS R/P 9T 1 DIAL	180
ANDREA F. A. D.	75	1929	AC	CONSOLE	95
ANDREA F. A. D.	77	1929	AC	CONSOLE RADIO/PHONO	
ANDREA F. A. D.	81	1930	DC	CONSOLE HB	135
ANDREA F. A. D.	82	1930	DC	CONSOLE LB	
ANDREA F. A. D.	84	1930	DC	CONSOLE LB	
ANDREA F. A. D.	86	1930	DC	CONSOLE HB	
ANDREA F. A. D.	104-A	1922	BAT	MANTLE 2 TUBE 1 DIAL	225
ANDREA F. A. D.	105-A	1922	BAT	TABLE 3 TUBE 1 DIAL	275
ANDREA F. A. D.	137-A	1922	BAT	TABLE LP 1 TUBE 2 DIAL	350
ANDREA F. A. D.	140-A	1922	BAT	TABLE LP 3 TUBE 2 DIAL	325
ANDREA F. A. D.	160	1923	BAT	TABLE LP 4 TUBE 3 DIAL	85
ANDREA F. A. D.	160-A	1924	BAT	TABLE LP 5 TUBE 3 DIAL	110
ANDREA F. A. D.	175-A	1924	BAT	TABLE NEUTROCEIVER 5 T 3 DIAL	185
ANDREA F. A. D.	175/90-A	1924	BAT	CONSOLE 5 TUBE 3 DIAL	90
ANDREA F. A. D.	185-A NEUTROLA	1924	BAT	TABLE HP SPEAKER 5 T 3 DL	140
ANDREA F. A. D.	185/90-A	1924	BAT	CONSOLE HB 5 TUBE 3 DIAL	100
ANDREA F. A. D.	192-A	1924	BAT	TABLE NEUTROLETTE 5 T 3 DIAL	120
ANDREA F. A. D.	192-BS	1924	BAT	PORTABLE	
ANDREA F. A. D.	192-S	1924	BAT	PORTABLE	
ANDREA F. A. D.	195 NEUTRO JR.	1924	BAT	TABLE LP 3 TUBE 2 DIAL	95
ANDREA F. A. D.	196-A	1924	BAT	CONSOLE R/P 4 TUBE 3 DIAL	
ANDREA F. A. D.	265-A	1927	BAT	TABLE 6 TUBE 3 DIAL	

MANUFACTURER	MODEL	YEAR	POWER	STYLE	PRICE
ANDREA F. A. D.	460-A	1926	BAT	TABLE LP LOOP ANT 6 T 2DL	
ANDREA F. A. D.	475-A	1927	BAT	TABLE LP LOOP ANT 7 T 2 DL	
ANDREA F. A. D.	480-B	1927	BAT	TABLE LP 8 TUBE 2 DIAL	260
ANDREA F. A. D.	SF-10/60	1927	BAT	TABLE DAVENPORT	
ANDREA F. A. D.	SF-10/70	1924	BAT	CONSOLE DAVENPORT 5 T 3 DIAL	
ANDREA F. A. D.	SF-20/60	1927	BAT	CONSOLE BEETHOVEN GRAND	
ANDREA F. A. D.	SF-20/70	1924	BAT	CONSOLE BEETHOVEN 5 T 3 DIAL	
ANDREA F. A. D.	SF-30/60	1927	BAT	CONSOLE QUEEN ANNE DESK	200
ANDREA F. A. D.	SF-30/70	1924	BAT	CONSOLE DESK QUEEN ANNE	200
ANDREA F. A. D.	SF-40/60	1927	BAT	CONSOLE	135
ANDREA F. A. D.	SF-45/77	1927	BAT	CONSOLE 7 TUBE 2 DIAL	135
ANDREA F. A. D.	SF-50/80	1927	BAT	CONSOLE 8 TUBE 1 DIAL	135
ANDREA F. A. D.	SPECIAL	1927	BAT	TABLE LP 6 TUBE 2 DIAL	75
ANDREA RADIO	2-D-5	1938	AC	TABLE BC/SW	
ANDREA RADIO	2-E-6	1938	AC	TABLE PB BC/SW/LW	25
ANDREA RADIO	2-E-8	1938	AC	TABLE PB BC/SW/LW	35
ANDREA RADIO	4-E-6	1938	AC	CONSOLE PB BC/SW/LW	
ANDREA RADIO	4-E-8	1938	AC	TABLE PB BC/SW/LW	
ANDREA RADIO	4-E-11	1938	AC	TABLE PB BC/SW/LW	40
ANDREA RADIO	5-E-11	1939	AC	CONSOLE PB BC/SW	100
ANDREA RADIO	6-D-5	1938	AC	TABLE PB BC/SW/LW	
ANDREA RADIO	6-E-6	1938	AC	CONSOLE R/P PB BC/SW/LW	45
ANDREA RADIO	6-E-8	1938	AC	CONSOLE PB BC/SW/LW	
ANDREA RADIO	6-E-11	1938	AC	CONSOLE PB BC/SW/LW	
ANDREA RADIO	8-E-8	1938	AC	CONSOLE R/P PB BC/SW/LW	
ANDREA RADIO	8-E-11	1938	AC	CONSOLE R/P PB BC/SW/LW	45
ANDREA RADIO	8-F-12	1940	AC	CONSOLE R/P W/ 33 TUBES	
ANDREA RADIO	10-E-8	1938	AC	CONSOLE R/P PB BC/SW	45
ANDREA RADIO	10-E-11	1938	AC	CONSOLE PB BC/SW/LW	50
ANDREA RADIO	12-FS	1940	AC	TABLE	30
ANDREA RADIO	14-E-6	1938	AC	TABLE PB BC/SW	25
ANDREA RADIO	21-F-5	1939	BAT	PORTABLE	15
ANDREA RADIO	62PX	1938	AC	TABLE BC/SW	
ANDREA RADIO	626	1939	BAT	CONSOLE R/P PB BC/SW/LW	
ANDREA RADIO	630	1938	AC	TABLE PB BC/SW/LW	
ANDREA RADIO	632	1938	AC	CONSOLE PB BC/SW/LW	
ANDREA RADIO	634	1938	AC	CONSOLE R/P PB BC/SW/LW	
ANDREA RADIO	1530	1938	AC	TABLE BC/SW	
ANDREA RADIO	1532	1938	AC	CONSOLE	
ANDREA RADIO	1534	1938	AC	CONSOLE R/P PB BC/SW/LW	
ANDREA RADIO	1536	1938	AC	CONSOLE R/P PB BC/SW/LW	
ANDREA RADIO	1538	1939	AC	CONSOLE R/P PB BC/SW/LW	
ANDREWS RADIO	1	1925	BAT	TABLE LP 5 TUBE 3 DIAL	65
ANDREWS RADIO	11	1925	BAT	TABLE LP 5 TUBE 3 DIAL	75
ANDREWS RADIO	111	1925	AC	TABLE HP SPEAKER 5 T 3 DL	
ANDREWS RADIO	A	1925	BAT	TABLE LP 5 TUBE 3 DIAL	95
ANDREWS RADIO	DERESNADYNE	1925	BAT	PORTABLE W/LOOP 2 DIAL	100
ANDREWS RADIO	P	1924	BAT	PORTABLE 6 TUBE 2 DIAL	175

MANUFACTURER	MODEL	YEAR	POWER	STYLE	PRICE
ANSLEY	17AX	1940	AC	CONSOLE RADIO/PHONO	
ANSLEY	21AX	1940	AC	CONSOLE RADIO/PHONO	
ANSLEY	25AX	1940	AC	CONSOLE RADIO/PHONO	
ANSLEY	B-1	1936	BAT	PORTABLE	
ANSLEY	D-3	1933	AC	COMPACT RADIO/PHONO	
ANSLEY	D-4	1933	AC	TABLE RADIO/PHONO	20
ANSLEY	D-6	1934	AC	TABLE RADIO/PHONO	
ANSLEY	D-7	1934	AC/DC	PORTABLE RADIO/PHONO	
ANSLEY	D-9	1935	AC/DC	PORTABLE RADIO/PHONO BC/SW	15
ANSLEY	D-10	1935	AC/DC	TABLE RADIO/PHONO BC/SW	15
ANSLEY	D-10-A	1941	AC	TABLE RADIO/PHONO	15
ANSLEY	D-14	1937	AC	CONSOLE RADIO/PHONO BC/SW	
ANSLEY	D-16	1937	AC	CONSOLE RADIO/PHONO	
ANSLEY	D-17	1937	AC	CONSOLE RADIO/PHONO BC/SW	
ANSLEY	D-18	1937	AC	CONSOLE RADIO/PHONO BC/SW	
ANSLEY	D-20	1937	AC	CONSOLE RADIO/PHONO AUTO.	
ANSLEY	D-21	1937	AC	CONSOLE RADIO/PHONO BC/SW	
ANSLEY	D-22	1937	AC	CONSOLE RADIO/PHONO BC/SW	
ANSLEY	D-23	1937	AC	CONSOLE BC/SW END TABLE	
ANSLEY	U-8	1932	AC/DC	PORTABLE	
ANSLEY	U-9	1932	AC/DC	PORTABLE	
ANSLEY	U-10	1935	AC	TABLE BC/SW	
ANSLEY	U-11	1938	AC	PORTABLE	
ANSLEY	Y-11	1937	AC/DC	PORTABLE BC/SW	
ANYLITE	KING COLE	1925	BAT	TABLE LP 4 TUBE 2 DIAL	
APEX ELECTRIC	5 SUPER	1925	BAT	TALE LP 5 TUBE 3 DIAL	80
APEX ELECTRIC	36	1928	AC	TABLE LP METAL 1 DIAL	50
APEX ELECTRIC	BABY GRAND	1925	BAT	CONSOLE LB 5 TUBE 3 DIAL	220
APEX ELECTRIC	DELUXE	1925	BAT	TABLE HP SPEAKER 5 T 3 DL	135
APEX ELECTRIC	MILAN	1927	BAT	CONSOLE LB DOORS 6 T 1 DL	130
APEX ELECTRIC	MINSTREL	1927	AC	CONSOLE HB 7 TUBE 1 DIAL	175
APEX ELECTRIC	TROUBADOUR	1927	AC	CONSOLE HB TECHNIDYNE	195
ARGUS RADIO	125-B	1927	AC	TABLE LP 6 TUBE 2 DIAL	
ARGUS RADIO	160	1925	BAT	TABLE LP 6 TUBE 2 DIAL	75
ARGUS RADIO	195-B	1927	AC	TABLE LP 6 TUBE 1 DIAL	165
ARGUS RADIO	250-B	1927	AC	CONSOLE LB 2 DIAL	
ARGUS RADIO	295-B	1927	AC	CONSOLE	
ARGUS RADIO	375-B	1926	AC	CONSOLE 6 TUBE 2 DIAL	
ARGUS RADIO	395-B	1927	AC	CONSOLE	
ARGUS RADIO	STANDARD	1926	BAT	TABLE LP 6 TUBE 2 DIAL	130
ARMLEY	KARRYADIO	1923	BAT	PORTABLE 6 TUBE 1 DIAL	
ATCHISON RADIO	12	1930	AC	CATHEDRAL	110

MANUFACTURER	MODEL	YEAR	POWER	STYLE	PRICE
ATLANTIC-PACIFIC	A-P SPECIAL	1922	BAT	PORTABLE HP 2 DIAL	195
ATLANTIC-PACIFIC	AR-10	1922	BAT	TABLE LP 3 TUBE 2 DIAL	250
ATLAS COLONIAL	16-FIVE	1924	BAT	TABLE LP METER 5 T 3 DIAL	85
ATLAS COLONIAL	16-SIX	1924	BAT	TABLE LP METER 6 T 3 DIAL	85
ATLAS COLONIAL	17	1924	BAT	TABLE LP METER 5 T 2 DIAL	85
ATWATER KENT	3925	1922	BAT	BREADBOARD 3 TUBE 1 DIAL	
ATWATER KENT	3945	1922	BAT	BREADBOARD 2 TUBE 1 DIAL	
ATWATER KENT	3955	1922	BAT	BREADBOARD 3 TUBE 2 DIAL	
ATWATER KENT	3975	1922	BAT	BREADBOARD 3 TUBE 2 DIAL	
ATWATER KENT	4052	1923	BAT	BREADBOARD 4 TUBE 1 DIAL	
ATWATER KENT	4066	1923	BAT	BREADBOARD 5 TUBE 1 DIAL	
ATWATER KENT	4205	1923	BAT	BREADBOARD 5 TUBE 1 DIAL	
ATWATER KENT	5 (4333)	1923	BAT	BREADBOARD 5 TUBE 1 DIAL	2500
ATWATER KENT	8 (4325)	1923	BAT	BREADBOARD 5 TUBE 2 DIAL	
ATWATER KENT	9 (4445)	1923	BAT	BREADBOARD 4 TUBE 2 DIAL	650
ATWATER KENT	9C (4660)	1923	BAT	BREADBOARD 4 TUBE 2 DIAL	575
ATWATER KENT	10 (4340)	1923	BAT	BREADBOARD RADIODYNE	900
ATWATER KENT	10 (4340)	1923	BAT	BREADBOARD BLK/GRN 5T 3 DIAL	400
ATWATER KENT	10 (4490)	1923	BAT	BREADBOARD 5 TUBE 3 DIAL	
ATWATER KENT	10 (4600)	1923	BAT	BREADBOARD BRN 5 TUBE 3 DIAL	400
ATWATER KENT	10A (4550)	1923	BAT	BREADBOARD BRN 5 TUBE 3 DIAL	450
ATWATER KENT	10A (4560)	1924	BAT	BREADBOARD BLK/GRN 5T 3 DIAL	450
ATWATER KENT	10B (4550)	1924	BAT	BREADBOARD BRN 5 TUBE 3 DIAL	350
ATWATER KENT	10B (4560)	1924	BAT	BREADBOARD BLK/GRN 5T 3 DIAL	450
ATWATER KENT	10C (4700)	1924	BAT	BREADBOARD COMPACT 5T 3 DIAL	450
ATWATER KENT	10C (4700)	1924	BAT	BREADBOARD STANDARD 5T 3 DIAL	400
ATWATER KENT	10C (4950)	1924	BAT	CONSOLE W/BREADBOARD	450
ATWATER KENT	12 (4375)	1923	BAT	BREADBOARD GRN RADIODYNE	
ATWATER KENT	12 (4620)	1923	BAT	BREADBOARD BRN 6 TUBE 3 DIAL	800
ATWATER KENT	12 (4910)	1923	BAT	BREADBOARD BRN 6 TUBE 3 DIAL	700
ATWATER KENT	19 (4880)	1924	BAT	TABLE LP 4 TUBE 2 DIAL	250
ATWATER KENT	20 (4640)	1924	BAT	TABLE LP 5 TUBE 3 DIAL	90
ATWATER KENT	20-C (7570)	1925	BAT	TABLE LP COMPACT 5 T 3 DL	90
ATWATER KENT	20C (7960)	1925	BAT	TABLE LP COMPACT 5 T 3 DIAL	100
ATWATER KENT	21 (7780)	1925	BAT	TABLE LP 5 TUBE 3 DIAL	300
ATWATER KENT	24/20D (4920)	1925	BAT	TABLE LP 5 TUBE 3 DIAL	150
ATWATER KENT	30 (7950)	1925	BAT	TABLE LP 6 TUBE 1 DIAL	120
ATWATER KENT	32 (8270)	1926	BAT	TABLE LP 7 TUBE 1 DIAL	85
ATWATER KENT	33 (8930)	1927	BAT	TABLE LP 6 TUBE 1 DIAL	75

MANUFACTURER	MODEL	YEAR	POWER	STYLE	PRICE
ATWATER KENT	35 (8100)	1926	BAT	TABLE LP METAL 6 TUBE 1 DIAL	95
ATWATER KENT	36	1927	AC	TABLE LP METAL 7 TUBE 1 DIAL	85
ATWATER KENT	37	1927	AC	TABLE LP METAL 7 TUBE 1 DIAL	85
ATWATER KENT	38	1927	AC	TABLE LP METAL 8 TUBE 1 DL	50
ATWATER KENT	40	1928	AC	TABLE LP METAL 7 TUBE 1 DIAL	80
ATWATER KENT	40	1928	AC	CONSOLE	200
ATWATER KENT	41	1928	AC	TABLE LP METAL 7 TUBE 1 DIAL	70
ATWATER KENT	42	1928	AC	TABLE LP METAL 7 TUBE 1 DIAL	70
ATWATER KENT	43	1928	AC	TABLE LP METAL 8 TUBE 1 DL	75
ATWATER KENT	44	1928	AC	TABLE LP METAL 8 TUBE 1 DL	65
ATWATER KENT	45	1929	AC	TABLE LP METAL	95
ATWATER KENT	46	1929	AC	TABLE LP METAL BLACK GREEN	110
ATWATER KENT	46 (8200)	1929	AC	CONSOLE HB #46 POOLEY CAB	175
ATWATER KENT	46 (8400)	1929	AC	CONSOLE HB #46 POOLEY CAB	195
ATWATER KENT	47	1929	AC	TABLE LP METAL	80
ATWATER KENT	48	1928	BAT	TABLE LP METAL 6 TUBE 1 DL	65
ATWATER KENT	49	1928	BAT	TABLE LP METAL 6 TUBE 1 DL	80
ATWATER KENT	50	1927	BAT	TABLE LP 7 TUBE 1 DIAL	175
ATWATER KENT	52	1928	AC	CONSOLE 7 TUBE 1 DIAL	125
ATWATER KENT	53	1929	AC	CONSOLE HB	125
ATWATER KENT	55	1929	AC	TABLE LP METAL	95
ATWATER KENT	55	1929	AC	CONSOLE	200
ATWATER KENT	56	1929	AC	CONSOLE	150
ATWATER KENT	57	1929	AC	CONSOLE METAL	140
ATWATER KENT	60	1929	AC	CONSOLE KIEL TABLE	165
ATWATER KENT	61	1929	DC	TABLE LP METAL	75
ATWATER KENT	66	1930	AC	CONSOLE KIEL TABLE	175
ATWATER KENT	67	1929	BAT	TABLE LP METAL	65
ATWATER KENT	70	1930	AC	CONSOLE LB	95
ATWATER KENT	70-Q	1930	BAT	CONSOLE LB	250
ATWATER KENT	72	1930	AC	CONSOLE HB	160
ATWATER KENT	74	1930	AC	CONSOLE LB	125
ATWATER KENT	75	1930	AC	CONSOLE LB RADIO/PHONO	175
ATWATER KENT	76	1930	AC	CONSOLE HB	150
ATWATER KENT	79	1929	AC	CONSOLE	280
ATWATER KENT	80	1931	AC	COMPACT 25 CYCLE	325

MANUFACTURER	MODEL	YEAR	POWER	STYLE	PRICE
ATWATER KENT	82	1931	AC	COMPACT 25 CYCLE	
ATWATER KENT	83	1931	AC	CONSOLE LB 25 CYCLE	160
ATWATER KENT	84	1931	AC	CATHEDRAL	270
ATWATER KENT	85	1931	AC	CONSOLE LB 25 CYCLE	
ATWATER KENT	85Q	1931	BAT	CONSOLE LB	
ATWATER KENT	86	1931	AC	CONSOLE	200
ATWATER KENT	87	1931	AC	CONSOLE LB	
ATWATER KENT	89	1931	AC	CONSOLE LB	200
ATWATER KENT	89P	1931	AC	CONSOLE RADIO/PHONO	
ATWATER KENT	90	1931	AC	CATHEDRAL	200
ATWATER KENT	112-N	1934	AC	CONSOLE BC/SW	230
ATWATER KENT	112-S	1934	AC	CONSOLE BC/SW	165
ATWATER KENT	135	1934	AC	TABLE BC/SW	
ATWATER KENT	145	1934	AC	TOMBSTONE BC/SW	85
ATWATER KENT	155	1933	AC	TABLE CHEST	50
ATWATER KENT	165	1933	AC	CATHEDRAL BC/SW	185
ATWATER KENT	184	1935	AC	TOMBSTONE COMPACT	100
ATWATER KENT	185	1934	AC	MIDGET	195
ATWATER KENT	188	1932	AC	CONSOLE LB	
ATWATER KENT	206	1934	AC	CATHEDRAL BC/SW	150
ATWATER KENT	215-E	1934	32/DC	CONSOLE BC/SW	
ATWATER KENT	217	1933	AC	CATHEDRAL BC/SW	200
ATWATER KENT	217-D	1933	DC	COMPACT BC/SW	
ATWATER KENT	237Q	1935	6/DC	COMPACT BC/SW	
ATWATER KENT	246	1933	AC	TABLE CHEST	200
ATWATER KENT	260	1932	AC	CONSOLE HB (SEMI)	250
ATWATER KENT	266	1933	AC	CONSOLE	
ATWATER KENT	275	1933	AC	TABLE BC/SW	75
ATWATER KENT	285Q	1935	2/DC	CONSOLE	
ATWATER KENT	286B	1935	AC	CONSOLE	
ATWATER KENT	305Z	1935	32/DC	COMPACT BC/SW	
ATWATER KENT	310-J	1933	AC	CONSOLE LB BC/SW	220
ATWATER KENT	310-T	1933	AC	CONSOLE HB BC/SW	220
ATWATER KENT	317	1935	AC	CONSOLE BC/SW	150
ATWATER KENT	318-C	1934	AC	CONSOLE	
ATWATER KENT	318-K	1934	AC	CONSOLE BC/SW	220
ATWATER KENT	318-N	1934	AC	CONSOLE BC/SW	
ATWATER KENT	325-E	1934	AC	CONSOLE BC/SW	165
ATWATER KENT	328	1935	AC	CONSOLE BC/SW	
ATWATER KENT	328-C	1935	AC	CONSOLE	
ATWATER KENT	337	1935	AC	TOMBSTONE BC/SW	120
ATWATER KENT	356	1935	AC	MIDGET	
ATWATER KENT	376-DE	1934	DC	CONSOLE BC/SW	
ATWATER KENT	376-E	1934	AC	CONSOLE BC/SW	
ATWATER KENT	376-KX	1934	AC	CONSOLE BC/SW	
ATWATER KENT	387	1933	BAT	COMPACT	
ATWATER KENT	412	1935	AC	CONSOLE BC/SW	
ATWATER KENT	415-Q	1935	2/DC	COMPACT	
ATWATER KENT	427	1933	AC	CONSOLE BC/SW	
ATWATER KENT	427-D	1933	DC	CONSOLE BC/SW	

MANUFACTURER	MODEL	YEAR	POWER	STYLE	PRICE
ATWATER KENT	427-Q	1933	BAT	CONSOLE	
ATWATER KENT	435	1935	AC	CONSOLE BC/SW	175
ATWATER KENT	435-T	1935	AC	CONSOLE	
ATWATER KENT	447	1934	AC	TABLE BC/SW	
ATWATER KENT	448	1933	AC	CONSOLE	195
ATWATER KENT	465-Q	1934	BAT	TABLE BC/SW	
ATWATER KENT	467-Q	1935	6/DC	CONSOLE BC/SW	
ATWATER KENT	469	1932	AC	CONSOLE LB	
ATWATER KENT	469-D	1932	BAT	CONSOLE LB	
ATWATER KENT	469-Q	1932	BAT	CONSOLE LB	
ATWATER KENT	475-B	1935	AC	CONSOLE	
ATWATER KENT	480	1932	AC	CONSOLE HB (SEMI) BC/SW	
ATWATER KENT	487-C	1935	AC	CONSOLE	
ATWATER KENT	509	1935	AC	CONSOLE BC/SW	
ATWATER KENT	510	1933	AC	CONSOLE BC/SW	90
ATWATER KENT	511-W	1934	AC	CONSOLE TUNE-O-MATIC BC/SW	165
ATWATER KENT	525	1934	AC	CONSOLE	
ATWATER KENT	545	1935	AC	COMPACT BC/SW	180
ATWATER KENT	555	1933	AC	TABLE CHEST	
ATWATER KENT	558	1932	AC	CATHEDRAL	300
ATWATER KENT	558-D	1932	DC	COMPACT	
ATWATER KENT	558-Q	1932	BAT	COMPACT	
ATWATER KENT	559-N	1934	AC	CONSOLE BC/SW	120
ATWATER KENT	559-S	1934	AC	CONSOLE BC/SW	
ATWATER KENT	565-Z	1935	32/DC	CONSOLE BC/SW	
ATWATER KENT	612	1932	AC	CONSOLE HB (SEMI)	160
ATWATER KENT	627	1932	AC	COMPACT	
ATWATER KENT	649	1935	AC	CONSOLE BC/SW	
ATWATER KENT	655-QE	1934	BAT	CONSOLE BC/SW	
ATWATER KENT	667	1933	AC	CONSOLE BC/SW	260
ATWATER KENT	667-D	1933	DC	CONSOLE BC/SW	
ATWATER KENT	708	1933	AC	CATHEDRAL BC/SW	265
ATWATER KENT	711-J	1933	AC	CONSOLE LB BC/SW	190
ATWATER KENT	711-R	1933	AC	CONSOLE BC/SW	190
ATWATER KENT	711-T	1933	AC	CONSOLE HB BC/SW	190
ATWATER KENT	735	1935	AC	MIDGET	105
ATWATER KENT	755	1935	AC	MIDGET	
ATWATER KENT	768-Q	1934	BAT	TABLE BC/SW	
ATWATER KENT	788-J	1934	AC	CONSOLE	
ATWATER KENT	808	1933	AC	CONSOLE	
ATWATER KENT	808-A	1933	AC	CONSOLE BC/SW	
ATWATER KENT	810	1935	AC	CONSOLE BC/SW	155
ATWATER KENT	812	1932	AC	CONSOLE HB	
ATWATER KENT	825	1934	AC	TABLE	
ATWATER KENT	854	1935	AC	MIDGET	60
ATWATER KENT	856	1935	AC	TOMBSTONE BC/SW	90
ATWATER KENT	875F	1935	AC	CONSOLE	
ATWATER KENT	944	1934	AC	CATHEDRAL	150
ATWATER KENT	976	1935	AC	CONSOLE BC/SW	

MANUFACTURER	MODEL	YEAR	POWER	STYLE	PRICE
ATWATER KENT	978-QK	1934	BAT	CONSOLE BC/SW	
AUDIOLA RADIO	4MB	1933	AC	MANTLE METAL	
AUDIOLA RADIO	5W	1933	AC	CHEST	
AUDIOLA RADIO	5WB	1933	AC	COMPACT	
AUDIOLA RADIO	6-C	1927	BAT	CONSOLE LB PIANO 6 T 1 DL	240
AUDIOLA RADIO	6-T	1927	BAT	TABLE LP 6 TUBE 1 DIAL	130
AUDIOLA RADIO	8-C	1927	BAT	CONSOLE LB PIANO 6 T 1·DL	240
AUDIOLA RADIO	8-T	1927	BAT	TABLE LP 8 TUBE 1 DIAL	135
AUDIOLA RADIO	60	1930	AC	CONSOLE LB	
AUDIOLA RADIO	70	1930	AC	CONSOLE LB	175
AUDIOLA RADIO	80	1930	AC	CONSOLE LB	165
AUDIOLA RADIO	416	1931	AC	TABLE	
AUDIOLA RADIO	506	1933	AC	TABLE	
AUDIOLA RADIO	516-V	1933	AC	MIDGET	
AUDIOLA RADIO	520	1933	AC	MIDGET	
AUDIOLA RADIO	527	1926	BAT	TABLE	
AUDIOLA RADIO	527-C	1926	BAT	CONSOLE	
AUDIOLA RADIO	627	1926	BAT	TABLE	
AUDIOLA RADIO	627-C	1926	BAT	CONSOLE LB PIANO 6 T 2 DL	225
AUDIOLA RADIO	710	1931	AC	TABLE	
AUDIOLA RADIO	712	1931	AC	CONSOLE LB	
AUDIOLA RADIO	714	1931	AC	CONSOLE LB	
AUDIOLA RADIO	716	1933	AC	COMPACT BC/SW	
AUDIOLA RADIO	811	1932	AC	TABLE	
AUDIOLA RADIO	816	1933	AC	COMPACT BC/SW	
AUDIOLA RADIO	816-32	1933	AC	COMPACT	
AUDIOLA RADIO	829	1928	AC	TABLE METAL	
AUDIOLA RADIO	843	1932	AC	CONSOLE LB	
AUDIOLA RADIO	868	1932	AC	CONSOLE HB	
AUDIOLA RADIO	914	1931	AC	CONSOLE LB	
AUDIOLA RADIO	1011	1932	AC	TABLE	
AUDIOLA RADIO	1016	1933	AC	TABLE BC/SW	
AUDIOLA RADIO	1050	1932	AC	CONSOLE LB	
AUDIOLA RADIO	1054	1933	AC	CONSOLE LB	
AUDIOLA RADIO	1068	1933	AC	CONSOLE HB	
AUDIOLA RADIO	10731	1933	AC	CONSOLE BC/SW	
AUDIOLA RADIO	11300DN	1932	AC	CONSOLE HB	
AUDIOLA RADIO	7330	1929	AC	TABLE METAL	
AUDIOLA RADIO	7558	1933	AC	CONSOLE BC/SW	
AUDIOLA RADIO	8430	1929	AC	TABLE METAL	
AUDIOLA RADIO	8557	1933	AC	CONSOLE BC/SW	
AUDIOLA RADIO	AUDIODYNE	1923	BAT	TABLE LP 4 TUBE 2 DIAL	
AUDIOLA RADIO	GRAND	1923	BAT	PORTABLE 3 TUBE 1 DIAL	
AUDIOLA RADIO	MIDGET	1923	BAT	PORTABLE 1 TUBE 1 DIAL	
AUDIOLA RADIO	SUPER	1923	BAT	TABLE LP 5 TUBE 2 DIAL	
AUDIOLA RADIO	WONDER	1923	BAT	PORTABLE 1 TUBE 1 DIAL	
AURORA	AURORA THREE	1925	BAT	TABLE LP 3 TUBE 2 DIAL	
AUTO INDICATOR	POCKET RADIO	1923	BAT	PORTABLE 1 TUBE 1 DIAL	
AUTOCRAT	4C	1935	AC	MANTLE	
AUTOCRAT	4M	1935	AC	MANTLE	

MANUFACTURER	MODEL	YEAR	POWER	STYLE	PRICE
AUTOCRAT	4S	1933	AC	TABLE BC/SW	
AUTOCRAT	5BL	1933	AC	TABLE BC/LW	
AUTOCRAT	5C	1935	AC	MANTLE	
AUTOCRAT	5SA	1933	AC	TABLE BC/SW	
AUTOCRAT	6	1933	AC	TABLE BC/SW	
AUTOCRAT	16-AC-S	1935	AC	TABLE BC/SW	
AUTOCRAT	25-AC-L	1935	AC	TABLE BC/SW	
AUTOCRAT	26-AC-L	1935	AC	TABLE BC/SW	
AUTOCRAT	29	1938	2/DC	TABLE	
AUTOCRAT	31	1938	6/DC	TABLE	
AUTOCRAT	34-B	1935	AC	TABLE BC/SW	
AUTOCRAT	40	1935	AC	MANTLE	
AUTOCRAT	40-M	1935	AC	MANTLE	
AUTOCRAT	40-SW	1935	AC	MANTLE BC/SW	
AUTOCRAT	42	1935	AC	MANTLE	
AUTOCRAT	42-SW	1935	AC	MANTLE BC/SW	
AUTOCRAT	45	1935	AC	MANTLE	
AUTOCRAT	45-SW	1935	AC	MANTLE BC/SW	
AUTOCRAT	51-UL	1935	AC	TABLE BC/SW	
AUTOCRAT	52	1935	AC	MANTLE	
AUTOCRAT	52-SW	1935	AC	MANTLE BC/SW	
AUTOCRAT	53	1935	AC	TABLE	
AUTOCRAT	53-SW	1935	AC	MANTLE BC/SW	
AUTOCRAT	56-SW	1935	AC	TABLE BC/SW	
AUTOCRAT	57-W	1933	AC	TABLE BC/SW	
AUTOCRAT	61-UL	1935	AC	TABLE BC/SW	
AUTOCRAT	66-US	1935	AC	TABLE BC/SW	
AUTOCRAT	80	1933	AC	TABLE BC/SW	
AUTOCRAT	90-SL	1932	AC	COMPACT BC/SW	
AUTOCRAT	101	1938	AC	MANTLE	
AUTOCRAT	260-UL	1935	AC	TABLE BC/SW/LW	
AUTOCRAT	661-US	1935	AC	TABLE BC/SW	
AUTOCRAT	682	1933	AC	CONSOLE	
AUTOCRAT	840	1933	AC	CONSOLE	
AUTOCRAT	900	1935	AC	CONSOLE BC/SW/LW	
AUTOCRAT	900	1935	AC	TABLE BC/SW/LW	
AUTOCRAT	910	1935	AC	CONSOLE RADIO/PHONO BC/SW	
AUTOCRAT	AA	1935	BAT	PORTABLE	
AUTOCRAT	BA-41	1935	2/DC	TABLE BC/SW	
AUTOMATIC RADIO	40	1938	AC	MANTLE	
AUTOMATIC RADIO	50	1938	AC	MANTLE	
AUTOMATIC RADIO	141	1941	AC	TABLE RADIO/PHONO	15
AUTOMATIC RADIO	325	1938	AC	TABLE	
AUTOMATIC RADIO	402	1939	AC	TABLE	25
AUTOMATIC RADIO	432	1940	AC	TABLE RADIO/PHONO	15
AUTOMATIC RADIO	458	1939	AC	TABLE	45
AUTOMATIC RADIO	480	1939	AC	CONSOLE BC/SW	150
AUTOMATIC RADIO	816-818	1938	AC	TABLE	
AUTOMATIC RADIO	840	1938	AC	TABLE	

MANUFACTURER	MODEL	YEAR	POWER	STYLE	PRICE
AUTOMATIC RADIO	850	1938	AC	TABLE BC/SW	
AUTOMATIC RADIO	855	1938	AC	TABLE BC/SW	
AUTOMATIC RADIO	892	1938	AC	CONSOLE BC/SW	
AUTOMATIC RADIO	905	1939	AC	MANTLE	
AUTOMATIC RADIO	911	1938	AC	TABLE	
AUTOMATIC RADIO	915	1938	AC	TABLE BC/SW	
AUTOMATIC RADIO	920	1938	AC	TABLE	
AUTOMATIC RADIO	929	1939	AC	MANTLE	
AUTOMATIC RADIO	933	1938	AC	MANTLE TOM THUMB	50
AUTOMATIC RADIO	935	1939	AC	MANTLE	
AUTOMATIC RADIO	939	1939	AC	TABLE BC/SW	
AUTOMATIC RADIO	949	1939	AC	TABLE BC/SW	
AUTOMATIC RADIO	950	1939	AC	MANTLE	
AUTOMATIC RADIO	979	1939	AC	TABLE BC/SW	
AUTOMATIC RADIO	986	1939	AC	TABLE PB BC/SW	
AUTOMATIC RADIO	988	1939	AC	TDABLE BC/SW	
AUTOMATIC RADIO	999	1939	AC	CONSOLE PB BC/SW	
AUTOMATIC RADIO	A 5	1935	6/DC	TABLE	
AUTOMATIC RADIO	A 6G	1935	AC	TABLE BC/SW	
AUTOMATIC RADIO	A.C.	1929	AC	PORTABLE TOM THUMB	
AUTOMATIC RADIO	A63	1935	AC	TABLE	
AUTOMATIC RADIO	A74	1935	AC	TABLE	
AUTOMATIC RADIO	AC 5	1935	AC	TABLE BC/SW	
AUTOMATIC RADIO	AC52	1935	AC	TABLE	
AUTOMATIC RADIO	AJ-400	1934	AC	TABLE	
AUTOMATIC RADIO	B	1930	BAT	PORTABLE TOM THUMB	180
AUTOMATIC RADIO	B DELUXE	1929	BAT	PORTABLE TOM THUMB	
AUTOMATIC RADIO	B10	1937	AC	TABLE FLAT	
AUTOMATIC RADIO	B10LW	1937	AC	TABLE BC/SW/LW FLAT	
AUTOMATIC RADIO	B10P	1937	AC	TABLE BC/SW FLAT	
AUTOMATIC RADIO	B15	1937	AC	TABLE FLAT	
AUTOMATIC RADIO	B20	1937	AC/DC	PORTABLE	
AUTOMATIC RADIO	B30	1937	AC	TABLE FLAT	
AUTOMATIC RADIO	B35	1937	AC	TABLE BC/SW FLAT	
AUTOMATIC RADIO	B40	1937	AC	TABLE BC/SW FLAT	
AUTOMATIC RADIO	B70	1937	AC	TABLE BC/SW FLAT	
AUTOMATIC RADIO	B80	1937	AC	TABLE BC/SW FLAT	
AUTOMATIC RADIO	B80LW	1937	AC	TABLE BC/SW/LW FLAT	
AUTOMATIC RADIO	B85	1937	AC	TABLE BC/SW FLAT	
AUTOMATIC RADIO	BL4	1935	AC	TABLE	
AUTOMATIC RADIO	BLUEBIRD	1925	BAT	TABLE LP 5 TUBE 3 DIAL	
AUTOMATIC RADIO	C- 4	1936	AC	MANTLE BC/SW	
AUTOMATIC RADIO	C-66	1934	AC	TABLE BC/SW	
AUTOMATIC RADIO	CA-50	1933	AC	CONSOLE	
AUTOMATIC RADIO	CA-66	1933	AC	CONSOLE BC/SW	
AUTOMATIC RADIO	D-500	1934	AC	TABLE	
AUTOMATIC RADIO	D.C.	1929	DC	PORTABLE TOM THUMB	
AUTOMATIC RADIO	E- 6	1936	AC	TABLE BC/SW	
AUTOMATIC RADIO	F-2-5	1934	AC	TABLE BC/SW	
AUTOMATIC RADIO	J25	1931	AC	TABLE	

MANUFACTURER	MODEL	YEAR	POWER	STYLE	PRICE
AUTOMATIC RADIO	P35	1931	AC	COMPACT	
AUTOMATIC RADIO	P43	1939	BAT	PORTABLE CLOTH	
AUTOMATIC RADIO	P45	1939	BAT	PORTABLE CLOTH	
AUTOMATIC RADIO	P45	1931	AC	COMPACT	
AUTOMATIC RADIO	P57	1939	BAT	PORTABLE LEATHERETTE	
AUTOMATIC RADIO	P57LW	1939	BAT	PORTABLE LEATHERETTE BC/LW	
AUTOMATIC RADIO	P58	1939	BAT	PORTABLE LEATHERETTE	
AUTOMATIC RADIO	P61	1939	BAT	PORTABLE CLOTH	
AUTOMATIC RADIO	P61LW	1939	BAT	PORTABLE CLOTHE BC/SW	
AUTOMATIC RADIO	S-6	1936	AC	TABLE BC/SW	
AUTOMATIC RADIO	S-6	1937	AC	TABLE BC/SW	
AUTOMATIC RADIO	S-50	1933	AC	TABLE BC/SW	
AUTOMATIC RADIO	S-66	1933	AC	TABLE BC/SW	
AUTOMATIC RADIO	T	1936	AC	TABLE BC/SW	
AUTOMATIC RADIO	T-5	1933	AC	TABLE BC/SW	
AUTOMATIC RADIO	T-5-3	1934	AC	TABLE BC/SW	
AUTOMATIC RADIO	TOM THUMB	1930	AC	CATHEDRAL	85
AUTOMATIC RADIO	V-45	1931	AC	TABLE	
AUTOMETAL	ARAGAIN	1925	BAT	TABLE LP 5 TUBE 3 DIAL	75
B					
BAKER-SMITH	565 SYLFAN	1925	BAT	TABLE LP SLOPE PAN. 5T 3DL	
BALDWIN	40	1930	AC	PORTABLE	
BALDWIN	50	1930	AC	CATHEDRAL	80
BALDWIN	51	1930	AC	CATHEDRAL	80
BALDWIN	70	1930	AC	CONSOLE LB	
BALDWIN	71	1930	AC	CONSOLE LB	
BALDWIN	75	1930	AC	PORTABLE	
BALKEIT RADIO	70	1934	AC	CATHEDRAL MODERN	65
BALKEIT RADIO	A-3	1928	AC	TABLE LP METAL	
BALKEIT RADIO	A-5	1928	AC	TABLE LP	135
BALKEIT RADIO	A-7	1928	AC	CONSOLE HB DOORS	145
BALKEIT RADIO	C	1929	AC	CONSOLE	
BALTIMORE HUB-WHEEL	STANDARD	1923	BAT	PORTABLE 5 TUBE 3 DIAL	
BAUMAN RADIO	55	1924	BAT	TABLE LP METER 5 TUBE 3 DIAL	
BEACON RADIO	TRINITY FIVE	1924	BAT	TABLE LP 5 TUBE 3 DIAL	75
BEACON RADIO	TRINITY SIX	1924	BAT	TABLE LP 6 TUBE 3 DIAL	
BEACON RADIO	TRINITY SIX	1924	BAT	CONSOLE HB 6 TUBE 3 DIAL	
BELMONT	41-A	1932	AC	TABLE	
BELMONT	45	1932	DC	TABLE	
BELMONT	51-C	1932	AC	TABLE	
BELMONT	71-A	1932	AC	TABLE	
BELMONT	100-A	1932	AC	CONSOLE	
BELMONT	401-M	1935	AC	TABLE	
BELMONT	403	1939	BAT	PORTABLE	15
BELMONT	404	1935	2/DC	TABLE	
BELMONT	415-19	1937	BAT	TABLE	
BELMONT	415-42	1937	BAT	CONSOLE	

MANUFACTURER	MODEL	YEAR	POWER	STYLE	PRICE
BELMONT	425	1933	AC	TABLE METAL BC/SW	
BELMONT	489	1938	2/DC	TABLE BC/SW	
BELMONT	510	1939	AC	MANTLE PLASTIC	25
BELMONT	519	1939	BAT	MANTLE PB	25
BELMONT	522-C	1936	2/DC	TABLE	
BELMONT	523-28	1937	2/DC	TABLE	
BELMONT	523-42	1937	2/DC	CONSOLE	
BELMONT	525	1933	AC	TABLE METAL	
BELMONT	526	1938	AC	TABLE	20
BELMONT	530	1933	AC	TABLE BC/SW	
BELMONT	531	1938	AC	TABLE	
BELMONT	533	1940	AC	TABLE RADIO/PHONO	15
BELMONT	533-D	1941	AC	TABLE RADIO/PHONO	15
BELMONT	534	1940	AC	MANTLE PLASTIC PB	25
BELMONT	540	1935	AC	TABLE	
BELMONT	545	1933	AC	TABLE BC/SW	
BELMONT	575	1934	AC	TOMBSTONE	40
BELMONT	582	1938	AC	TABLE BC/SW	
BELMONT	585	1935	AC	TABLE BC/SW	
BELMONT	585-R	1935	AC	TABLE	
BELMONT	586-J	1936	AC	TABLE BC/SW	
BELMONT	588	1937	AC	TABLE BC/SW	
BELMONT	601-F	1936	AC	MANTLE	
BELMONT	602	1937	AC	TABLE	
BELMONT	625	1933	AC	TABLE BC/SW	
BELMONT	671	1940	AC	CONSOLE RADIO/PHONO	
BELMONT	675	1934	AC	TABLE BC/SW	
BELMONT	675-E	1934	AC	CONSOLE BC/SW	65
BELMONT	686-B	1934	AC	TABLE BC/SW	
BELMONT	740	1937	AC	TABLE BC/SW	
BELMONT	761	1938	AC	TABLE BC/SW	
BELMONT	775-T	1934	AC	MIDGET	
BELMONT	777-K	1935	AC	CONSOLE	
BELMONT	777-L	1935	AC	CONSOLE BC/SW	
BELMONT	777-T	1935	AC	TABLE BC/SW	
BELMONT	778-A	1934	AC	TABLE BC/SW	
BELMONT	778-M	1934	AC	CONSOLE BC/SW	
BELMONT	792	1939	AC	CONSOLE PB	85
BELMONT	793	1940	AC	CONSOLE RADIO/PHONO	40
BELMONT	796	1940	AC	CONSOLE RADIO/PHONO	
BELMONT	797	1940	AC	CONSOLE R/P RECORDER BC/SW	45
BELMONT	800	1940	AC	CONSOLE PB BC/SW	110
BELMONT	801	1940	AC	CONSOLE RADIO/PHONO	
BELMONT	804-41	1937	AC	CONSOLE BC/SW	
BELMONT	840	1937	AC	CONSOLE BC/SW	
BELMONT	860	1938	AC	CONSOLE BC/SW	
BELMONT	1070-N	1936	AC	CONSOLE BC/SW	
BELMONT	1077	1935	AC	CONSOLE BC/SW	
BELMONT	1170-B	1937	AC	CONSOLE BC/SW	

MANUFACTURER	MODEL	YEAR	POWER	STYLE	PRICE
BELMONT	1175	1938	AC	CONSOLE BC/SW	
BENSEN	4-X	1925	BAT	TABLE 4 TUBE 1 DIAL	
BENSEN	6-C	1925	BAT	CONSOLE 6 TUBE 3 DIAL	
BENSEN	A- 4	1927	BAT	TABLE LP 4 TUBE 1 DIAL	
BENSEN	A- 4	1927	BAT	CONSOLE 4 TUBE 1 DIAL	
BENSEN	SUPERFLEX	1924	BAT	PORTABLE 1 TUBE 2 DIAL	
BETTER RADIO	DELUXE MELODY	1924	BAT	TABLE LP SLOPE PAN. 4T 2DL	
BETTER RADIO	DELUXE MELODY	1924	BAT	CONSOLE HB 4 TUBE 2 DIAL	
BETTER RADIO	STANDARD	1925	BAT	TABLE LP 4 TUBE 2 DIAL	
BETTER RADIO	SUPERIOR	1924	BAT	TABLE LP 4 TUBE 2 DIAL	
BETTS & BETTS	T-8	1923	BAT	TABLE HP 8 TUBE 2 DIAL	
BIG THREE RADIO	WIDE WORLD	1924	BAT	TABLE LP 5 TUBE 2 DIAL	
BLAIR RADIO	6	1926	BAT	TABLE LP SLOPE PAN. 6T 3DL	85
BLAIR RADIO	11	1925	BAT	TABLE LP SLOPE PAN. 6T 3DL	95
BLAIR RADIO	11 BLAIRCO	1923	BAT	TABLE LP TRIANON 1 DIAL	
BLAIR RADIO	L'ELE'GANT	1925	BAT	CONSOLE 6 TUBE 3 DIAL	
BOISSONAULT	1926	1925	BAT	TABLE LP SLOPE PAN. 5T 3DL	75
BOSWORTH ELECTRIC	B-1	1925	BAT	TABLE LP SLOPE PAN 5 T 2DL	130
BOSWORTH ELECTRIC	B-2	1926	BAT	TABLE LP SLOPE PAN 5 T 2DL	110
BOSWORTH ELECTRIC	B-3	1925	BAT	TABLE LP SLOPE PAN 6 T 2DL	125
BOSWORTH ELECTRIC	B-5	1927	AC	TABLE LP 1 DIAL	
BOSWORTH ELECTRIC	B-6	1927	BAT	TABLE LP 2 DIAL	
BOSWORTH ELECTRIC	B-7	1927	AC	CONSOLE 1 DIAL	
BRANDEIS RADIO	B-10	1929	AC	TABLE	
BRANDEIS RADIO	B-15	1929	AC	CONSOLE	
BRANDEIS RADIO	B-16	1929	AC	CONSOLE	
BRANDEIS RADIO	BC-150	1924	BAT	CONSOLE HB 5 TUBE 1 DIAL	
BRANDEIS RADIO	BG-200	1924	BAT	CONSOLE LB 5 TUBE 1 DIAL	
BRANDEIS RADIO	BRANDOLA	1924	BAT	TABLE LP 6 TUBE 1 DIAL	175
BRANDEIS RADIO	VT-100	1924	BAT	TABLE LP SLOPE PAN. 5T 1DL	
BRANSTON	R-45	1925	BAT	TABLE LP 5 TUBE 3 DIAL	65
BRANSTON	R-47	1925	BAT	TABLE HP SPEAKER 5 T 2 DL	
BRANSTON	V HETROLA	1925	BAT	TABLE LP 2 DIAL	85
BREMER TULLY	5	1925	BAT	TABLE LP COUNTERPHASE 2 DIAL	75
BREMER TULLY	6	1925	BAT	TABLE LP COUNTERPHASE 2 DIAL	80
BREMER TULLY	6-35	1927	BAT	TABLE LP 6 TUBE 2 DIAL	110
BREMER TULLY	6-37	1927	BAT	CONSOLE 6 TUBE 2 DIAL	130
BREMER TULLY	6-40	1928	AC	TABLE LP 7 TUBE 1 DIAL	110
BREMER TULLY	6-41	1928	AC	CONSOLE	
BREMER TULLY	7-70	1928	AC	TABLE	
BREMER TULLY	7-71	1928	AC	CONSOLE	
BREMER TULLY	8	1926	BAT	TABLE LP COUNTERPHASE 1 DIAL	150
BREMER TULLY	8-20	1928	AC	TABLE LP 9 TUBE 1 DIAL	130
BREMER TULLY	8-21	1928	AC	CONSOLE	
BREMER TULLY	80	1929	BAT	CONSOLE	
BREMER TULLY	81	1929	AC	CONSOLE	
BREMER TULLY	82	1929	AC	CONSOLE	

MANUFACTURER	MODEL	YEAR	POWER	STYLE	PRICE
BRENARD	HERALDYNE	1926	BAT	CONSOLE LB 3 DIAL	
BRISTOL	HH-5	1923	BAT	PORTABLE 6 TUBE 2 DIAL	
BRISTOL	SINGLE CONTROL	1923	BAT	TABLE INVERSE DUPLEX 4T 1 DIAL	175
BROWN & MANHART	44	1930	AC	CATHEDRAL	90
BROWNING-DRAKE	5-A	1927	BAT	TABLE LP DRUM DIAL	90
BROWNING-DRAKE	5-R	1926	BAT	TABLE LP 5 TUBE 2 DIAL	75
BROWNING-DRAKE	6-A	1927	BAT	TABLE LP 6 TUBE 1 DIAL	80
BROWNING-DRAKE	7-A	1927	BAT	TABLE LP 7 TUBE 1 DIAL	95
BROWNING-DRAKE	7-A	1927	BAT	CONSOLE 7 TUBE 1 DIAL	
BROWNING-DRAKE	30	1928	AC	TABLE LP METAL	145
BROWNING-DRAKE	32	1928	AC	CONSOLE LB	175
BROWNING-DRAKE	34	1928	AC	TABLE LP 8 TUBE 1 DIAL	90
BROWNING-DRAKE	53	1929	AC	TABLE	80
BROWNING-DRAKE	54	1929	AC	CONSOLE	120
BROWNING-DRAKE	56	1929	AC	CONSOLE	
BROWNING-DRAKE	57	1929	AC	CONSOLE	
BROWNING-DRAKE	63	1929	AC	TABLE	80
BROWNING-DRAKE	64	1929	AC	CONSOLE	
BROWNING-DRAKE	66	1929	AC	CONSOLE	
BROWNING-DRAKE	67	1929	AC	CONSOLE	
BROWNING-DRAKE	68	1930	AC	TABLE	
BROWNING-DRAKE	69	1930	AC	CONSOLE LB	
BROWNING-DRAKE	70	1930	AC	CONSOLE LB	
BROWNING-DRAKE	70-R	1930	AC	CONSOLE LB	
BROWNING-DRAKE	71	1930	AC	CONSOLE HB	
BROWNING-DRAKE	71-R	1930	AC	CONSOLE HB	
BROWNING-DRAKE	72	1930	AC	CONSOLE HB RADIO/PHONO	
BROWNING-DRAKE	B-D JUNIOR	1925	BAT	TABLE LP 5 TUBE 2 DIAL	80
BROWNING-DRAKE	B-D SENIOR	1925	BAT	TABLE HP SPEAKER 6 T 2 DL	175
BROWNING-DRAKE	B-D STANDARD	1925	BAT	TABLE LP 5 TUBE 2 DIAL	120
BRUNSWICK	3-NC-8	1928	AC	CONSOLE HB RADIO/PHONO	
BRUNSWICK	3-NW-8	1929	AC	CONSOLE RADIO/PHONO	
BRUNSWICK	5-KR	1928	AC	TABLE	95
BRUNSWICK	10	1931	AC	TABLE	
BRUNSWICK	11	1931	AC	TABLE	
BRUNSWICK	12	1931	AC	CONSOLE HB	
BRUNSWICK	14	1929	AC	CONSOLE LB	
BRUNSWICK	15	1930	AC	CONSOLE LB	160
BRUNSWICK	16	1931	AC	CONSOLE LB	
BRUNSWICK	17	1931	AC	CONSOLE LB	
BRUNSWICK	21	1929	AC	CONSOLE HB	
BRUNSWICK	22	1930	AC	CONSOLE HB	
BRUNSWICK	24	1931	AC	CONSOLE HB	
BRUNSWICK	31	1929	AC	CONSOLE RADIO/PHONO	
BRUNSWICK	33	1931	AC	CONSOLE RADIO/PHONO	
BRUNSWICK	42	1930	AC	CONSOLE RADIO/PHONO	
BRUNSWICK	275	1940	AC	CONSOLE RADIO/PHONO	
BRUNSWICK	276	1940	AC	CONSOLE RADIO/PHONO	45

MANUFACTURER	MODEL	YEAR	POWER	STYLE	PRICE
BRUNSWICK	277	1940	AC	CONSOLE R/P PB BC/SW	60
BRUNSWICK	280	1940	AC	CONSOLE RADIO/PHONO	
BRUNSWICK	282	1940	AC	CONSOLE RADIO/PHONO	
BRUNSWICK	284	1940	AC	CONSOLE RADIO/PHONO	
BRUNSWICK	285	1940	AC	CONSOLE COMMODE R/P BC/SW	50
BRUNSWICK	286	1940	AC	CONSOLE R/P PB BC/SW	45
BRUNSWICK	299-FM	1941	AC	CONSOLE R/P BC/FM	85
BRUNSWICK	1680	1939	AC	CONSOLE PB SW/LW	
BRUNSWICK	1700	1939	AC	CONSOLE PB BC/SW/LW	
BRUNSWICK	2660	1939	AC	CONSOLE PB SW/LW	
BRUNSWICK	2689	1939	AC	CONSOLE PB SW/LW	
BRUNSWICK	3580	1939	AC	CONSOLE	
BRUNSWICK	3689	1939	AC	CONSOLE PB SW/LW	60
BRUNSWICK	4580	1939	AC	CONSOLE	
BRUNSWICK	4580.5	1939	AC	CONSOLE MAHOGANY	
BRUNSWICK	A-1020	1939	AC	CONSOLE PB BC/SW/LW	
BRUNSWICK	A-2020	1939	AC	CONSOLE PB BC/SW/LW	
BRUNSWICK	A-2600	1939	AC	CONSOLE PB SW/LW	
BRUNSWICK	A-2700	1939	AC	CONSOLE PB BC/SW/LW	
BRUNSWICK	A-3720	1939	AC	CONSOLE PB BC/SW/LW	
BRUNSWICK	P-5580	1939	AC	CONSOLE RADIO/PHONO	
BRUNSWICK	PR-138-C	1927	AC	CONSOLE LB RADIO/PHONO 2 DIAL	400
BRUNSWICK	PR-148-C	1926	BAT	CONSOLE RADIO/PHONO	
BRUNSWICK	R- 1	1929	AC	CONSOLE HB	
BRUNSWICK	T-1580	1939	AC	TABLE	
BRUNSWICK	T-2580	1939	AC	TABLE	
BUCKINGHAM RADIO	1	1928	AC	CONSOLE 8 TUBE 1 DIAL	150
BUCKINGHAM RADIO	2	1928	AC	CONSOLE 8 TUBE 1 DIAL	
BUCKINGHAM RADIO	3	1928	AC	CONSOLE 8 TUBE 1 DIAL	
BUCKINGHAM RADIO	4	1928	AC	TABLE 8 TUBE 1 DIAL	
BUCKWALTER RADIO	BURAD	1926	BAT	TABLE	
BUCKWALTER RADIO	COMRADE	1927	AC	CONSOLE	
BUCKWALTER RADIO	ENCHANTER	1927	BAT	TABLE	
BUCKWALTER RADIO	GRANDEE	1927	AC	CONSOLE	
BUCKWALTER RADIO	MUSE	1927	BAT	CONSOLE	
BUCKWALTER RADIO	SUPERTONE	1925	BAT	TABLE	
BUELL	PORTABLE	1925	BAT	PORTABLE 3 TUBE 2 DIAL	
BUELL	REGULAR	1925	BAT	MANTLE 3 TUBE 2 DIAL	
BUSH & LANE PIANO	10-C	1929	AC	CONSOLE	
BUSH & LANE PIANO	11-C	1928	AC	CONSOLE LB	
BUSH & LANE PIANO	12-C	1928	AC	CONSOLE RADIO/PHONO	
BUSH & LANE PIANO	20	1929	AC	TABLE	
BUSH & LANE PIANO	21	1929	AC	CONSOLE	
BUSH & LANE PIANO	30	1929	AC	CONSOLE	
BUSH & LANE PIANO	32	1929	AC	CONSOLE	
BUSH & LANE PIANO	34	1929	AC	CONSOLE	
BUSH & LANE PIANO	40	1929	AC	CONSOLE	
BUSH & LANE PIANO	50	1929	AC	CONSOLE	

MANUFACTURER	MODEL	YEAR	POWER	STYLE	PRICE
BUSH & LANE PIANO	55	1929	AC	CONSOLE LB	175
BUSH & LANE PIANO	60	1929	AC	CONSOLE	
BUSH & LANE PIANO	70	1929	AC	CONSOLE	
BUSH & LANE PIANO	90	1929	AC	CONSOLE LB	
C					
CADILLAC RADIO	CADILLAC	1926	BAT	TABLE LP SLOPE PANEL 3 DL	95
CADILLAC RADIO	CADILLAC	1926	BAT	TABLE LP SLOPE PANEL 2 DL	110
CAPEHART	1	1932	AC	CONSOLE LB RADIO/PHONO	
CAPEHART	11	1932	AC	CONSOLE HB RADIO/PHONO BC/SW	
CAPEHART	12	1932	AC	CONSOLE HB RADIO/PHONO BC/SW	
CAPEHART	14	1932	AC	CONSOLE HB RADIO/PHONO BC/SW	
CAPEHART	15	1931	AC	CONSOLE RADIO/PHONO	
CAPEHART	17-K- 3	1940	AC	CONSOLE	60
CAPEHART	20	1931	AC	CONSOLE RADIO/PHONO	
CAPEHART	21	1931	AC	CONSOLE RADIO/PHONO	
CAPEHART	28-F	1929	AC	CONSOLE RADIO/PHONO	
CAPEHART	101-C	1934	AC	CONSOLE RADIO/PHONO	
CAPEHART	101-CAW	1934	AC	CONSOLE R/P BC/SW/LW	
CAPEHART	110	1929	AC	CONSOLE HB	200
CAPEHART	200	1932	AC	CONSOLE HB RADIO/PHONO	
CAPEHART	202	1936	AC	CONSOLE RADIO/PHONO BC/SW	
CAPEHART	300	1932	AC	CONSOLE HB RADIO/PHONO	
CAPEHART	301-C	1934	AC	CONSOLE RADIO/PHONO	
CAPEHART	301-CAW	1934	AC	CONSOLE R/P BC/SW/LW	
CAPEHART	302-D	1935	AC	CONSOLE RADIO/PHONO BC/SW	
CAPEHART	400	1931	AC	CONSOLE RADIO/PHONO	
CAPEHART	400-A	1932	AC	CONSOLE HB RADIO/PHONO	
CAPEHART	400-B	1933	AC	CONSOLE RADIO/PHONO	
CAPEHART	400-BRC	1933	AC	CONSOLE RADIO/PHONO	
CAPEHART	401	1931	AC	CONSOLE RADIO/PHONO	
CAPEHART	402	1931	AC	CONSOLE RADIO/PHONO	
CAPEHART	402-A	1932	AC	CONSOLE HB RADIO/PHONO	
CAPEHART	402-B	1933	AC	CONSOLE RADIO/PHONO	
CAPEHART	402-BRC	1933	AC	CONSOLE RADIO/PHONO	
CAPEHART	402-CAW	1934	AC	CONSOLE R/P BC/SW/LW	
CAPEHART	404-B	1933	AC	CONSOLE RADIO/PHONO	
CAPEHART	404-BRC	1933	AC	CONSOLE RADIO/PHONO	
CAPEHART	404-C	1934	AC	CONSOLE RADIO/PHONO	
CAPEHART	404-CAW	1934	AC	CONSOLE R/P BC/SW/LW	
CAPEHART	404-D	1935	AC	CONSOLE RADIO/PHONO BC/SW	
CAPEHART	404-DR	1935	AC	CONSOLE RADIO/PHONO RC BC/SW	
CAPEHART	405-CAW	1934	AC	CONSOLE R/P BC/SW/LW	

MANUFACTURER	MODEL	YEAR	POWER	STYLE	PRICE
CAPEHART	405-D	1935	AC	CONSOLE RADIO/PHONO BC/SW	
CAPEHART	405-DR	1935	AC	CONSOLE RADIO/PHONO RC B/S	
CAPEHART	406-D	1935	AC	CONSOLE RADIO/PHONO BC/SW	
CAPEHART	406-DR	1935	AC	CONSOLE RADIO/PHONO RC B/S	
CARDINAL RADIO	70	1930	AC	TOMBSTONE RADIO/PHONO	60
CARDINAL RADIO	80	1930	AC	CATHEDRAL MODERN	50
CARDWELL	112-A	1922	BAT	TABLE HP 2 DIAL	195
CARDWELL	125-A	1922	BAT	TABLE LP 3 TUBE 2 DIAL	250
CARLOYD ELECTRIC	MARK II	1923	BAT	TABLE HP SLOPE PAN. 5T 3DL	
CARTERET RADIO	AC-7	1930	AC	CATHEDRAL	
CARTERET RADIO	AC-8	1930	AC	CATHEDRAL	
CARTERET RADIO	DC-8	1930	DC	CATHEDRAL	
CASE ELECTRIC	500	1935	AC	TABLE BC/SW	
CASE ELECTRIC	501	1935	AC	CONSOLE BC/SW	
CASE ELECTRIC	700	1935	AC	TABLE BC/SW	
CASE ELECTRIC	701	1935	AC	CONSOLE BC/SW	
CASE ELECTRIC	1000	1935	AC	TABLE BC/SW	
CASE ELECTRIC	1001	1935	AC	CONSOLE BC/SW	
CASE ELECTRIC	1003	1935	AC	CONSOLE BC/SW	
CHAMPION RADIO	S-2	1930	AC	CATHEDRAL W/CLOCK	95
CHELSEA RADIO	5 SUPER	1925	BAT	TABLE LP SLOPE PAN 5 T 3DL	125
CHELSEA RADIO	6 SUPER	1925	BAT	TABLE LP 6 TUBE 3 DIAL	90
CHELSEA RADIO	6 TRUPHONIC	1926	BAT	TABLE LP 6 TUBE 3 DIAL	100
CHELSEA RADIO	102	1923	BAT	TABLE HP 3 TUBE 2 DIAL	120
CHELSEA RADIO	107	1925	BAT	TABLE 4 TUBE 2 DIAL	95
CHELSEA RADIO	122	1925	BAT	TABLE	85
CHELSEA RADIO	130	1925	BAT	CONSOLE SUPER 5	110
CHELSEA RADIO	140	1926	BAT	TABLE BEARCAT 3 TUBE 2 DL	100
CHELSEA RADIO	REGENERATIVE	1922	BAT	TABLE HP 3 DIAL	225
CHICAGO NIPPLE	6-A	1927	AC	TABLE	
CHICAGO NIPPLE	8-A	1927	AC	TABLE W/LOOP	
CHICAGO NIPPLE	9-A	1927	AC	CONSOLE DESK	
CHICAGO RADIO LABORATORIES	Z-NITH	1921	BAT	TABLE LP 3 DIAL	450
CHISHOLM-BARFIELD	RADIO KNIGHT	1925	BAT	TABLE LP 3 DIAL	90
CLAIRE	AERIAL	1924	BAT	TABLE LP 5 TUBE 3 DIAL	
CLAPP-EASTHAM	BABY EMERSON	1927	BAT	TABLE LP 1 MULTIVALVE 1 DIAL	400
CLAPP-EASTHAM	C- 3	1923	BAT	TABLE HP 3 TUBE 2 DIAL	
CLAPP-EASTHAM	C-23	1923	BAT	TABLE HP 4 TUBE 2 DIAL	
CLAPP-EASTHAM	C-64	1923	BAT	TABLE HP 5 TUBE 2 DIAL	
CLAPP-EASTHAM	DD	1924	BAT	TABLE LP 3 TUBE 2 DIAL	
CLAPP-EASTHAM	H R RADAK	1922	BAT	TABLE HP 1 TUBE 2 DIAL	350
CLAPP-EASTHAM	HR/HZ RADAK	1922	BAT	TABLE	600
CLAPP-EASTHAM	R Z RADAK	1922	BAT	TABLE HP 3 TUBE 2 DIAL	450
CLAPP-EASTHAM	R- 4 RADAK	1923	BAT	TABLE HP 1 TUBE 1 DIAL	300

MANUFACTURER	MODEL	YEAR	POWER	STYLE	PRICE
CLAPP-EASTHAM	RHM	1924	BAT	TABLE 2 DIAL	
CLAPP-EASTHAM	SUPER-RADAK	1924	BAT	TABLE LP 4 TUBE 2 DIAL	
CLAPP-EASTHAM	T5-A	1925	BAT	TABLE LP 5 TUBE 3 DIAL	
CLAPP-EASTHAM	T5-B	1925	BAT	TABLE LP 5 TUBE 3 DIAL	
CLARION	60	1931	AC	MIDGET	
CLARION	61	1933	AC	MIDGET	
CLARION	70	1937	AC	TABLE BC/SW	170
CLARION	71	1937	AC	TABLE BC/SW	
CLARION	400	1937	3/DC	TABLE	
CLARION	401	1937	3/DC	TABLE	
CLARION	407	1937	3/DC	TABLE BC/SW	
CLARION	500	1937	6/DC	TABLE BC/SW	
CLARION	506	1937	AC	TABLE	
CLARION	600	1937	AC	TABLE BC/SW	
CLARION	601	1937	AC	TABLE BC/SW	
CLARION	750	1937	AC	TABLE BC/SW	
CLARION	751	1937	AC	TABLE BC/SW	
CLEARTONE RADIO	3	1923	BAT	TABLE	250
CLEARTONE RADIO	60	1923	BAT	TABLE LP 4 TUBE 3 DIAL	
CLEARTONE RADIO	62	1923	BAT	CONSOLE LB 4 TUBE 3 DIAL	
CLEARTONE RADIO	70	1923	BAT	TABLE LP 4 TUBE 3 DIAL	
CLEARTONE RADIO	72	1923	BAT	CONSOLE LB 4 TUBE 2 DIAL	
CLEARTONE RADIO	80	1925	BAT	PORTABLE 5 TUBE 3 DIAL	
CLEARTONE RADIO	82	1925	BAT	CONSOLE 5 TUBE 3 DIAL	
CLEARTONE RADIO	90	1925	BAT	TABLE 7 TUBE 2 DIAL	
CLEARTONE RADIO	91	1925	BAT	TABLE 7 TUBE 2 DIAL	
CLEARTONE RADIO	92	1925	BAT	CONSOLE LB 7 TUBE 2 DIAL	
CLEARTONE RADIO	110 COMPACT	1927	AC	TABLE LP 5 TUBE 1 DIAL	
CLEARTONE RADIO	110 STANDARD	1926	AC	TABLE LP 5 TUBE 2 DIAL	
CLEARTONE RADIO	110 STANDARD	1927	AC	TABLE LP 5 TUBE 1 DIAL	
CLEARTONE RADIO	110C CONSOLE	1927	AC	CONSOLE 5 TUBE 1 DIAL	
CLEARTONE RADIO	112	1930	AC	TABLE	
CLEARTONE RADIO	112	1930	AC	CONSOLE HB	
CLEARTONE RADIO	MAYFLOWER	1927	AC	CONSOLE LB 5 TUBE 1 DIAL	130
CLEARTONE RADIO	MAYFLOWER	1927	AC	TABLE LP 5 TUBE 1 DIAL	120
CLEVELAND AUTO. ACCESSORIES	A	1924	BAT	TABLE LP GLASS PAN. 5T 3DL	
CLEVELAND PRODUCTS	A-5	1923	BAT	TABLE LP 5 TUBE 3 DIAL	75
CLIMAX RADIO & TELEVISION	2-VOLTER	1937	2/DC	TABLE BC/SW	
CLIMAX RADIO & TELEVISION	6-VOLTER	1937	6/DC	TABLE BC/SW	
CLIMAX RADIO & TELEVISION	7	1936	AC	MANTLE	
CLIMAX RADIO & TELEVISION	10	1936	AC	MANTLE	
CLIMAX RADIO & TELEVISION	12	1936	AC	TABLE	
CLIMAX RADIO & TELEVISION	30	1936	AC	TABLE BC/SW	
CLIMAX RADIO & TELEVISION	35	1936	AC	TABLE BC/SW	
CLIMAX RADIO & TELEVISION	40	1936	AC	TABLE BC/SW	
CLIMAX RADIO & TELEVISION	41	1936	AC	TABLE BC/SW	
CLIMAX RADIO & TELEVISION	90	1936	AC	TABLE BC/SW	
CLIMAX RADIO & TELEVISION	91	1936	AC	TABLE BC/SW	
CLIMAX RADIO & TELEVISION	92E	1936	AC	TABLE BC/SW	

MANUFACTURER	MODEL	YEAR	POWER	STYLE	PRICE
CLIMAX RADIO & TELEVISION	92E358	1936	AC	CONSOLE BC/SW	
CLIMAX RADIO & TELEVISION	A-64	1937	AC	TABLE BC/SW	
CLIMAX RADIO & TELEVISION	A-65	1937	AC	TABLE BC/SW	
CLIMAX RADIO & TELEVISION	DIAMOND	1937	AC	MANTLE BC/SW	
CLIMAX RADIO & TELEVISION	DUKE	1937	AC	MANTLE	
CLIMAX RADIO & TELEVISION	ESQUIRE	1937	AC	CONSOLE BC/SW	
CLIMAX RADIO & TELEVISION	ESQUIRE JR.	1937	AC	CONSOLE BC/SW	
CLIMAX RADIO & TELEVISION	GARNET	1937	AC	MANTLE BC/SW	
CLIMAX RADIO & TELEVISION	MONA	1937	AC	TABLE BC/SW	
CLIMAX RADIO & TELEVISION	OPAL	1937	AC	MANTLE	
CLIMAX RADIO & TELEVISION	U-64	1937	AC	TABLE BC/SW	
CLIMAX RADIO & TELEVISION	U-65	1937	AC	TABLE BC/SW	
CLIMAX RADIO & TELEVISION	V-56	1937	AC	TABLE BC/SW	
CLIMAX RADIO & TELEVISION	VISION	1937	AC	TABLE BC/SW	
COLONIAL RADIO	16-5	1925	BAT	TABLE	
COLONIAL RADIO	16-6	1925	BAT	TABLE	
COLONIAL RADIO	17-5	1925	BAT	TABLE	
COLONIAL RADIO	21	1925	BAT	CONSOLE	
COLONIAL RADIO	26	1926	BAT	CONSOLE 6 TUBE 3 DIAL	
COLONIAL RADIO	28 AC	1927	AC	TABLE	
COLONIAL RADIO	28 DC	1927	DC	TABLE	
COLONIAL RADIO	31 AC	1928	DC	CONSOLE 7 TUBE 1 DIAL	
COLONIAL RADIO	31 DC	1928	BAT	CONSOLE 6 TUBE 1 DIAL	
COLONIAL RADIO	32	1929	DC	CONSOLE CAVALIER	
COLONIAL RADIO	32	1929	DC	CONSOLE PICADILLY	
COLONIAL RADIO	32	1929	DC	CONSOLE MODERNE	
COLONIAL RADIO	32	1929	AC	CONSOLE CAVALIER	
COLONIAL RADIO	32	1929	AC	CONSOLE PICADILLY	
COLONIAL RADIO	32	1929	AC	CONSOLE MODERNE	
COLONIAL RADIO	33	1930	DC	CONSOLE HB MAYFLOWER	
COLONIAL RADIO	33	1930	DC	CONSOLE HB WINDSOR	
COLONIAL RADIO	33	1930	AC	CONSOLE LB PRINCESS	
COLONIAL RADIO	33	1930	AC	CONSOLE HB WINDSOR	
COLONIAL RADIO	33	1930	AC	CONSOLE HB MAYFLOWER	
COLONIAL RADIO	34	1930	DC	CONSOLE HB LAFAYETTE	
COLONIAL RADIO	34	1930	AC	CONSOLE HB LAFAYETTE	
COLONIAL RADIO	36C	1931	AC	CLOCK	
COLONIAL RADIO	47H	1931	AC	CONSOLE HB	
COLONIAL RADIO	47L	1931	AC	CONSOLE LB	
COLONIAL RADIO	47M	1931	AC	CONSOLE HB	
COLONIAL RADIO	48	1931	AC	TABLE	
COLONIAL RADIO	48A	1931	AC	TABLE	
COLONIAL RADIO	136	1933	AC	TABLE BC/SW	
COLONIAL RADIO	250	1933	AC	TABLE BC/SW	
COLONIAL RADIO	300	1933	AC	TABLE BC/SW	
COLONIAL RADIO	301	1933	AC	TABLE BC/SW	
COLONIAL RADIO	400	1933	AC	TABLE BC/SW	
COLONIAL RADIO	500	1933	AC	CONSOLE END TABLE BC/SW	
COLONIAL RADIO	501	1933	AC	CONSOLE BOOKCASE BC/SW	
COLONIAL RADIO	603	1934	AC	CONSOLE BC/SW	

MANUFACTURER	MODEL	YEAR	POWER	STYLE	PRICE
COLONIAL RADIO	604	1934	AC	CONSOLE BC/SW	
COLONIAL RADIO	605	1934	AC	CONSOLE BC/SW	
COLONIAL RADIO	651	1934	AC	TABLE BC/SW	
COLONIAL RADIO	652	1934	AC	TABLE BC/SW	
COLONIAL RADIO	653	1934	AC	TABLE	
COLONIAL RADIO	654	1934	AC	TABLE	
COLONIAL RADIO	655	1934	AC	TABLE BC/SW	
COLONIAL RADIO	656	1934	AC	TABLE BC/SW	
COLONIAL RADIO	657	1934	AC	TABLE BC/SW	
COLONIAL RADIO	658	1934	AC	TABLE BC/SW	
COLONIAL RADIO	659	1934	BAT	TABLE BC/SW	
COLUMBIA PHONOGRAPH	920	1929	AC	CONSOLE LB RADIO/PHONO	175
COLUMBIA PHONOGRAPH	940	1929	AC	CONSOLE RADIO/PHONO PB	300
COLUMBIA PHONOGRAPH	950	1929	AC	CONSOLE RADIO/PHONO	
COLUMBIA PHONOGRAPH	960	1928	AC	CONSOLE RADIO/PHONO	
COLUMBIA PHONOGRAPH	961	1929	AC	CONSOLE RADIO/PHONO	
COLUMBIA PHONOGRAPH	C-1	1928	AC	TABLE	
COLUMBIA PHONOGRAPH	C-2	1928	AC	TABLE	
COLUMBIA PHONOGRAPH	C-3	1928	AC	CONSOLE HB	
COLUMBIA PHONOGRAPH	C-4	1928	AC	CONSOLE HB	
COLUMBIA PHONOGRAPH	C-5	1928	AC	CONSOLE HB	
COLUMBIA PHONOGRAPH	C-6	1928	BAT	TABLE	
COLUMBIA PHONOGRAPH	C-7	1929	BAT	CONSOLE HB	
COLUMBIA PHONOGRAPH	C-11	1929	AC	CONSOLE HB PUSHBUTTON	300
COLUMBIA PHONOGRAPH	C-20	1930	AC	CONSOLE LB	
COLUMBIA PHONOGRAPH	C-21	1930	AC	CONSOLE HB	
COLUMBIA PHONOGRAPH	C-22	1931	AC	CONSOLE	
COLUMBIA PHONOGRAPH	C-31	1931	AC	COMPACT	
COLUMBIA PHONOGRAPH	C-32	1931	AC	CONSOLE	
COLUMBIA PHONOGRAPH	C-33	1931	AC	COMPACT	
COLUMBIA PHONOGRAPH	C-34	1931	AC	CONSOLE	
COLUMBIA PHONOGRAPH	C-81	1932	AC	COMPACT	
COLUMBIA PHONOGRAPH	C-83	1932	AC	CONSOLE LB	
COLUMBIA PHONOGRAPH	C-84	1932	AC	CONSOLE HB	
COLUMBIA PHONOGRAPH	C-85	1932	AC	CONSOLE RADIO/PHONO	
COLUMBIA PHONOGRAPH	C-93	1932	AC	CONSOLE HB	
COLUMBIA PHONOGRAPH	C-95	1932	AC	CONSOLE	
COLUMBIA PHONOGRAPH	C-103	1932	AC	CONSOLE LB	
COLUMBIA PHONOGRAPH	C-123	1932	BAT	CONSOLE	
COLUMBIA RADIO	6 SENTINEL	1927	BAT	TABLE LP 2 DIAL	110
CONCERT RADIOPHONE	CONCERT GRAND	1923	BAT	TABLE LP 4 TUBE 3 DIAL	75
CONCERT RADIOPHONE	CONCERT SENIOR	1923	BAT	PORTABLE 2 TUBE 2 DIAL	150
CONSOLIDATED	25 ARBORPHONE	1927	BAT	TABLE 6 TUBE 1 DIAL	
CONSOLIDATED	27 ARBORPHONE	1927	BAT	TABLE 5 TUBE 2 DIAL	90
CONSOLIDATED	37-AC ARBORP.	1928	AC	TABLE LP 8 TUBE 1 DIAL	140
CONSOLIDATED	45 ARBORPHONE	1928	AC	TABLE LP	60
CONSOLIDATED	55 ARBORPHONE	1928	AC	TABLE	
CONSOLIDATED	252 ARBORPHONE	1927	BAT	TABLE 6 TUBE 1 DIAL	
CONSOLIDATED	253 ARBORPHONE	1927	BAT	CONSOLE HB 6 TUBE 1 DIAL	200
CONSOLIDATED	255 ARBORPHONE	1927	BAT	CONSOLE RADIO/PHONO 6T 1DL	

MANUFACTURER	MODEL	YEAR	POWER	STYLE	PRICE
CONSOLIDATED	271 ARBORPHONE	1927	BAT	CONSOLE 5 TUBE 2 DIAL	
CONSOLIDATED	272 ARBORPHONE	1927	BAT	CONSOLE 5 TUBE 2 DIAL	
CONSUMERS RADIO	6 SUPER	1927	BAT	TABLE LP 1 DIAL	95
CONSUMERS RADIO	CON-RAD	1927	BAT	TABLE LP 1 DIAL	85
CONTINENTAL	12-B- 5	1940	AC	MANTLE PLASTIC EBONY BC/SW	20
CONTINENTAL	13-C- 5	1940	AC	MANTLE PLASTIC MAHOGANY	20
CONTINENTAL	14-B-4	1940	AC	MANTLE PLASTIC IVORY BC/SW	25
CONTINENTAL	14-C-5	1940	AC	MANTLE PLASTIC IVORY	20
CONTINENTAL	15-B-5	1940	AC	MANTLE PLASTIC WAL BC/SW	20
CONTINENTAL	15-D-5	1940	AC	MANTLE PLAS STREAMLINED	20
CONTINENTAL	16-B-5	1940	AC	MANTLE PLASTIC IVORY BC/SW	20
CONTINENTAL	16-D-5	1940	AC	MANTLE PLAS IV STREAMLINED	25
CONTINENTAL	17-B-5	1940	AC	TABLE BC/SW	25
CONTINENTAL	18-B-5	1940	AC	TABLE BC/SW	25
CONTINENTAL	20-A-6	1940	AC	MANTLE PLASTIC WALNUT	25
CONTINENTAL	21-A-6	1940	AC	MANTLE PLASTIC IVORY	25
CONTINENTAL	22-A-6	1940	AC	TABLE	30
CONTINENTAL	23-A-6	1940	AC	TABLE	35
CONTINENTAL	24-A-5	1940	AC	CONSOLE	80
CONTINENTAL	25-Q-5	1940	AC	TABLE PB	25
CONTINENTAL	27-G-4	1940	BAT	PORTABLE BANTAM CAMERA	20
CONTINENTAL	33-F-5	1940	AC	PORTABLE	20
CONTINENTAL	34-F-5	1940	AC	PORTABLE LEATHERETTE	20
CONTINENTAL	35-G-6	1940	AC	PORTABLE LEATHERETTE	15
CONTINENTAL	37-G-6	1940	AC	PORTABLE W/WOOD PANELS	25
CONTINENTAL	43-B-4	1940	2/DC	TABLE	25
CONTINENTAL	47-J-55	1940	AC	TABLE BC/SW	35
CONTINENTAL	48-J-6	1940	AC	TABLE PLASTIC PB BC/SW	20
CONTINENTAL	49-J-6	1940	AC	TABLE PLASTIC IV. PB BC/SW	25
CONTINENTAL	50-J-6	1940	AC	TABLE PLASTIC PB BC/SW	35
CONTINENTAL	51-J-55	1940	AC	MANTLE PLASTIC MAHOG BC/SW	20
CONTINENTAL	51-K-6	1940	AC	MANTLE PLASTIC BC/SW	20
CONTINENTAL	52-J-55	1940	AC	MANTLE PLASTIC IVORY BC/SW	25
CONTINENTAL	52-K-6	1940	AC	MANTLE PLASTIC IVORY BC/SW	25
CONTINENTAL	53-K-6	1940	AC	TABLE BC/SW	35
CONTINENTAL	54-XJ-55	1940	AC	TABLE R/P BC/SW	15
CONTINENTAL	55-A-7	1940	AC	TABLE PB BC/SW	40
CONTINENTAL	56-A-77	1940	AC	CONSOLE PB BC/SW	90
CONTINENTAL	57-B-7	1940	AC	CONSOLE R/P PB BC/SW	50
CONTINENTAL	58-A-11	1940	AC	CONSOLE R/P PB TE BC/SW	50
CONTINENTAL	59-A-11	1940	AC	CONSOLE R/P PB TE BC/SW	50
CONTINENTAL	61-K-7	1940	AC	TABLE R/P W/RECORDER	20

MANUFACTURER	MODEL	YEAR	POWER	STYLE	PRICE
CONTINENTAL	62-B-7	1940	AC	CONSOLE RADIO/PHONO	45
CONTINENTAL	63-A-11	1940	AC	CONSOLE PB TUN EYE BC/SW	115
CONTINENTAL	64-K-5	1940	AC	TABLE RADIO/PHONO	
CONTINENTAL	69-M-5	1941	AC	TABLE RADIO/PHONO	15
CONTINENTAL	70-K-5	1941	AC	TABLE RADIO/PHONO	15
CONTINENTAL	72-D-5	1941	AC	TABLE	30
CONTINENTAL	102-6B	1938	AC	TABLE BC/SW	30
CONTINENTAL	103-6B	1938	AC	TABLE BC/SW	30
CONTINENTAL	104-4A	1939	BAT	TABLE	35
CONTINENTAL	113-5A	1938	AC	MANTLE PLASTIC BLACK	
CONTINENTAL	113-5A	1938	AC	TABLE PLASTIC	
CONTINENTAL	114-5A	1938	AC	TABLE PLASTIC	
CONTINENTAL	115-5A	1938	AC	MANTLE PLASTIC IVORY	
CONTINENTAL	115-5A	1938	AC	MANTLE PLASTIC WALNUT	
CONTINENTAL	115-5A	1938	AC	TABLE PLASTIC	
CONTINENTAL	123-5E	1938	AC	TABLE PLASTIC BLACK	15
CONTINENTAL	124-5E	1938	AC	TABLE PLASTIC WALNUT	15
CONTINENTAL	125-5E	1938	AC	TABLE PLASTIC IVORY	20
CONTINENTAL	126-5E	1938	AC	TABLE PLASTIC RED	20
CONTINENTAL	148-6K	1939	AC	TABLE PB	55
CONTINENTAL	153-5L	1939	AC	TABLE RADIO/PHONO	15
CONTINENTAL	159-5L	1939	AC	TABLE RADIO/PHONO	15
CONTINENTAL	162-5L	1939	AC	TABLE PB	25
CONTINENTAL	305-7C	1939	AC	TABLE PB BC/SW	
CONTINENTAL	311-4D	1939	BAT	PORTABLE RADIO/PHONO	15
CONTINENTAL	335-4Z	1939	BAT	PORTABLE LUGGAGE STYLE	15
CONTINENTAL	336-5N	1939	AC	PORTABLE LUGGAGE STYLE	15
CONTINENTAL	351-4A	1939	BAT	CONSOLE	70
CONTINENTAL	352-5R	1939	AC	TABLE PB	35
CONTINENTAL	361-5Q	1939	AC	MANTLE PLASTIC PB	
CONTINENTAL	362-5Q	1939	AC	MANTLE PLASTIC PB	
CONTINENTAL	366-6J	1939	AC	MANTLE PLASTIC WALNUT PB	20
CONTINENTAL	367-6J	1939	AC	MANTLE PLASTIC IVORY PB	20
CONTINENTAL	368-6J	1939	AC	MANTLE PLASTIC ONYX PB	65
CONTINENTAL	369-6J	1939	AC	MANTLE PB	35
CONTINENTAL	371-5R	1939	AC	MANTLE PLASTIC PB	20
CONTINENTAL	372-5R	1939	AC	MANTLE PLASTIC PB	25
CONTINENTAL	373-5R	1939	AC	MANTLE PLASTIC PB	65
CONTINENTAL	380-7H	1939	AC	CONSOLE PB BC/SW	90
CONTINENTAL	381-7H	1939	AC	CONSOLE PB BC/SW	90
CONTINENTAL	382-7H	1939	AC	CONSOLE R/P PB BC/SW	45
CONTINENTAL	383-7H	1939	AC	CONSOLE R/P PB BC/SW	45
CONTINENTAL	384-5S	1939	AC	TABLE	30
CONTINENTAL	392-7C	1940	AC	CONSOLE	65
CONTINENTAL	394-11B	1939	AC	CONSOLE PB BC/SW	100
CONTINENTAL	395-11B	1939	AC	CONSOLE R/P PB BC/SW	45
CONTINENTAL	396-6M	1940	AC	MANTLE PLASTIC WAL PB	20
CONTINENTAL	397-6M	1940	AC	MANTLE PLASTIC IVORY PB	20
CONTINENTAL	398-6M	1940	AC	MANTLE PLASTIC ONYX PB	20
CONTINENTAL	399-6M	1940	AC	TABLE PB	25

MANUFACTURER	MODEL	YEAR	POWER	STYLE	PRICE
CONTINENTAL	510-6B	1938	AC	TABLE RADIO/PHONO BC/SW	
CONTINENTAL	512-6D	1938	AC	TABLE BC/SW	20
CONTINENTAL	516-5C	1938	AC	MANTLE PLASTIC	
CONTINENTAL	516-5G	1938	AC	MANTLE PLASTIC	
CONTINENTAL	516-5G	1938	AC	MANTLE PLASTIC WALNUT	
CONTINENTAL	517-5C	1938	AC	MANTLE PLASTIC	
CONTINENTAL	517-5G	1938	AC	MANTLE PLASTIC WALNUT	
CONTINENTAL	517-5G	1938	AC	MANTLE PLASTIC IVORY	
CONTINENTAL	521-5C	1938	AC	TABLE RADIO/PHONO	15
CONTINENTAL	521-5G	1938	AC	TABLE	
CONTINENTAL	527-5X	1938	AC	TABLE	
CONTINENTAL	915-9G	1938	AC	CONSOLE BC/SW	
CONTINENTAL	920-6Q	1937	BAT	TABLE BC/SW UPRIGHT	
CONTINENTAL	925-16R	1937	AC	CONSOLE BC/SW TILT TUNER	
CONTINENTAL	930-16R	1937	AC	CONSOLE BC/SW TILT TUNER	180
CONTINENTAL	935-11S	1937	AC	CONSOLE BC/SW TILT TUNER	
CONTINENTAL	940-11S	1937	AC	CONSOLE BC/SW	
CONTINENTAL	945-8K	1937	AC	CONSOLE BC/SW	
CONTINENTAL	945-8T	1937	AC	CONSOLE BC/SW	
CONTINENTAL	950-6A	1938	6/DC	CONSOLE BC/SW	
CONTINENTAL	950-6P	1937	BAT	CONSOLE BC/SW	
CONTINENTAL	955-8K	1937	AC	CONSOLE BC/SW ARMCHAIR	
CONTINENTAL	955-8T	1937	AC	CONSOLE BC/SW ARMCHAIR	
CONTINENTAL	960-8K	1937	AC	TABLE BC/SW LAYDOWN	
CONTINENTAL	960-8T	1937	AC	TABLE BC/SW LAYDOWN	
CONTINENTAL	965-6A	1938	6/DC	TABLE BC/SW	
CONTINENTAL	965-6P	1937	BAT	TABLE BC/SW LAYDOWN	
CONTINENTAL	965-7M	1937	AC	TABLE BC/SW LAYDOWN	
CONTINENTAL	975-6W	1937	AC	TABLE BC/SW LAYDOWN	
CONTINENTAL	980-5X	1937	AC	TABLE BC/SW LAYDOWN	
CONTINENTAL	985-5Z	1937	AC	TABLE LAYDOWN	
CONTINENTAL	985-6Y	1937	AC	TABLE BC/SW LAYDOWN	
CONTINENTAL	990-5Z	1937	AC	TABLE LAYDOWN	
CONTINENTAL	990-6Y	1937	AC	TABLE BC/SW LAYDOWN	
CONTINENTAL	39411-B	1939	AC	CONSOLE PB BC/SW	130
CONTINENTAL	A126	1936	AC	TABLE SEMI-UPRIGHT BC/SW	
CONTINENTAL	AM387	1936	AC	CONSOLE BC/SW/LW	
CONTINENTAL	AM688	1936	AC	CONSOLE TILT TUNER BC/SW	
CONTINENTAL	AM786	1936	AC	CONSOLE BC/SW	140
CONTINENTAL	AM787	1936	AC	CONSOLE TILT-TUNE BC/SW	190
CONTINENTAL	AM889	1936	AC	CONSOLE TILT TUNER BC/SW	
CONTINENTAL	AZ393	1936	AC	CONSOLE BC/SW	
CONTINENTAL	AZ593	1936	6/DC	CONSOLE BC/SW	
CONTINENTAL	B125	1936	AC	TABLE BC/SW	
CONTINENTAL	B225	1936	AC	TABLE BC/SW	
CONTINENTAL	L215A	1936	AC	CONSOLE BC/SW	
CONTINENTAL	M551	1936	6/DC	TABLE UPRIGHT BC/SW/LW	
CONTINENTAL	M587	1936	6/DC	CONSOLE BC/SW/LW	
CONTINENTAL	ML156	1936	AC	TABLE BC/SW	
CONTINENTAL	ML266	1935	AC	TABLE BC/SW	

MANUFACTURER	MODEL	YEAR	POWER	STYLE	PRICE
CONTINENTAL	ML477	1936	AC	TABLE BC/SW	
CONTINENTAL	ML481	1936	AC	CONSOLE BC/SW	
CONTINENTAL	MX218	1936	AC	CONSOLE BC/SW	
CONTINENTAL	MX239	1936	AC	TABLE BC/SW	
CONTINENTAL	MX241	1936	AC	TABLE BC/SW	
CONTINENTAL	N351	1936	AC	TABLE UPRIGHT BC/SW/LW	
CONTINENTAL	R 58-B-11	1940	AC	CONSOLE R/P W/REC.PB BC/SW	60
CONTINENTAL	R 59-B-11	1940	AC	CONSOLE R/P W/REC.PB BC/SW	60
CONTINENTAL	R-20	1929	AC	CONSOLE 11 TUBE 2 DIAL	
CONTINENTAL	R-25	1929	AC	CONSOLE	
CONTINENTAL	R-30	1929	AC	CONSOLE RADIO/PHONO 11T 2 DIAL	
CONTINENTAL	R-105	1929	AC	CONSOLE RADIO/PHONO CUSTOM	
CONTINENTAL	RP-40	1929	AC	CONSOLE RADIO/PHONO 11T 2 DIAL	
CONTINENTAL	RP-115	1929	AC	CONSOLE RADIO/PHONO CUSTOM	
CONTINENTAL	X118	1936	AC	CONSOLE BC/SW	
CONTINENTAL	X140	1936	AC	TABLE BC/SW	
CONTINENTAL	X141	1936	AC	TABLE BC/SW	
CONTINENTAL	X341	1936	AC	TABLE BC/SW	
CONTINENTAL	X541	1936	AC	TABLE BC/SW	
CONTINENTAL	X641	1936	AC	TABLE BC/SW	
CONTINENTAL	X718	1936	AC	CONSOLE BC/SW	
CONTINENTAL	Z344	1936	AC	TABLE BC/SW	
CONTINENTAL	Z544	1936	6/DC	TABLE BC/SW	
COSMOPOLITAN	COSMOPOLITAN 5	1924	BAT	TABLE LP 5 TUBE 3 DIAL	65
COTO-COIL	COTO SYMPHONIC	1924	BAT	TABLE LP 4 TUBE 2 DIAL	
CROSLEY RADIO	2C1	1934	AC	MANTLE (SAMPLER)	
CROSLEY RADIO	4	1934	BAT	TABLE	
CROSLEY RADIO	4-29	1926	BAT	TABLE LP SLOPE PAN 4 T 2DL	140
CROSLEY RADIO	4-29-P	1926	BAT	PORTABLE 4 TUBE 2 DIAL	100
CROSLEY RADIO	4C1	1934	BAT	PORTABLE	
CROSLEY RADIO	5	1933	BAT	CONSOLE LB	
CROSLEY RADIO	5	1933	BAT	TABLE	
CROSLEY RADIO	5-38	1926	BAT	TABLE LP SLOPE PAN 5 T 3DL	200
CROSLEY RADIO	5-50	1926	BAT	TABLE LP SLOPE PAN 5 T 1DL	60
CROSLEY RADIO	5-75	1926	BAT	CONSOLE HB 5 TUBE 1 DIAL	175
CROSLEY RADIO	5B2	1934	BAT	CONSOLE	
CROSLEY RADIO	5B2	1934	BAT	TABLE	
CROSLEY RADIO	5C2	1934	BAT	PORTABLE	
CROSLEY RADIO	5DL	1934	AC	TABLE	
CROSLEY RADIO	5DL-LB	1934	AC	CONSOLE LB	
CROSLEY RADIO	5H1	1934	AC	CONSOLE BC/SW	
CROSLEY RADIO	5H1	1934	AC	TABLE BC/SW	
CROSLEY RADIO	5M3	1934	AC	TABLE	
CROSLEY RADIO	5V1	1934	AC	CONSOLE BC/SW	

MANUFACTURER	MODEL	YEAR	POWER	STYLE	PRICE
CROSLEY RADIO	5V1	1934	AC	TABLE BC/SW	
CROSLEY RADIO	6	1936	32/DC	TABLE BC/SW	
CROSLEY RADIO	6-60	1927	BAT	TABLE	160
CROSLEY RADIO	6-85	1927	BAT	CONSOLE	
CROSLEY RADIO	6H2	1934	AC	CONSOLE BC/SW	
CROSLEY RADIO	6H2	1934	AC	TABLE BC/SW	60
CROSLEY RADIO	7H2	1934	AC	CONSOLE BC/SW	
CROSLEY RADIO	7H2	1934	AC	TABLE BC/SW	
CROSLEY RADIO	7V2	1934	AC	CONSOLE BC/SW	85
CROSLEY RADIO	7V2	1934	AC	TABLE BC/SW	
CROSLEY RADIO	8	1933	BAT	COMPACT	
CROSLEY RADIO	8	1934	BAT	CONSOLE	
CROSLEY RADIO	8AF	1934	AC	CONSOLE	
CROSLEY RADIO	8AF	1934	AC	TABLE	
CROSLEY RADIO	8B2	1934	BAT	CONSOLE	
CROSLEY RADIO	8B2	1934	BAT	TABLE	
CROSLEY RADIO	8H1	1934	AC	CONSOLE BC/SW	
CROSLEY RADIO	8H1	1934	AC	TABLE BC/SW	
CROSLEY RADIO	9	1933	BAT	CONSOLE LB	
CROSLEY RADIO	10	1934	AC	CONSOLE	
CROSLEY RADIO	10-LB	1934	AC	CONSOLE LB	
CROSLEY RADIO	10P3	1934	AC	CONSOLE BC/SW	
CROSLEY RADIO	12	1934	AC	CONSOLE	
CROSLEY RADIO	16-AL	1940	AC	TABLE PB	35
CROSLEY RADIO	20	1929	AC	TABLE CHEST	
CROSLEY RADIO	21	1929	AC	TABLE LP METAL 5 TUBE 1 DL	
CROSLEY RADIO	22	1929	AC	CONSOLE 5 TUBE 1 DIAL	
CROSLEY RADIO	22-CB	1941	AC	CONSOLE PB BC/SW/FM	160
CROSLEY RADIO	22-CP	1941	AC	CONSOLE BC/SW/FM	55
CROSLEY RADIO	25-AY	1940	AC	CONSOLE PB BC/SW	120
CROSLEY RADIO	26-H	1930	BAT	TABLE	
CROSLEY RADIO	26-J	1930	BAT	CONSOLE HB CRONY	
CROSLEY RADIO	26-K	1930	BAT	CONSOLE HB PARTNER	
CROSLEY RADIO	28-AZ	1940	AC	CONSOLE R/P W/RECORDER	45
CROSLEY RADIO	28N	1931	BAT	TABLE	
CROSLEY RADIO	29-AT	1940	AC	CONSOLE RADIO/PHONO	
CROSLEY RADIO	29-BA	1940	AC	CONSOLE RADIO/PHONO	
CROSLEY RADIO	30-BC	1940	AC	CONSOLE R/P PB BC/SW	45
CROSLEY RADIO	30-S	1929	AC	TABLE CHEST	
CROSLEY RADIO	31	1929	AC	TABLE 6 TUBE 1 DIAL	
CROSLEY RADIO	31-BF	1940	AC	CONSOLE RADIO/PHONO	
CROSLEY RADIO	31-S	1929	AC	TABLE METAL	
CROSLEY RADIO	32	1933	DC	CONSOLE LB	
CROSLEY RADIO	32	1933	DC	TABLE	
CROSLEY RADIO	32	1929	AC	CONSOLE 6 TUBE 1 DIAL	
CROSLEY RADIO	33-BG	1940	AC	TABLE RADIO/PHONO	15
CROSLEY RADIO	33-S	1929	AC	CONSOLE LB	145
CROSLEY RADIO	34-BH	1940	AC	CONSOLE RADIO/PHONO	
CROSLEY RADIO	34-S	1929	AC	CONSOLE	
CROSLEY RADIO	35-AK	1941	AC	TABLE RADIO/PHONO	15

MANUFACTURER	MODEL	YEAR	POWER	STYLE	PRICE
CROSLEY RADIO	40	1933	AC	MIDGET	
CROSLEY RADIO	40-S	1929	AC	CONSOLE 7 TUBE 1 DIAL	
CROSLEY RADIO	41	1934	AC	COMPACT	
CROSLEY RADIO	41	1929	AC	TABLE METAL 7 TUBE 1 DIAL	
CROSLEY RADIO	41-A	1929	AC	TABLE METAL 8 TUBE 1 DIAL	
CROSLEY RADIO	41-S	1929	AC	TABLE 7 TUBE 1 DIAL	
CROSLEY RADIO	42	1929	AC	CONSOLE 7 TUBE 1 DIAL	
CROSLEY RADIO	42-S	1929	AC	CONSOLE 7 TUBE 1 DIAL	
CROSLEY RADIO	46	1936	6/DC	TABLE	
CROSLEY RADIO	48	1931	AC	CATHEDRAL (REPWOOD)	
CROSLEY RADIO	50	1924	BAT	MANTLE 1 TUBE 1 DIAL	130
CROSLEY RADIO	50-P	1924	BAT	PORTABLE 1 TUBE 1 DIAL	200
CROSLEY RADIO	50AF	1934	AC	COMPACT	
CROSLEY RADIO	50AF	1934	AC	CONSOLE LB	
CROSLEY RADIO	51	1924	BAT	MANTLE 2 TUBE 1 DIAL	90
CROSLEY RADIO	51	1934	AC	COMPACT	
CROSLEY RADIO	51-P	1924	BAT	PORTABLE LEATHERETTE CASE	175
CROSLEY RADIO	51-S	1924	BAT	TABLE SLOPE PAN. 2 T 1 DL	120
CROSLEY RADIO	51/51-A	1924	BAT	TABLE & AMPLIFIER	225
CROSLEY RADIO	52	1924	BAT	TABLE LP 3 TUBE 1 DIAL	125
CROSLEY RADIO	52	1934	AC	COMPACT	
CROSLEY RADIO	52-P	1924	BAT	PORTABLE LEATHERETTE 3T 1 DIAL	160
CROSLEY RADIO	52-S	1924	BAT	TABLE SLOPE PAN. 3 T 1 DL	145
CROSLEY RADIO	52-SD	1924	BAT	TABLE SPECIAL DELUXE 3T 1 DIAL	125
CROSLEY RADIO	52-TA	1941	AC	TABLE	30
CROSLEY RADIO	52-TQ	1941	AC	TABLE RADIO/PHONO	15
CROSLEY RADIO	53-E	1930	AC	CONSOLE LB MATE	140
CROSLEY RADIO	53-F	1930	AC	CONSOLE LB PAL	
CROSLEY RADIO	53-M	1930	AC	CONSOLE DESK	
CROSLEY RADIO	54-G	1930	AC	TOMBSTONE NEW BUDDY	130
CROSLEY RADIO	60	1933	AC	CONSOLE LB	
CROSLEY RADIO	60	1933	AC	TABLE	
CROSLEY RADIO	60LB	1934	AC	CONSOLE LB	
CROSLEY RADIO	61	1929	DC	TABLE LP METAL 8 TUBE 1 DIAL	50
CROSLEY RADIO	61AF	1934	AC	TOMBSTONE BC/SW	125
CROSLEY RADIO	61DL-LB	1934	AC	CONSOLE LB	
CROSLEY RADIO	61LB	1934	AC	CONSOLE LB	
CROSLEY RADIO	62	1929	DC	CONSOLE 8 TUBE 1 DIAL	
CROSLEY RADIO	63-S	1929	DC	CONSOLE	
CROSLEY RADIO	64MD	1934	AC	CONSOLE LB	95
CROSLEY RADIO	65	1931	AC	CATHEDRAL (REPWOOD)	
CROSLEY RADIO	70	1934	AC	TABLE	
CROSLEY RADIO	70-LB	1934	AC	CONSOLE LB	
CROSLEY RADIO	72AF	1934	AC	TABLE	100
CROSLEY RADIO	72LB	1934	AC	CONSOLE LB	
CROSLEY RADIO	76-A	1930	DC	CONSOLE LB DIRECTOR	
CROSLEY RADIO	77-A	1930	AC	CONSOLE LB DIRECTOR	110

MANUFACTURER	MODEL	YEAR	POWER	STYLE	PRICE
CROSLEY RADIO	77-B	1930	AC	CONSOLE LB R/P ARBITER	110
CROSLEY RADIO	77-L	1930	AC	CONSOLE HB RADIO/PHONO	
CROSLEY RADIO	80	1934	AC	TABLE	
CROSLEY RADIO	80AW	1934	AC	CONSOLE LB BC/SW	
CROSLEY RADIO	82	1929	AC	CONSOLE HB 8 TUBE 1 DIAL	150
CROSLEY RADIO	82-H	1929	AC	CONSOLE 7 TUBE 1 DIAL	
CROSLEY RADIO	82-S	1929	AC	CONSOLE 7 TUBE 1 DIAL	
CROSLEY RADIO	82-S	1929	AC	CONSOLE HB DOORS 8 T 1 DL	175
CROSLEY RADIO	83	1929	DC	CONSOLE 8 TUBE 1 DIAL	
CROSLEY RADIO	84-C	1930	AC	CONSOLE LB RONDEAU	
CROSLEY RADIO	84-D	1930	AC	CONSOLE LB R/P SONDO	
CROSLEY RADIO	117	1936	AC	CONSOLE BC/SW	
CROSLEY RADIO	121-1C	1931	AC	CONSOLE HB	
CROSLEY RADIO	121-1D	1931	AC	CONSOLE HB	
CROSLEY RADIO	122	1930	BAT	CATHEDRAL SUPER BUDDY BOY	90
CROSLEY RADIO	124H	1931	AC	CATHEDRAL	110
CROSLEY RADIO	124J	1931	AC	CONSOLE HB	
CROSLEY RADIO	124K	1931	AC	CONSOLE HB	
CROSLEY RADIO	124L	1931	AC	CONSOLE HB	
CROSLEY RADIO	124M	1931	AC	CLOCK GRANDFATHER	
CROSLEY RADIO	125N	1931	AC	COMPACT	75
CROSLEY RADIO	126-1	1931	AC	CHASSIS	
CROSLEY RADIO	128-1-LC	1932	BAT	CONSOLE	
CROSLEY RADIO	128-2-EA	1932	BAT	TABLE	
CROSLEY RADIO	129-1-2-K	1932	AC	TABLE CHEST	
CROSLEY RADIO	129-2-BA	1932	AC	TABLE	
CROSLEY RADIO	129-2-L	1932	AC	CONSOLE	
CROSLEY RADIO	132-1-2-MB	1932	AC	CONSOLE	
CROSLEY RADIO	132-1-2-W	1932	AC	CONSOLE	
CROSLEY RADIO	136-1-2-PA	1932	AC	CONSOLE BC/SW	
CROSLEY RADIO	136-1-2-S	1932	AC	TABLE BC/SW CHEST	
CROSLEY RADIO	137	1936	AC	CONSOLE BC/SW	
CROSLEY RADIO	140	1934	AC	CONSOLE	
CROSLEY RADIO	141-2-GA	1932	AC	TABLE CHEST	
CROSLEY RADIO	141-2-T	1932	AC	TABLE CHEST	75
CROSLEY RADIO	146-1-2-PA	1932	AC	CONSOLE	
CROSLEY RADIO	146-2-E	1932	AC	TABLE	
CROSLEY RADIO	146-2-Q	1932	AC	CONSOLE	
CROSLEY RADIO	148 FIVER-D	1933	AC	MIDGET	85
CROSLEY RADIO	154 LEADER-D	1933	AC	MIDGET	
CROSLEY RADIO	157	1933	AC	MIDGET	
CROSLEY RADIO	160 TWELVE	1933	AC	CONSOLE LB	
CROSLEY RADIO	167	1936	AC	CONSOLE BC/SW	125
CROSLEY RADIO	169	1934	AC	CATHEDRAL	180
CROSLEY RADIO	250	1936	AC	TABLE BC/SW	
CROSLEY RADIO	251	1936	BAT	PORTABLE BC/SW	
CROSLEY RADIO	295	1936	AC	TABLE BC/SW	
CROSLEY RADIO	299	1936	BAT	PORTABLE BC/SW	
CROSLEY RADIO	318-F	1940	AC	CONSOLE R/P RECORDER PB	50

MANUFACTURER	MODEL	YEAR	POWER	STYLE	PRICE
CROSLEY RADIO	349	1936	AC	TABLE BC/SW	
CROSLEY RADIO	395	1936	AC	TABLE BC/SW	
CROSLEY RADIO	400	1927	AC	CONSOLE W/601 BANDBOX	
CROSLEY RADIO	401 BANDBOX JR	1928	BAT	TABLE LP METAL 4 TUBE 1 DL	80
CROSLEY RADIO	401-A BAND BOX JR	1928	BAT	TABLE LP METAL 5 TUBE 1 DL	
CROSLEY RADIO	405	1927	AC	CONSOLE W/601 BANDBOX	
CROSLEY RADIO	410	1927	AC	CONSOLE DELUXE W/601	
CROSLEY RADIO	415-AA	1935	2/DC	TABLE	
CROSLEY RADIO	425-H	1935	BAT	PORTABLE NEW TRAVO	25
CROSLEY RADIO	449	1936	AC	TABLE BC/SW	
CROSLEY RADIO	495	1936	AC	TABLE BC/SW	
CROSLEY RADIO	499	1936	AC	CONSOLE BC/SW	
CROSLEY RADIO	505-MK	1935	AC	CONSOLE GALLEON BC/SW	
CROSLEY RADIO	515-AC	1935	AC	TOMBSTONE FIVER BC/SW	90
CROSLEY RADIO	516	1934	AC	TOMBSTONE BC/SW	50
CROSLEY RADIO	517-A	1937	AC	TABLE BC/SW	
CROSLEY RADIO	517-B	1937	AC	TABLE BC/SW	
CROSLEY RADIO	517-M	1937	AC	CONSOLE BC/SW	
CROSLEY RADIO	519-A	1939	AC	MANTLE PLASTIC	25
CROSLEY RADIO	525-B	1935	AC	TABLE GALLEON BC/SW	45
CROSLEY RADIO	529-A	1939	AC	MANTLE PLASTIC PB	
CROSLEY RADIO	529-D	1939	AC	TABLE PB	
CROSLEY RADIO	539-M	1939	AC	TABLE PB	
CROSLEY RADIO	545-AD	1935	AC	TABLE PRIVATEER	
CROSLEY RADIO	547-A	1937	AC	TABLE BC/SW	50
CROSLEY RADIO	547-B	1937	AC	TABLE BC/SW	
CROSLEY RADIO	547-M	1937	AC	CONSOLE BC/SW	
CROSLEY RADIO	555-AH	1935	AC	TABLE BC/SW	50
CROSLEY RADIO	555-KC	1935	2/DC	TABLE	30
CROSLEY RADIO	555-SA	1935	2/DC	CONSOLE	
CROSLEY RADIO	567-M	1937	AC	CONSOLE CHAIRSIDE BC/SW	
CROSLEY RADIO	567-N	1937	AC	CONSOLE CHAIRSIDE BC/SW	
CROSLEY RADIO	567-P	1937	AC	CONSOLE CHAIRSIDE BC/SW	
CROSLEY RADIO	599-A	1939	AC	MANTLE PLASTIC	
CROSLEY RADIO	601 BANDBOX	1927	BAT	TABLE LP METAL 6 TUBE 1 DIAL	80
CROSLEY RADIO	602 BANDBOX	1927	AC	TABLE LP METAL 7 TUBE 1 DIAL	110
CROSLEY RADIO	608 GEMBOX	1928	AC	TABLE LP METAL 6 TUBE 1 DIAL	95
CROSLEY RADIO	609 GEMCHEST	1928	AC	CONSOLE METAL CHINESE CAB.	120
CROSLEY RADIO	610 GEMCHEST	1928	AC	TABLE METAL	
CROSLEY RADIO	614EH	1934	AC	MIDGET	125
CROSLEY RADIO	614PG	1934	AC	CONSOLE	100
CROSLEY RADIO	615-C	1935	AC/DC	TABLE CRUISER BC/SW	50
CROSLEY RADIO	615-M	1935	AC	CONSOLE CRUISER BC/SW	
CROSLEY RADIO	625-E	1935	DC	TABLE BC/SW	
CROSLEY RADIO	625-NB	1935	DC	CONSOLE BC/SW	
CROSLEY RADIO	629-M	1939	AC	CONSOLE RADIO/PHONO PB	

MANUFACTURER	MODEL	YEAR	POWER	STYLE	PRICE
CROSLEY RADIO	635-C	1935	AC	TABLE BUCCANEER BC/SW	50
CROSLEY RADIO	635-M	1935	AC	CONSOLE BUCCANEER BC/SW	
CROSLEY RADIO	637-A	1937	AC	TABLE BC/SW	
CROSLEY RADIO	639	1940	AC	CONSOLE	50
CROSLEY RADIO	645-CB	1935	32/DC	TABLE BC/SW	40
CROSLEY RADIO	645-MB	1935	DC	CONSOLE BC/SW	45
CROSLEY RADIO	648-A	1939	AC	MANTLE PLASTIC PB	
CROSLEY RADIO	648-D	1939	AC	TABLE PB	
CROSLEY RADIO	655-C	1935	AC	TABLE OLYMPIA BC/SW	55
CROSLEY RADIO	655-M	1935	AC	CONSOLE OLYMPIA BC/SW	
CROSLEY RADIO	699	1936	AC	CONSOLE BC/SW	
CROSLEY RADIO	704 JEWELBOX	1928	AC	TABLE LP 7 TUBE 1 DIAL	100
CROSLEY RADIO	704-A JEWELBOX	1928	AC	TABLE LP METAL 8 TUBE 1 DIAL	
CROSLEY RADIO	704-B JEWELBOX	1928	AC	TABLE LP METAL 9 TUBE 1 DIAL	
CROSLEY RADIO	705 SHOWBOX	1928	DC	TABLE LP METAL 8 TUBE 1 DIAL	
CROSLEY RADIO	706 SHOWBOX	1928	AC	TABLE LP METAL 9 TUBE 1 DIAL	85
CROSLEY RADIO	714GA	1934	AC	MIDGET	100
CROSLEY RADIO	714NA	1934	AC	CONSOLE	95
CROSLEY RADIO	715-D	1935	AC	TABLE CORSAIR BC/SW	55
CROSLEY RADIO	715-N	1935	AC	CONSOLE CORSAIR BC/SW	50
CROSLEY RADIO	719-A	1939	AC	MANTLE PLASTIC PB	
CROSLEY RADIO	719-D	1939	AC	TABLE PB	
CROSLEY RADIO	725-F	1935	AC	TABLE VIKING BC/SW	55
CROSLEY RADIO	725-P	1935	AC	CONSOLE VIKING BC/SW	60
CROSLEY RADIO	729-A	1939	AC	TABLE PB BC/SW	
CROSLEY RADIO	729-D	1939	AC	TABLE PB BC/SW	
CROSLEY RADIO	739-A	1939	AC	TABLE PB BC/SW	
CROSLEY RADIO	804 JEWELBOX	1928	AC	TABLE LP METAL 8 TUBE 1 DIAL	
CROSLEY RADIO	814-FA	1934	AC	MIDGET	100
CROSLEY RADIO	814-QB	1934	AC	CONSOLE LB	95
CROSLEY RADIO	815-EC	1935	2/DC	TABLE BC/SW	40
CROSLEY RADIO	815-NC	1935	2/DC	CONSOLE BC/SW	45
CROSLEY RADIO	817-A	1937	AC	TABLE BC/SW	
CROSLEY RADIO	819-M	1939	AC	CONSOLE PB BC/SW	130
CROSLEY RADIO	855-D	1935	AC	TABLE MERRIMAC BC/SW	
CROSLEY RADIO	855-N	1935	AC	CONSOLE MERRIMAC BC/SW	
CROSLEY RADIO	865-F	1935	AC	TABLE MONITOR BC/SW	
CROSLEY RADIO	865-P	1935	AC	CONSOLE BC/SW MONITOR	
CROSLEY RADIO	899	1936	AC	CONSOLE BC/SW	
CROSLEY RADIO	915-EK	1935	AC	TABLE CLIPPER BC/SW	
CROSLEY RADIO	915-RA	1935	AC	CONSOLE CLIPPER BC/SW	
CROSLEY RADIO	955-RC	1936	AC	CONSOLE	70
CROSLEY RADIO	1014	1935	AC	CONSOLE LB	
CROSLEY RADIO	1014	1935	AC	TABLE	
CROSLEY RADIO	1055-EK	1935	AC	CONSOLE CONSTITUTION BC/SW	

MANUFACTURER	MODEL	YEAR	POWER	STYLE	PRICE
CROSLEY RADIO	1055-EK	1935	AC	TABLE CONSTITUTION BC/SW	
CROSLEY RADIO	1117-M	1937	AC	CONSOLE BC/SW	
CROSLEY RADIO	1123	1923	BAT	TABLE LP SLOPE PAN. 3T 2DL	
CROSLEY RADIO	5515-MH	1936	AC	CONSOLE	65
CROSLEY RADIO	5539-M	1939	AC	CONSOLE PB	
CROSLEY RADIO	5555-MR	1936	AC	CONSOLE BC/SW	65
CROSLEY RADIO	5648-A	1939	AC	MANTLE PB	
CROSLEY RADIO	5648-D	1939	AC	MANTLE PB	
CROSLEY RADIO	6625-NA	1936	AC	CONSOLE	70
CROSLEY RADIO	7739-M	1939	AC	CONSOLE PB BC/SW	
CROSLEY RADIO	AC-7	1927	AC	TABLE LP SLOPE PAN 7 T 1DL	65
CROSLEY RADIO	AC-7-C	1927	AC	CONSOLE LB 7 TUBE 1 DIAL	135
CROSLEY RADIO	ACE 3-B	1923	BAT	TABLE LP (V/2B) 3 T 1 DL	125
CROSLEY RADIO	ACE 3-C	1923	BAT	TABLE LP 3 TUBE 1 DIAL	150
CROSLEY RADIO	ACE TRU	1922	BAT	MANTLE 1 TUBE 1 DIAL	150
CROSLEY RADIO	ACE TRU/AV-2	1922	BAT	TABLE LP 2 BOXES 3 T 1 DL	225
CROSLEY RADIO	ACE V (V-C)	1922	BAT	MANTLE 1 TUBE 1 DIAL	175
CROSLEY RADIO	ACE V (V-D)	1922	BAT	MANTLE OVERHANG LID 1T 1 DIAL	185
CROSLEY RADIO	ACE V/2B	1923	BAT	TABLE 2 BOXES 3 T 1 DIAL	
CROSLEY RADIO	ACE V/AV-2(V-C)	1922	BAT	TABLE LP 2 BOXES 3 T 1 DL	225
CROSLEY RADIO	ACE V/AV-2(V-D)	1922	BAT	TABLE LP 2 BOXES 3 T 1 DL	240
CROSLEY RADIO	B-339A	1939	2/DC	PORTABLE CLOTH	20
CROSLEY RADIO	B-345	1936	2/DC	TABLE BC/SW	
CROSLEY RADIO	B-375	1936	6/DC	TABLE	
CROSLEY RADIO	B-425	1936	2/DC	TABLE BC/SW	
CROSLEY RADIO	B-429A	1939	BAT	PORTABLE CLOTH	
CROSLEY RADIO	B-445	1936	6/DC	TABLE BC/SW	
CROSLEY RADIO	B-449A	1939	BAT	MANTLE PLASTIC PB	
CROSLEY RADIO	B-459A	1939	BAT	TABLE PB	
CROSLEY RADIO	B-495	1936	2/DC	TABLE BC/SW	
CROSLEY RADIO	B-499	1936	2/DC	CONSOLE BC/SW	
CROSLEY RADIO	B-527-A	1937	2/DC	TABLE	
CROSLEY RADIO	B-549A	1939	2/DC	PORTABLE CLOTH	20
CROSLEY RADIO	B-557-A	1937	2/DC	TABLE BC/SW	
CROSLEY RADIO	B-557-M	1937	2/DC	CONSOLE BC/SW	
CROSLEY RADIO	B-579A	1939	BAT	TABLE PB	45
CROSLEY RADIO	B-589A	1939	BAT	TABLE PB BC/SW	
CROSLEY RADIO	B-599	1936	2/DC	TABLE BC/SW	
CROSLEY RADIO	B-675	1936	6/DC	CONSOLE BC/SW	
CROSLEY RADIO	B-695	1936	6/DC	TABLE BC/SW	
CROSLEY RADIO	B-699	1936	2/DC	CONSOLE BC/SW	
CROSLEY RADIO	B-899	1936	6/DC	CONSOLE BC/SW	
CROSLEY RADIO	B-5549A	1939	2/DC	PORTABLE RADIO/PHONO	15
CROSLEY RADIO	B-5579M	1939	BAT	CONSOLE PB	
CROSLEY RADIO	B-5589M	1939	BAT	CONSOLE PB BC/SW	
CROSLEY RADIO	B-6579A	1939	BAT	TABLE PB	
CROSLEY RADIO	BAT. 5	1933	BAT	TABLE	
CROSLEY RADIO	BAT. 8	1933	BAT	COMPACT	
CROSLEY RADIO	BAT. 9	1933	BAT	CONSOLE LB	

MANUFACTURER	MODEL	YEAR	POWER	STYLE	PRICE
CROSLEY RADIO	BUDDY	1929	AC	CONSOLE METAL	
CROSLEY RADIO	BUDDY	1929	AC	TABLE METAL	
CROSLEY RADIO	C-647-A	1937	AC	TABLE BC/SW	
CROSLEY RADIO	CASA	1933	AC	MANTLE BC/SW	
CROSLEY RADIO	COMPANION	1933	AC	TABLE BC/SW	
CROSLEY RADIO	DUAL 10 MODERNE	1933	AC	CONSOLE	
CROSLEY RADIO	DUAL 12 MODERNE	1933	AC	CONSOLE	
CROSLEY RADIO	FIRE-SCREEN	1933	AC	CONSOLE FIRE-SCREEN BC/SW	
CROSLEY RADIO	FIVE	1933	AC	COMPACT SHERATON BC/SW	
CROSLEY RADIO	FIVE	1933	AC	CONSOLE CABRIOLET BC/SW	
CROSLEY RADIO	FIVE	1933	AC	CONSOLE LB BC/SW	
CROSLEY RADIO	FIVE	1933	AC	TABLE BC/SW	
CROSLEY RADIO	FIVER JR	1934	AC	CONSOLE LB	
CROSLEY RADIO	FIVER JR.	1934	AC	TABLE	85
CROSLEY RADIO	FOUR	1933	AC	TABLE BC/SW	
CROSLEY RADIO	FOURTEEN	1933	AC	CONSOLE BC/SW	
CROSLEY RADIO	HARKO	1921	BAT	TABLE LP 1 TUBE 1 DIAL	240
CROSLEY RADIO	HARKO SENIOR	1922	BAT	TABLE LP 1 TUBE 1 DIAL	200
CROSLEY RADIO	HARKO SR./AMP.	1922	BAT	TABLE LP 2 BOXES 3 T 1 DL	350
CROSLEY RADIO	PUP	1924	BAT	MANTLE 1 TUBE	260
CROSLEY RADIO	REPOSE	1933	AC	CONSOLE BC/SW	
CROSLEY RADIO	REPOSE JR.	1933	AC	CONSOLE RC BC/SW	
CROSLEY RADIO	RFL-60	1926	BAT	TABLE LP 5 TUBE 3 DIAL	175
CROSLEY RADIO	RFL-75	1926	BAT	TABLE LP 5 TUBE 3 DIAL	180
CROSLEY RADIO	RFL-90	1926	BAT	CONSOLE HB 6 TUBE 2 DIAL	
CROSLEY RADIO	SEVEN	1933	AC	CONSOLE LB BC/SW	
CROSLEY RADIO	SEVEN	1933	AC	TABLE BC/SW	
CROSLEY RADIO	SEVETTE	1933	AC	CONSOLE FLOOR BC/SW	
CROSLEY RADIO	SIX	1933	AC	TABLE BC/SW	
CROSLEY RADIO	SUPER-TRIRDYN	1925	BAT	TABLE LP REGULAR 3 T 2 DL	95
CROSLEY RADIO	SUPER-TRIRDYN	1925	BAT	TABLE LP SPECIAL 3 T 2 DL	100
CROSLEY RADIO	SUPER-TRIRDYN	1925	BAT	CONSOLE DELUXE 3 TUBE 2 DIAL	100
CROSLEY RADIO	TEN	1933	AC	COMPACT BC/SW	
CROSLEY RADIO	TEN	1933	AC	CONSOLE LB BC/SW	
CROSLEY RADIO	TOTEM	1933	BAT	PORTABLE	
CROSLEY RADIO	TOTEM DELUXE	1933	BAT	PORTABLE CHEST	
CROSLEY RADIO	TRAVETTE	1933	AC	MANTLE METAL BC/SW	30
CROSLEY RADIO	TRAVO	1933	AC	MANTLE METAL	
CROSLEY RADIO	TRIRDYN 3R3	1924	BAT	TABLE LP REGULAR 3 T 2 DL	115
CROSLEY RADIO	TRIRDYN 3R3	1924	BAT	TABLE LP SPECIAL 3 T 2 DL	100
CROSLEY RADIO	TWELVE	1933	AC	COMPACT BC/SW	
CROSLEY RADIO	TWELVE	1933	AC	CONSOLE LB BC/SW	
CROSLEY RADIO	V HARKO SENIOR	1922	BAT	TABLE LP 1 TUBE 1 DIAL	230
CROSLEY RADIO	VI	1922	BAT	TABLE LP 2 TUBE 2 DIAL	235
CROSLEY RADIO	VI	1923	BAT	PORTABLE 2 TUBE 2 DIAL	150
CROSLEY RADIO	VI SUPER	1924	BAT	TABLE LP 2 TUBE 2 DIAL	95
CROSLEY RADIO	VIII	1923	BAT	TABLE LP 3 TUBE 2 DIAL	170
CROSLEY RADIO	VIII	1923	BAT	PORTABLE 3 TUBE 2 DIAL	195

MANUFACTURER	MODEL	YEAR	POWER	STYLE	PRICE
CROSLEY RADIO	X	1922	BAT	TABLE LP 4 TUBE 2 DIAL	280
CROSLEY RADIO	X-J	1923	BAT	TABLE LP 4 TUBE 2 DIAL	240
CROSLEY RADIO	X-L	1923	BAT	CONSOLE 4 TUBE 2 DIAL	200
CUTTING & WASHINGTON	11	1922	BAT	TABLE LP 2 DIAL	250
CUTTING & WASHINGTON	12-A	1923	BAT	TABLE HP 2 DIAL	275
CUTTING & WASHINGTON	TELEDYNE	1924	BAT	TABLE LP 3 DIAL	275
D					
DALCO RADIO	AIRTROLA	1924	BAT	TABLE LP 5 TUBE 3 DIAL	75
DALCO RADIO	DALCOFONE	1924	BAT	MANTLE 1 TUBE 1 DIAL	
DAVISON HAYNES	69-A	1930	AC	CATHEDRAL	90
DAYFAN ELECTRIC	66	1929	AC	TABLE	
DAYFAN ELECTRIC	67	1929	AC	TABLE	
DAYFAN ELECTRIC	68	1929	AC	CONSOLE	
DAYFAN ELECTRIC	69	1929	AC	CONSOLE	
DAYFAN ELECTRIC	72	1929	AC	CONSOLE	
DAYFAN ELECTRIC	73	1929	BAT	TABLE	
DAYFAN ELECTRIC	74	1929	AC	CONSOLE 7 TUBE 1 DIAL	
DAYTON FAN & MOTOR				See also GENERAL MOTORS	
DAYTON FAN & MOTOR	DAYOLA	1924	BAT	TABLE LP SLOPE PANEL 3 DL	175
DeFOREST RADIO	CS-5	1930	BAT	MANTLE METAL BC/SW COILS	300
DeFOREST RADIO	D-4	1923	BAT	TABLE 1 TUBE 2 DIAL	350
DeFOREST RADIO	D-4/D-5	1923	BAT	TABLE/AMPLIFIER	
DeFOREST RADIO	D-6	1922	BAT	TABLE LP 3 TUBE 2 DIAL	900
DeFOREST RADIO	D-7	1922	BAT	TABLE HP LOOP ANT 3 T 2DL	400
DeFOREST RADIO	D-7-A	1923	BAT	TABLE HP LOOP ANT 3 T 2 DIAL	490
DeFOREST RADIO	D-10	1923	BAT	TABLE LOOP ANT 4 TUBE 2DL	550
DeFOREST RADIO	D-12	1924	BAT	TABLE HP LOOP ANT 4 T 2 DIAL	600
DeFOREST RADIO	D-14	1924	BAT	TABLE HP SLOPE PAN 5 T 2DL	500
DeFOREST RADIO	D-17	1925	BAT	TABLE HP LOOP ANT 5 T 2DL	340
DeFOREST RADIO	D-17-A	1925	BAT	TABLE 5 TUBE 2 DIAL	300
DeFOREST RADIO	DT-900	1922	BAT	TABLE HP 2 DIAL	
DeFOREST RADIO	F-5	1925	BAT	TABLE HP 5 TUBE 3 DIAL	275
DeFOREST RADIO	F-5-AL	1925	BAT	TABLE HP 5 TUBE 3 DIAL	
DeFOREST RADIO	F-5-AW	1925	BAT	TABLE HP 5 TUBE 3 DIAL	300
DeFOREST RADIO	F-5-L	1925	BAT	TABLE HP 5 TUBE 3 DIAL	
DeFOREST RADIO	F-5-M	1925	BAT	TABLE HP 5 TUBE 3 DIAL	300
DeFOREST RADIO	FIFTEEN PANEL	1919	BAT	TABLE HP 3 TUBE 2 DIAL	1500
DeFOREST RADIO	INTERPANEL	1921	BAT	TABLE HP 3 TUBE 2 DIAL	950
DeFOREST RADIO	INTERPANEL MS-1	1921	BAT	TABLE HP 3 TUBE 2 DIAL	800
DeFOREST RADIO	MR-6	1923	BAT	TABLE INTERPANEL	750
DeFOREST RADIO	RADIOCRAFT	1920	BAT	TABLE SW	700
DeFOREST RADIO	RADIOCRAFT	1920	BAT	TABLE LW-SW	700
DeFOREST RADIO	RADIOHOME	1922	BAT	TABLE 1 TUBE 2 DIAL	400
DeFOREST RADIO	T-200/P-300	1920	BAT	TABLE 3 TUBE 1 DIAL	500
DeFOREST RADIO	W-5-F	1925	BAT	CONSOLE 6 TUBE 2 DIAL	300
DeFOREST RADIO	W-6-F	1925	BAT	CONSOLE RENAISSANCE 6T 1 DIAL	450

MANUFACTURER	MODEL	YEAR	POWER	STYLE	PRICE
DeFOREST RADIO	W-6-T	1925	BAT	TABLE SPEAKER 5 TUBE 2 DL	320
DETROLA RADIO	2B1	1935	2/DC	TABLE	
DETROLA RADIO	4F1	1934	AC	TABLE BC/SW	
DETROLA RADIO	4WG2	1935	AC	TABLE BC/SW	
DETROLA RADIO	4WM3	1935	AC	TABLE BC/SW	
DETROLA RADIO	4XG1	1935	AC	TABLE BC/SW	
DETROLA RADIO	4XM2	1935	AC	TABLE BC/SW	
DETROLA RADIO	5B3	1934	AC	TABLE BC/SW	
DETROLA RADIO	5XG1	1935	AC	TABLE BC/SW	
DETROLA RADIO	5XM1	1935	AC	TABLE BC/SW	
DETROLA RADIO	5XM4	1935	AC	CONSOLE BC/SW	
DETROLA RADIO	5XM9	1935	AC	TABLE BC/SW	
DETROLA RADIO	6B1	1935	6/DC	TABLE	
DETROLA RADIO	6WG1	1935	AC	TABLE BC/SW	
DETROLA RADIO	6WM1	1935	AC	TABLE BC/SW	
DETROLA RADIO	6WM3	1935	AC	CONSOLE BC/SW	
DETROLA RADIO	6XM1	1935	AC	TABLE BC/SW	
DETROLA RADIO	6XM5	1935	AC	CONSOLE BC/SW	
DETROLA RADIO	6XM99	1935	AC	TABLE BC/SW	
DETROLA RADIO	6Z1	1935	AC	TABLE	
DETROLA RADIO	6Z3	1935	AC	CONSOLE	
DETROLA RADIO	7A3	1934	AC	TABLE BC/SW	
DETROLA RADIO	7Z1	1935	AC	TABLE	
DETROLA RADIO	7Z3	1935	AC	CONSOLE	
DETROLA RADIO	10Z1	1935	AC	TABLE BC/SW	
DETROLA RADIO	10Z3	1935	AC	CONSOLE BC/SW	
DETROLA RADIO	32V1	1935	32/DC	TABLE BC/SW	
DETROLA RADIO	100X	1937	AC	TABLE BC/SW	
DETROLA RADIO	134AB	1937	AC	TABLE BC/SW	
DETROLA RADIO	134AZ	1937	AC	TABLE BC/SW	
DETROLA RADIO	135E	1937	AC	TABLE BC/SW	
DETROLA RADIO	137A	1937	AC	TABLE BC/SW	
DETROLA RADIO	139A	1937	AC	TABLE BC/SW	
DETROLA RADIO	139B	1937	AC	TABLE BC/SW	
DETROLA RADIO	139EA	1937	AC	TABLE BC/SW	
DETROLA RADIO	139EAB	1937	AC	TABLE BC/SW WHITE	
DETROLA RADIO	139EAE	1937	AC	TABLE BC/SW EBONY	
DETROLA RADIO	140A	1937	AC	TABLE BC/SW	
DETROLA RADIO	140C	1937	AC	CONSOLE BC/SW	
DETROLA RADIO	146A1	1937	AC	TABLE BC/SW	
DETROLA RADIO	146C1	1937	AC	CONSOLE BC/SW	
DETROLA RADIO	146EA1	1937	AC	TABLE BC/SW	
DETROLA RADIO	147A	1937	AC	TABLE BC/SW	
DETROLA RADIO	147EA1	1937	AC	TABLE BC/SW	
DETROLA RADIO	148B	1937	AC	TABLE BC/SW	
DETROLA RADIO	148EA	1937	AC/DC	TABLE BC/SW	
DETROLA RADIO	150A	1937	AC/DC	TABLE BC/SW	
DETROLA RADIO	150B	1937	AC	TABLE BC/SW	
DETROLA RADIO	154A	1937	AC	TABLE BC/SW	
DETROLA RADIO	154EA	1937	AC	TABLE BC/SW	

MANUFACTURER	MODEL	YEAR	POWER	STYLE	PRICE
DETROLA RADIO	162AK	1937	AC	TABLE	
DETROLA RADIO	163E	1937	AC	CONSOLE BC/SW	
DETROLA RADIO	165ECR	1937	AC	CONSOLE BC/SW	
DETROLA RADIO	169EA	1937	AC/DC	TABLE BC/SW	
DETROLA RADIO	172A	1937	AC	TABLE BC/SW	
DETROLA RADIO	173EC	1937	AC	CONSOLE BC/SW	
DETROLA RADIO	174EC	1937	AC/DC	CONSOLE BC/SW	
DETROLA RADIO	197	1939	AC	MANTLE PLASTIC	
DETROLA RADIO	197	1938	AC	TABLE	
DETROLA RADIO	199	1938	AC	TABLE	
DETROLA RADIO	208	1938	AC	TABLE	
DETROLA RADIO	209-EA	1939	AC	TABLE PB BC/SW	
DETROLA RADIO	211	1938	AC	TABLE BC/SW	
DETROLA RADIO	212	1938	AC	TABLE BC/SW	
DETROLA RADIO	212-EA	1939	AC	TABLE BC/SW	
DETROLA RADIO	218	1939	AC	MANTLE PLASTIC BC/SW	
DETROLA RADIO	220-A	1939	AC	TABLE PB BC/SW	
DETROLA RADIO	221-A	1939	AC	TABLE PB BC/SW	
DETROLA RADIO	222	1939	AC	TABLE PB BC/SW	
DETROLA RADIO	225-A	1939	AC	TABLE PB BC/SW	
DETROLA RADIO	228-AP	1939	AC	TABLE RADIO/PHONO BC/SW	
DETROLA RADIO	231-A	1939	AC	TABLE PB BC/SW	
DETROLA RADIO	233-AP	1939	AC	TABLE RADIO/PHONO BC/SW	
DETROLA RADIO	235-AP	1939	AC	TABLE RADIO/PHONO	
DETROLA RADIO	248-A	1939	AC	TABLE BC/SW	
DETROLA RADIO	249-A	1939	AC	TABLE BC/SW	
DETROLA RADIO	250-A	1939	AC	TABLE BC/SW	
DETROLA RADIO	251-A	1939	AC	TABLE BC/SW	
DETROLA RADIO	259-EPC	1939	AC	CONSOLE R/P PB BC/SW	
DETROLA RADIO	270-EPC	1939	AC	CONSOLE BC/SW	
DETROLA RADIO	274	1939	AC	MANTLE PLASTIC	
DETROLA RADIO	276	1939	AC	MANTLE PLASTIC	
DETROLA RADIO	280-U	1939	AC	MANTLE PLASTIC	
DETROLA RADIO	282	1939	BAT	PORTABLE	
DETROLA RADIO	283-B	1939	AC	MANTLE PLASTIC	
DETROLA RADIO	284	1939	AC	MANTLE PLASTIC	
DETROLA RADIO	286	1939	BAT	PORTABLE	
DETROLA RADIO	288	1939	BAT	PORTABLE	
DETROLA RADIO	295	1939	AC	PORTABLE	
DETROLA RADIO	302	1939	AC	TABLE CATHEDRAL W/CLOCK	70
DETROLA RADIO	318	1939	AC	CONSOLE RADIO/PHONO BC/SW	45
DETROLA RADIO	378	1941	AC	PORTABLE MARBLE MINIATURE	25
DETROLA RADIO	383	1941	AC	PORTABLE AEROPLANE CLOTH	15
DETROLA RADIO	2741	1939	AC	TABLE RADIO/PHONO	
DETROLA RADIO	2742	1939	AC	TABLE RADIO/PHONO	
DETROLA RADIO	2743	1939	AC	PORTABLE RADIO/PHONO	
DETROLA RADIO	2744	1939	AC	TABLE RADIO/PHONO	
DETROLA RADIO	2831	1939	AC	TABLE RADIO/PHONO	

MANUFACTURER	MODEL	YEAR	POWER	STYLE	PRICE
DETROLA RADIO	2832	1939	AC	TABLE RADIO/PHONO	
DETROLA RADIO	3051	1939	AC	TABLE PB BC/SW	30
DETROLA RADIO	3861	1941	AC	TABLE RADIO/PHONO	15
DeWITT-LaFRANCE	SUPERADIO	1925	BAT	TABLE LP 5 TUBE 2 DIAL	95
DEXTER RADIO	5 DEXTER	1925	BAT	TABLE LP 3 DIAL	70
DIAMOND-T RADIO	AMERICAN BEAUTY	1928	BAT	TABLE	
DIAMOND-T RADIO	BABY GRAND	1927	AC	CONSOLE 6 TUBE 1 DIAL	
DIAMOND-T RADIO	CHIEF	1927	AC	CONSOLE 7 TUBE 1 DIAL	
DIAMOND-T RADIO	D-15	1924	BAT	TABLE LP 5 TUBE 3 DIAL	75
DIAMOND-T RADIO	DIAMOND SPECIAL	1927	BAT	TABLE 6 TUBE 1 DIAL	
DIAMOND-T RADIO	SUPER SPECIAL	1927	BAT	TABLE 6 TUBE 1 DIAL	
DISTANTONE	C	1926	BAT	TABLE HP 1 DIAL	90
DISTANTONE RADIOS	DISTANTONE FIVE	1925	BAT	TABLE LP 5 TUBE 3 DIAL	85
DORON BROTHERS	R-5	1924	BAT	TABLE LP 5 TUBE 3 DIAL	70
DOTSON-HIEBERT	DOTSON SIX	1924	BAT	TABLE LP 5 TUBE 3 DIAL	
DYNAMOTIVE RADIO	RC-250	1923	BAT	TABLE LP 5 TUBE 3 DIAL	
E					
EAGLE RADIO	B	1923	BAT	TABLE LP 5 TUBE 3 DIAL	125
EAGLE RADIO	B-S	1925	BAT	CONSOLE LB 5 TUBE 3 DIAL	
EAGLE RADIO	C	1925	BAT	TABLE SLOPE PAN. 3 T 2 DL	
EAGLE RADIO	C-1	1925	BAT	CONSOLE LB 5 TUBE 3 DIAL	
EAGLE RADIO	D	1925	BAT	TABLE LP 5 TUBE 3 DIAL	65
EAGLE RADIO	F	1926	BAT	TABLE LP 5 TUBE 3 DIAL	85
ECHOPHONE RADIO				See also RADIO SHOP	
ECHOPHONE RADIO	4	1932	AC	TABLE	140
ECHOPHONE RADIO	5	1932	AC	CATHEDRAL	120
ECHOPHONE RADIO	10	1932	AC	TABLE BC/SW	
ECHOPHONE RADIO	12	1932	AC	TABLE	
ECHOPHONE RADIO	14	1932	AC	TABLE	
ECHOPHONE RADIO	15	1932	AC	TABLE BC/SW	
ECHOPHONE RADIO	16	1932	AC	TABLE	
ECHOPHONE RADIO	17	1932	AC	TABLE	
ECHOPHONE RADIO	18	1932	AC	CONSOLE LB	
ECHOPHONE RADIO	20	1932	AC	CONSOLE LB BC/SW	
ECHOPHONE RADIO	35	1932	AC	CONSOLE LB BC/SW	
ECHOPHONE RADIO	40	1931	AC	CATHEDRAL ECHOETTE	
ECHOPHONE RADIO	40 ECHOETTE	1927	AC	TABLE	
ECHOPHONE RADIO	44	1932	AC	TABLE	
ECHOPHONE RADIO	60	1931	AC	CATHEDRAL	
ECHOPHONE RADIO	70	1931	AC	COMPACT	
ECHOPHONE RADIO	80	1931	AC	CATHEDRAL	
ECHOPHONE RADIO	90	1931	AC	CONSOLE LB	
ECHOPHONE RADIO	110	1934	AC	MANTLE BC/SW	
ECHOPHONE RADIO	111	1934	AC	MANTLE	
ECHOPHONE RADIO	112	1934	AC	TABLE BC/SW	
ECHOPHONE RADIO	113	1934	AC	TABLE BC/SW	
ECHOPHONE RADIO	115	1934	AC	TABLE BC/SW	
ECHOPHONE RADIO	119	1934	AC	TABLE BC/SW	
ECHOPHONE RADIO	124	1934	AC	CONSOLE BC/SW	

MANUFACTURER	MODEL	YEAR	POWER	STYLE	PRICE
ECHOPHONE RADIO	126	1934	AC	TABLE	
ECHOPHONE RADIO	128	1934	AC	MANTLE	
ECHOPHONE RADIO	143	1934	AC	TABLE BC/SW	
ECHOPHONE RADIO	ECHOPHONE	1930	AC	CATHEDRAL	130
ECODYNE	R- 5	1923	BAT	TABLE LP 5 TUBE 4 DIAL	
ECODYNE	RT-13	1923	BAT	TABLE LP 5 TUBE 4 DIAL	
EDISON	C-1	1928	AC	CONSOLE RADIO/PHONO 7 T 1 DIAL	
EDISON	C-2	1928	AC	CONSOLE RADIO/PHONO 7 T 1 DIAL	
EDISON	C-4	1929	AC	CONSOLE RADIO/PHONO 8T 1 DIAL	
EDISON	R-1	1928	AC	CONSOLE DESK 7 TUBE 1 DIAL	300
EDISON	R-2	1928	AC	CONSOLE HB 7 TUBE 1 DIAL	
EDISON	R-4	1929	AC	CONSOLE 8 TUBE 1 DIAL	
EDISON	R-5	1929	AC	CONSOLE 8 TUBE 1 DIAL	
EDISON	R-6	1930	AC	CONSOLE 9 TUBE 1 DIAL	
EDISON	R-7	1930	AC	CONSOLE HB 9 TUBE 1 DIAL	
EISEMANN MAGNETO	6-D	1924	BAT	TABLE LP 5 TUBE 3 DIAL	75
EISEMANN MAGNETO	8-D	1924	BAT	TABLE LP 5 TUBE 2 DIAL	75
ELECTRICAL RESEARCH	419-3	1924	BAT	CONSOLE HB 3 TUBE 1 DIAL	
ELECTRICAL RESEARCH	P-3	1924	BAT	TABLE LP SPEAKER 3 T 1 DL	
ELECTRICAL RESEARCH	S-3	1924	BAT	TABLE LP 3 TUBE 1 DIAL	
ELECTRICAL RESEARCH LABS	5 DELUXE	1925	BAT	TABLE LP 5 TUBE 3 DIAL	75
ELECTRICAL RESEARCH LABS	5 PEARSON	1925	BAT	TABLE LP 5 TUBE 3 DIAL	80
ELECTRICAL RESEARCH LABS	5 STANDARD	1925	BAT	TABLE LP 5 TUBE 3 DIAL	70
ELECTRICAL RESEARCH LABS	21P	1931	AC	TABLE	
ELECTRICAL RESEARCH LABS	22P	1931	AC	TABLE	
ELECTRICAL RESEARCH LABS	30	1929	AC	CONSOLE HB	145
ELECTRICAL RESEARCH LABS	30-248	1931	AC	CONSOLE	
ELECTRICAL RESEARCH LABS	31	1929	AC	CONSOLE HB	
ELECTRICAL RESEARCH LABS	32	1929	AC	CONSOLE HB	
ELECTRICAL RESEARCH LABS	33	1929	AC	CONSOLE RADIO/PHONO	
ELECTRICAL RESEARCH LABS	35	1930	AC	CONSOLE LB	
ELECTRICAL RESEARCH LABS	37	1930	AC	CONSOLE HB	
ELECTRICAL RESEARCH LABS	38	1930	AC	CONSOLE HB RADIO/PHONO	
ELECTRICAL RESEARCH LABS	39	1930	AC	CONSOLE HB RADIO/PHONO	
ELECTRICAL RESEARCH LABS	41	1931	AC	TABLE	
ELECTRICAL RESEARCH LABS	60BC	1936	2/DC	CONSOLE	
ELECTRICAL RESEARCH LABS	60BT	1936	2/DC	TABLE	
ELECTRICAL RESEARCH LABS	61P	1931	AC	TABLE	
ELECTRICAL RESEARCH LABS	62P	1931	AC	TABLE	
ELECTRICAL RESEARCH LABS	63BC	1936	6/DC	CONSOLE	
ELECTRICAL RESEARCH LABS	63BT	1936	6/DC	TABLE	
ELECTRICAL RESEARCH LABS	65BC	1936	2/DC	CONSOLE BC/SW	
ELECTRICAL RESEARCH LABS	65BT	1936	2/DC	TABLE BC/SW	
ELECTRICAL RESEARCH LABS	66BC	1936	8/DC	CONSOLE BC/SW	
ELECTRICAL RESEARCH LABS	66BT	1936	8/DC	TABLE BC/SW	
ELECTRICAL RESEARCH LABS	67LC	1936	32/DC	CONSOLE	

MANUFACTURER	MODEL	YEAR	POWER	STYLE	PRICE
ELECTRICAL RESEARCH LABS	67LT	1936	32/DC	TABLE	
ELECTRICAL RESEARCH LABS	67P	1931	BAT	COMPACT	
ELECTRICAL RESEARCH LABS	68BC	1936	6/DC	CONSOLE BC/SW	
ELECTRICAL RESEARCH LABS	68BCE	1936	6/DC	CONSOLE BC/SW	
ELECTRICAL RESEARCH LABS	68BT	1936	6/DC	TABLE BC/SW	
ELECTRICAL RESEARCH LABS	C-60	1927	BAT	CONSOLE 7 TUBE 1 DIAL	
ELECTRICAL RESEARCH LABS	C-61	1927	BAT	CONSOLE 7 TUBE 1 DIAL	
ELECTRICAL RESEARCH LABS	C-62	1927	AC	CONSOLE HB DOORS 7 T 1 DIAL	145
ELECTRICAL RESEARCH LABS	DELUXE CONSOLE	1925	BAT	CONSOLE LB 5 TUBE 3 DIAL	160
ELECTRICAL RESEARCH LABS	ERLA	1924	BAT	TABLE LP 5 TUBE 3 DIAL	60
ELECTRICAL RESEARCH LABS	S-51	1927	BAT	TABLE LP 6 TUBE 1 DIAL	
ELECTRICAL RESEARCH LABS	S-52	1927	BAT	TABLE LP 6 TUBE 1 DIAL	70
ELECTRICAL RESEARCH LABS	SUPERFLEX	1925	BAT	TABLE LP 3 TUBE 2 DIAL	120
ELVIN RADIO	DELUXE	1923	BAT	TABLE LP METER 3 TUBE 2 DIAL	
EMERSON RADIO	19	1934	AC	COMPACT	
EMERSON RADIO	20A	1933	AC	COMPACT	
EMERSON RADIO	23	1934	AC	COMPACT	
EMERSON RADIO	25A	1933	AC	COMPACT	
EMERSON RADIO	26	1934	AC	MIDGET	
EMERSON RADIO	28	1934	AC	TABLE UPRIGHT BC/SW	
EMERSON RADIO	30-AW	1933	AC	MANTLE BC/SW	40
EMERSON RADIO	30-LW	1933	AC	MANTLE BC/LW	
EMERSON RADIO	31-AW	1934	AC	COMPACT BC/SW	
EMERSON RADIO	32	1934	AC	COMPACT	
EMERSON RADIO	33-AW	1933	AC	TABLE BC/SW	
EMERSON RADIO	33-LW	1933	AC	TABLE BC/LW	
EMERSON RADIO	34-C	1935	AC	TABLE BC/SW	
EMERSON RADIO	34-F7	1935	BAT	TABLE BC/SW	
EMERSON RADIO	35	1933	AC	MIDGET	
EMERSON RADIO	36	1935	AC	TABLE BC/SW	50
EMERSON RADIO	38	1934	AC	TABLE BC/SW	40
EMERSON RADIO	38-LW	1934	AC	TABLE UPRIGHT BC/SW	
EMERSON RADIO	39	1934	AC	MIDGET	
EMERSON RADIO	40	1933	AC	CONSOLETTE	
EMERSON RADIO	45	1934	AC	TABLE UPRIGHT BC/SW	
EMERSON RADIO	45-LW	1934	AC	TABLE UPRIGHT BC/SW	
EMERSON RADIO	49	1934	AC	TABLE	
EMERSON RADIO	50	1933	AC	CONSOLETTE	
EMERSON RADIO	50-L	1933	AC	CONSOLETTE BC/LW	
EMERSON RADIO	50-S	1933	AC	CONSOLETTE BC/SW	
EMERSON RADIO	59	1934	AC	CONSOLE	
EMERSON RADIO	69	1934	AC	CONSOLE	
EMERSON RADIO	71	1934	AC	TABLE UPRIGHT BC/SW	45
EMERSON RADIO	77	1934	AC	CONSOLE	
EMERSON RADIO	100	1934	AC	CONSOLE	
EMERSON RADIO	101	1935	AC	CONSOLE BC/SW	
EMERSON RADIO	102	1935	AC	CONSOLE BC/SW	
EMERSON RADIO	102-LW	1935	AC	CONSOLE BC/SW/LW	

MANUFACTURER	MODEL	YEAR	POWER	STYLE	PRICE
EMERSON RADIO	103	1935	BAT	TABLE BC/SW	
EMERSON RADIO	104	1935	AC	TABLE BC/SW	
EMERSON RADIO	104-LW	1935	AC	TABLE BC/SW/LW	
EMERSON RADIO	105	1935	AC	CONSOLE BC/SW	
EMERSON RADIO	105-LW	1935	AC	CONSOLE BC/SW/LW	
ELECTRICAL RESEARCH LABS	68BTE	1936	6/DC	TABLE BC/SW	
ELECTRICAL RESEARCH LABS	71	1930	AC	TABLE	
ELECTRICAL RESEARCH LABS	72	1930	AC	TABLE	
ELECTRICAL RESEARCH LABS	72AT	1937	AC	TABLE BC/SW	
ELECTRICAL RESEARCH LABS	72ATE	1937	AC	CONSOLE BC/SW	
ELECTRICAL RESEARCH LABS	73	1930	AC	CONSOLE LB	
ELECTRICAL RESEARCH LABS	73BT	1937	6/DC	TABLE	
ELECTRICAL RESEARCH LABS	75	1928	AC	TABLE 7 TUBE 1 DIAL	
ELECTRICAL RESEARCH LABS	75	1928	AC	CONSOLE 7 TUBE 1 DIAL	
ELECTRICAL RESEARCH LABS	76AC	1937	AC	CONSOLE BC/SW	
ELECTRICAL RESEARCH LABS	77	1930	BAT	TABLE	
ELECTRICAL RESEARCH LABS	78BCE	1937	6/DC	CONSOLE BC/SW	
ELECTRICAL RESEARCH LABS	78BTE	1937	6/DC	TABLE BC/SW	
ELECTRICAL RESEARCH LABS	80BCT	1937	2/DC	CONSOLE	
ELECTRICAL RESEARCH LABS	80BT	1937	2/DC	TABLE	
ELECTRICAL RESEARCH LABS	81P	1931	AC	COMPACT	
ELECTRICAL RESEARCH LABS	82AC	1937	AC	CONSOLE BC/SW	
ELECTRICAL RESEARCH LABS	82AT	1937	AC	TABLE BC/SW	
ELECTRICAL RESEARCH LABS	82P	1931	AC	COMPACT	
ELECTRICAL RESEARCH LABS	86ACE	1937	AC	CONSOLE BC/SW	
ELECTRICAL RESEARCH LABS	86ATE	1937	AC	TABLE BC/SW	
ELECTRICAL RESEARCH LABS	90BCT	1937	2/DC	CONSOLE BC/SW	
ELECTRICAL RESEARCH LABS	90BT	1937	2/DC	TABLE BC/SW	
ELECTRICAL RESEARCH LABS	90P	1931	BAT	CONSOLE	
ELECTRICAL RESEARCH LABS	93LC	1937	32/DC	CONSOLE BC/SW	
ELECTRICAL RESEARCH LABS	93LT	1937	32/DC	TABLE BC/SW	
ELECTRICAL RESEARCH LABS	95BC	1937	2/DC	CONSOLE	
ELECTRICAL RESEARCH LABS	95BT	1937	2/DC	TABLE BC/SW	
ELECTRICAL RESEARCH LABS	96BCE	1937	6/DC	CONSOLE BC/SW	
ELECTRICAL RESEARCH LABS	96BTE	1937	6/DC	TABLE BC/SW	
ELECTRICAL RESEARCH LABS	450	1929	AC	CONSOLE RADIO/PHONO	
ELECTRICAL RESEARCH LABS	4143	1934	AC	TABLE	
ELECTRICAL RESEARCH LABS	4536	1934	AC	TABLE BC/SW	
ELECTRICAL RESEARCH LABS	5211	1934	AC	TABLE BC/SW	
ELECTRICAL RESEARCH LABS	5628	1934	32/DC	TABLE	
ELECTRICAL RESEARCH LABS	5721	1934	AC	TABLE BC/SW	
ELECTRICAL RESEARCH LABS	6232	1934	BAT	CONSOLE	
ELECTRICAL RESEARCH LABS	6241	1934	BAT	TABLE	
ELECTRICAL RESEARCH LABS	6315	1934	AC	TABLE BC/SW	
ELECTRICAL RESEARCH LABS	6317	1934	AC	CONSOLE BC/SW	
ELECTRICAL RESEARCH LABS	6323	1934	AC	TABLE BC/SW	
ELECTRICAL RESEARCH LABS	7732	1934	BAT	CONSOLE BC/SW	
ELECTRICAL RESEARCH LABS	7741	1934	BAT	TABLE BC/SW	
ELECTRICAL RESEARCH LABS	C- 4-F	1929	AC	CONSOLE HB RADIO/PHONO	
ELECTRICAL RESEARCH LABS	C- 5-F	1929	AC	CONSOLE HB	

MANUFACTURER	MODEL	YEAR	POWER	STYLE	PRICE
ELECTRICAL RESEARCH LABS	C-12	1927	BAT	TABLE LP 5 TUBE 3 DIAL	
ELECTRICAL RESEARCH LABS	C-13	1927	BAT	TABLE	
ELECTRICAL RESEARCH LABS	C-22	1927	BAT	CONSOLE 6 TUBE 1 DIAL	
ELECTRICAL RESEARCH LABS	C-50	1927	BAT	CONSOLE 5 TUBE 2 DIAL	
ELECTRICAL RESEARCH LABS	C-51	1927	BAT	CONSOLE 6 TUBE 1 DIAL	
EMERSON RADIO	106	1935	AC	TABLE BC/SW	
EMERSON RADIO	107	1935	AC	TABLE BC/SW	80
EMERSON RADIO	107-LW	1935	AC	TABLE BC/SW/LW	
EMERSON RADIO	108	1935	AC	TABLE BC/SW	40
EMERSON RADIO	108-LW	1935	AC	MANTLE BC/LW	
EMERSON RADIO	109	1935	AC	MANTLE	35
EMERSON RADIO	110	1935	AC	TABLE BC/SW	
EMERSON RADIO	110-LW	1935	AC	TABLE BC/LW	
EMERSON RADIO	111	1935	AC	TABLE BC/SW	
EMERSON RADIO	111-LW	1935	AC	TABLE BC/SW/LW	
EMERSON RADIO	119	1936	AC	TABLE BC/SW	
EMERSON RADIO	126	1936	AC	MANTLE	
EMERSON RADIO	126-LW	1936	AC	MANTLE BC/LW	
EMERSON RADIO	250-AW	1933	AC	MANTLE BC/SW	45
EMERSON RADIO	250-LW	1933	AC	TABLE BC/LW	
EMERSON RADIO	300	1933	AC	COMPACT	
EMERSON RADIO	321-AW	1933	AC	MANTLE BC/SW	
EMERSON RADIO	321-LW	1933	AC	MANTLE BC/LW	
EMERSON RADIO	330	1940	AC	MANTLE PLASTIC	20
EMERSON RADIO	331	1940	AC	TABLE BC/SW	35
EMERSON RADIO	332	1940	AC	TABLE BC/SW	30
EMERSON RADIO	336	1940	AC	MANTLE PLASTIC IVORY	25
EMERSON RADIO	343	1940	AC	MANTLE PLASTIC WAL BC/SW	20
EMERSON RADIO	344	1941	2/DC	TABLE PLASTIC	20
EMERSON RADIO	345	1940	AC	TABLE RADIO/PHONO	15
EMERSON RADIO	350	1940	AC	TABLE BC/SW	35
EMERSON RADIO	350-AW	1933	AC	TABLE BC/SW	
EMERSON RADIO	350-LW	1933	AC	TABLE BC/LW	
EMERSON RADIO	352	1940	AC	TABLE BC/SW	35
EMERSON RADIO	357	1940	AC	PORTABLE	15
EMERSON RADIO	364	1940	AC	TABLE RADIO/PHONO	15
EMERSON RADIO	365	1940	AC	TABLE TUNING EYE	45
EMERSON RADIO	368	1940	AC	CONSOLE PB BC/SW	90
EMERSON RADIO	369	1940	AC	CONSOLE PB BC/SW	130
EMERSON RADIO	370	1940	AC	CONSOLE R/P PB TE BC/SW	45
EMERSON RADIO	371	1940	AC	CONSOLE RADIO/PHONO	
EMERSON RADIO	372	1940	AC	CONSOLE R/P PB TE BC/SW	50
EMERSON RADIO	374	1940	AC	CONSOLE RADIO/PHONO	
EMERSON RADIO	375	1940	AC	MANTLE ONYX	85
EMERSON RADIO	375	1933	AC	TABLE	
EMERSON RADIO	380	1940	BAT	PORTABLE PERSONAL	20
EMERSON RADIO	382	1940	AC	PORTABLE R/P LEATHERETTE	15
EMERSON RADIO	383	1940	AC	TABLE RADIO/PHONO	15
EMERSON RADIO	384	1940	AC	TABLE RADIO/PHONO	30
EMERSON RADIO	385	1940	AC	PORTABLE AEROCLOTH	15

MANUFACTURER	MODEL	YEAR	POWER	STYLE	PRICE
EMERSON RADIO	397	1940	AC	CONSOLE RADIO/PHONO BC/SW	40
EMERSON RADIO	400-1,2,3	1940	AC	MANTLE PLASTIC	30
EMERSON RADIO	405	1940	AC	TABLE CHEST	60
EMERSON RADIO	406	1940	AC	TABLE PLASTIC	30
EMERSON RADIO	409	1933	AC	MANTLE "MICKEY MOUSE"	275
EMERSON RADIO	410	1933	AC	MANTLE "MICKEY MOUSE"	275
EMERSON RADIO	411	1933	AC	MANTLE "MICKEY MOUSE"	275
EMERSON RADIO	412	1941	AC	TABLE RADIO/PHONO	15
EMERSON RADIO	414	1933	AC	MANTLE "MICKEY MOUSE"	275
EMERSON RADIO	416	1934	AC	COMPACT	
EMERSON RADIO	420	1933	AC	MANTLE PLASTIC	
EMERSON RADIO	421	1941	AC	TABLE PLASTIC	20
EMERSON RADIO	423	1941	AC	TABLE RADIO/PHONO	15
EMERSON RADIO	424	1941	AC	PORTABLE LEATHERETTE	15
EMERSON RADIO	426	1941	BAT	PORTABLE STIPPLED PLASTIC	25
EMERSON RADIO	427	1941	AC	PORTABLE LEATHERETTE	15
EMERSON RADIO	442	1941	AC	TABLE	35
EMERSON RADIO	447	1941	AC	TABLE RADIO/PHONO	15
EMERSON RADIO	450	1934	BAT	TABLE UPRIGHT BC/SW	
EMERSON RADIO	456	1941	AC	TABLE BC/SW	30
EMERSON RADIO	459	1941	AC	TABLE BC/SW	35
EMERSON RADIO	520-J	1940	AC	TABLE PB TUNING EYE BC/SW	40
EMERSON RADIO	755-L	1933	AC	COMPACT BC/LW	
EMERSON RADIO	755-M	1933	AC	COMPACT	
EMERSON RADIO	755-S	1933	AC	COMPACT BC/SW	
EMERSON RADIO	770	1934	AC	CONSOLE BC/SW	
EMERSON RADIO	A-130	1936	AC	COMPACT	
EMERSON RADIO	A-132	1936	AC	COMPACT	
EMERSON RADIO	AA-204	1938	AC	TABLE BC/SW	
EMERSON RADIO	AA-207	1938	AC	TABLE BC/SW	
EMERSON RADIO	AB-182	1937	AC	CONSOLE BC/SW	
EMERSON RADIO	AB-184	1937	AC	CONSOLE RADIO/PHONO BC/SW	
EMERSON RADIO	AC-149	1937	AC	TABLE WAL,BLK,IVORY BC/SW	
EMERSON RADIO	AD-108	1937	AC	MANTLE WAL OR IVORY BC/SW	
EMERSON RADIO	AF-176	1937	2/DC	TABLE BC/SW	
EMERSON RADIO	AF-179	1937	2/DC	TABLE FARM BC/SW	
EMERSON RADIO	AJ-130	1937	2/DC	TABLE	
EMERSON RADIO	AJ-137	1937	2/DC	TABLE FARM	
EMERSON RADIO	AJ-149	1937	2/DC	TABLE FARM	
EMERSON RADIO	AJ-192	1938	2/DC	TABLE FARM	
EMERSON RADIO	AL-149-LW	1937	AC	TABLE WAL,BLK,IVORY BC/LW	
EMERSON RADIO	AL-164	1937	AC	PORTABLE RADIO/PHONO BC/SW	
EMERSON RADIO	AL-164-LW	1937	AC	PORTABLE R/P BC/LW	
EMERSON RADIO	AL-168	1937	AC	TABLE BC/SW	
EMERSON RADIO	AL-168-LW	1937	AC	TABLE BC/LW	
EMERSON RADIO	AM-131	1937	AC	TABLE BC/SW	

MANUFACTURER	MODEL	YEAR	POWER	STYLE	PRICE
EMERSON RADIO	AM-131-LW	1937	AC	TABLE BC/SW/LW	
EMERSON RADIO	AM-169	1937	AC	TABLE BC/SW	
EMERSON RADIO	AM-169-LW	1937	AC	TABLE BC/SW/LW	
EMERSON RADIO	AM-187	1937	AC	TABLE BC/SW	
EMERSON RADIO	AM-187-LW	1937	AC	TABLE BC/SW/LW	
EMERSON RADIO	AP-165	1937	AC	CONSOLE CHAIRSIDE R/P	
EMERSON RADIO	AP-166	1937	AC	TABLE RADIO/PHONO BC/SW	
EMERSON RADIO	AP-174	1937	AC	CONSOLE CHAIRSIDE BC/SW	
EMERSON RADIO	AP-176	1937	AC	TABLE BC/SW/LW	
EMERSON RADIO	AP-177	1937	AC	TABLE RADIO/PHONO BC/SW/LW	
EMERSON RADIO	AP-185	1937	AC	TABLE BC/SW/LW	
EMERSON RADIO	AR-165	1937	AC	CONSOLE CHAIRSIDE R/P	95
EMERSON RADIO	AR-166	1937	AC	TABLE RADIO/PHONO BC/SW	
EMERSON RADIO	AR-171	1937	AC	TABLE BC/SW	
EMERSON RADIO	AR-173	1937	AC	TABLE BC/SW	
EMERSON RADIO	AR-174	1937	AC	CONSOLE CHAIRSIDE BC/SW	
EMERSON RADIO	AR-176	1937	AC	TABLE BC/SW	
EMERSON RADIO	AR-177	1937	AC	TABLE RADIO/PHONO BC/SW	
EMERSON RADIO	AR-180	1937	AC	CONSOLE BC/SW	
EMERSON RADIO	AS-179	1937	6/DC	TABLE BC/SW	
EMERSON RADIO	AS-179	1938	6/DC	TABLE FARM BC/SW	
EMERSON RADIO	AT-170	1937	AC	TABLE BC/SW	
EMERSON RADIO	AT-172	1937	AC	TABLE BC/SW	35
EMERSON RADIO	AT-181	1937	AC	CONSOLE BC/SW	
EMERSON RADIO	AU-190	1938	AC	TABLE BC/SW	130
EMERSON RADIO	AU-213	1938	AC	TABLE	
EMERSON RADIO	AW- 55	1932	AC	TABLE BC/SW	
EMERSON RADIO	AX-211	1938	AC	MANTLE PLASTIC GEMLOID DIAL	35
EMERSON RADIO	AX-212	1938	AC	MANTLE PLASTIC GEMLOID DIAL	35
EMERSON RADIO	AX-217	1938	AC	MANTLE PLASTIC GEMLOID DIAL	35
EMERSON RADIO	AX-219	1938	AC	TABLE RADIO/PHONO	
EMERSON RADIO	AX-219	1938	AC	TABLE RADIO/PHONO	
EMERSON RADIO	AX-221	1938	AC	TABLE RADIO/PHONO	
EMERSON RADIO	AX-222	1938	AC	PORTABLE RADIO/PHONO	
EMERSON RADIO	AX-232	1938	AC	PORTABLE RADIO/PHONO	
EMERSON RADIO	AX-235	1938	AC	MANTLE	70
EMERSON RADIO	AZ-196	1938	AC	TABLE	25
EMERSON RADIO	B-131	1936	AC	TABLE BC/SW	
EMERSON RADIO	BA-199	1938	AC	TABLE	30
EMERSON RADIO	BA-201	1938	AC	TABLE	20
EMERSON RADIO	BB-208	1938	AC	TABLE	20
EMERSON RADIO	BB-209	1938	AC	TABLE	
EMERSON RADIO	BD-197	1938	AC	TABLE BC/SW ART DECO	95
EMERSON RADIO	BE-198	1938	AC	TABLE	20
EMERSON RADIO	BF-169	1938	AC	TABLE BC/SW	
EMERSON RADIO	BF-204	1938	AC	TABLE BC/SW	

MANUFACTURER	MODEL	YEAR	POWER	STYLE	PRICE
EMERSON RADIO	BF-207	1938	AC	TABLE BC/SW	
EMERSON RADIO	BJ-200	1938	AC	TABLE BC/SW	
EMERSON RADIO	BJ-210	1938	AC	TABLE BC/SW	
EMERSON RADIO	BJ-214	1938	AC	TABLE BC/SW	
EMERSON RADIO	BJ-218	1938	AC	TABLE RADIO/PHONO BC/SW	
EMERSON RADIO	BJ-220	1938	AC	TABLE RADIO/PHONO BC/SW	
EMERSON RADIO	BL-200	1938	AC	TABLE BC/SW	
EMERSON RADIO	BL-210	1938	AC	TABLE BC/SW	
EMERSON RADIO	BL-214	1938	AC	TABLE BC/SW	
EMERSON RADIO	BL-218	1938	AC	TABLE RADIO/PHONO BC/SW	
EMERSON RADIO	BL-220	1938	AC	TABLE RADIO/PHONO BC/SW	
EMERSON RADIO	BM-206	1938	AC	MANTLE	
EMERSON RADIO	BM-215	1938	AC	TABLE	
EMERSON RADIO	BM-216	1938	AC	TABLE RADIO/PHONO	20
EMERSON RADIO	BM-242	1939	AC	TABLE RADIO/PHONO	15
EMERSON RADIO	BQ-223	1938	AC	CONSOLE RADIO/PHONO BC/SW	
EMERSON RADIO	BQ-225	1938	AC	CONSOLE BC/SW	
EMERSON RADIO	BQ-228	1938	AC	TABLE BC/SW	30
EMERSON RADIO	BR-224	1938	AC	CONSOLE RADIO/PHONO BC/SW	
EMERSON RADIO	BR-226	1938	AC	CONSOLE BC/SW	165
EMERSON RADIO	BS-227	1938	AC	CONSOLE BC/SW	
EMERSON RADIO	BT-245	1939	AC	TABLE HIGH PROFILE ONYX	95
EMERSON RADIO	BU-229	1938	AC	TABLE BC/SW	
EMERSON RADIO	BU-230	1938	AC	CONSOLE BC/SW	
EMERSON RADIO	BW-231	1938	AC	TABLE BC/SW	35
EMERSON RADIO	BY-233	1938	AC	MANTLE	
EMERSON RADIO	C-134	1936	AC	TABLE BC/SW	
EMERSON RADIO	C-134-LW	1936	AC	TABLE BC/SW/LW	
EMERSON RADIO	C-138	1936	AC	CONSOLE BC/SW	
EMERSON RADIO	C-138-LW	1936	AC	CONSOLE BC/SW/LW	
EMERSON RADIO	C-142	1936	AC	CONSOLE BC/SW	
EMERSON RADIO	C-142-LW	1936	AC	CONSOLE BC/SW/LW	
EMERSON RADIO	C-145	1936	AC	CONSOLE RADIO/PHONO BC/SW	
EMERSON RADIO	CA-234	1938	AC	TABLE BC/SW	
EMERSON RADIO	CE-259	1939	AC	PORTABLE	15
EMERSON RADIO	CF-255	1939	AC	MANTLE PLASTIC	
EMERSON RADIO	CG-268	1939	AC	MANTLE PLASTIC BC/SW	
EMERSON RADIO	CG-276	1939	AC	TABLE BC/SW	
EMERSON RADIO	CG-293	1939	AC	CONSOLE RADIO/PHONO BC/SW	
EMERSON RADIO	CG-294	1939	AC	CONSOLE RADIO/PHONO BC/SW	
EMERSON RADIO	CG-318	1939	AC	TABLE BC/SW	40
EMERSON RADIO	CH-246	1939	AC	MANTLE PLASTIC IVORY	30
EMERSON RADIO	CH-253	1939	AC	TABLE	25
EMERSON RADIO	CH-256	1939	AC	TABLE	30
EMERSON RADIO	CJ-248	1939	AC	TABLE CHEST	

MANUFACTURER	MODEL	YEAR	POWER	STYLE	PRICE
EMERSON RADIO	CJ-257	1939	AC	TABLE	
EMERSON RADIO	CM 266	1939	AC	PORTABLE RADIO/PHONO	20
EMERSON RADIO	CQ-269	1939	AC	TABLE PB BC/SW	
EMERSON RADIO	CQ-271	1939	AC	TABLE PB BC/SW	
EMERSON RADIO	CQ-273	1939	AC	TABLE PB BC/SW	
EMERSON RADIO	CR-1-303	1939	AC	PORTABLE RADIO/PHONO	
EMERSON RADIO	CR-261	1939	AC	MANTLE PLASTIC	
EMERSON RADIO	CR-261-LW	1939	AC	TABLE WALNUT BC/LW	
EMERSON RADIO	CR-262	1939	AC	MANTLE PLASTIC	
EMERSON RADIO	CR-262-LW	1939	AC	TABLE WALNUT BC/LW	
EMERSON RADIO	CR-274	1939	AC	MANTLE PLASTIC	
EMERSON RADIO	CR-274-LW	1939	AC	MANTLE PLASTIC BC/LW	
EMERSON RADIO	CR-297	1939	AC	CONSOLE RADIO/PHONO	
EMERSON RADIO	CR-303	1939	AC	PORTABLE RADIO/PHONO	
EMERSON RADIO	CS-268	1939	AC	TABLE PLASTIC BC/SW	20
EMERSON RADIO	CS-268-LW	1939	AC	TABLE PLASTIC BC/SW/LW	
EMERSON RADIO	CS-270	1939	AC	TABLE BC/SW	
EMERSON RADIO	CS-270-LW	1939	AC	TABLE WALNUT BC/SW/LW	
EMERSON RADIO	CS-272	1939	AC	TABLE BC/SW	
EMERSON RADIO	CS-272-LW	1939	AC	TABLE WALNUT BC/SW/LW	
EMERSON RADIO	CS-317	1939	AC	TABLE BC/SW	35
EMERSON RADIO	CS-320	1939	AC	TABLE BC/SW	40
EMERSON RADIO	CU-265	1939	AC	MANTLE PLASTIC	
EMERSON RADIO	CU-265-LW	1939	AC	MANTLE PLASTIC BC/LW	
EMERSON RADIO	CV-1-290	1939	AC	PORTABLE RADIO/PHONO	
EMERSON RADIO	CV-1-291	1939	AC	TABLE RADIO/PHONO	
EMERSON RADIO	CV-264	1939	AC	MANTLE PLASTIC	
EMERSON RADIO	CV-268	1939	AC	TABLE PLASTIC BC/SW	
EMERSON RADIO	CV-280	1939	AC	PORTABLE	
EMERSON RADIO	CV-289	1939	AC	TABLE RADIO/PHONO	
EMERSON RADIO	CV-290	1939	AC	PORTABLE RADIO/PHONO	
EMERSON RADIO	CV-291	1939	AC	TABLE RADIO/PHONO	
EMERSON RADIO	CV-295	1939	AC	MANTLE	
EMERSON RADIO	CV-298	1939	AC	TABLE	
EMERSON RADIO	CV-313	1939	AC	TABLE	35
EMERSON RADIO	CV-316	1939	AC	TABLE	40
EMERSON RADIO	CW-279	1939	AC	MANTLE PLASTIC PB	
EMERSON RADIO	CX-263	1939	BAT	PORTABLE CLOTH	
EMERSON RADIO	CX-283	1939	BAT	PORTABLE	
EMERSON RADIO	CX-284	1939	BAT	PORTABLE	
EMERSON RADIO	CX-285	1940	2/DC	TABLE	30
EMERSON RADIO	CX-292	1939	AC	PORTABLE RADIO/PHONO	
EMERSON RADIO	CX-305	1939	BAT	PORTABLE WALNUT	
EMERSON RADIO	CY-269	1939	AC	TABLE PLASTIC PB BC/SW	
EMERSON RADIO	CY-286	1939	AC	TABLE PB BC/SW	
EMERSON RADIO	CY-288	1939	AC	TABLE PB BC/SW	
EMERSON RADIO	CZ-282	1939	AC	TABLE PB	
EMERSON RADIO	D-139	1936	AC	CONSOLE BC/SW	
EMERSON RADIO	D-139-LW	1936	AC	CONSOLE BC/SW/LW	
EMERSON RADIO	D-140	1936	AC	TABLE BC/SW	

MANUFACTURER	MODEL	YEAR	POWER	STYLE	PRICE
EMERSON RADIO	D-140-LW	1936	AC	TABLE BC/SW/LW	
EMERSON RADIO	DA-287	1939	AC	TABLE BC/SW	
EMERSON RADIO	DB-247	1939	AC	MANTLE PLASTIC	
EMERSON RADIO	DB-269	1939	AC	MANTLE PLASTIC	
EMERSON RADIO	DB-301	1939	AC	MANTLE PLASTIC	
EMERSON RADIO	DB-315	1939	AC	MANTLE	35
EMERSON RADIO	DC-308	1939	BAT	PORTABLE	
EMERSON RADIO	DF-302	1939	AC	PORTABLE WAL CARRYING CASE	25
EMERSON RADIO	DF-306	1939	AC	PORTABLE	15
EMERSON RADIO	DH-264	1939	BAT	MANTLE	
EMERSON RADIO	DJ-310	1939	BAT	PORTABLE PIGSKIN FABRIKOID	15
EMERSON RADIO	DQ-333	1940	AC	MANTLE PLASTIC	20
EMERSON RADIO	EMERSONETTE	1939	AC	MANTLE--IV, GRN, WAL, RED	20
EMERSON RADIO	F-133	1936	AC	TABLE BC/SW	
EMERSON RADIO	F-133-LW	1936	AC	BC/SW/LW	
EMERSON RADIO	G-127	1936	AC	PORTABLE R/P BC/SW	
EMERSON RADIO	H-130	1936	2/DC	COMPACT	
EMERSON RADIO	H-137	1936	2/DC	PORTABLE	
EMERSON RADIO	K-121	1936	AC	TABLE BC/SW	
EMERSON RADIO	K-123	1936	AC	TABLE BC/SW	
EMERSON RADIO	L-117	1936	AC	TABLE BC/SW	
EMERSON RADIO	L-117-LW	1936	AC	TABLE BC/LW	
EMERSON RADIO	L-122	1936	AC	TABLE BC/SW	
EMERSON RADIO	L-122-LW	1936	AC	TABLE BC/LW	
EMERSON RADIO	L-135	1936	AC	CONSOLE BC/SW	
EMERSON RADIO	L-135-LW	1936	AC	CONSOLE BC/LW	
EMERSON RADIO	L-141	1936	AC	TABLE BC/SW	
EMERSON RADIO	L-141-LW	1936	AC	TABLE BC/LW	
EMERSON RADIO	L-143	1936	AC	TABLE RADIO/PHONO BC/SW	
EMERSON RADIO	L-144	1936	AC	CONSOLE RADIO/PHONO BC/SW	
EMERSON RADIO	L-456	1932	DC	TABLE	
EMERSON RADIO	L-457	1932	DC	COMPACT	
EMERSON RADIO	L-459	1932	DC	TABLE CHEST	
EMERSON RADIO	L-556	1932	AC	TABLE	
EMERSON RADIO	L-557	1932	AC	COMPACT	
EMERSON RADIO	L-559	1932	AC	TABLE CHEST	
EMERSON RADIO	L-755	1933	AC	MIDGET	
EMERSON RADIO	LB-458	1932	AC	COMPACT BARONET	
EMERSON RADIO	LB-460	1932	AC	TABLE	
EMERSON RADIO	P-117	1936	6/DC	TABLE	
EMERSON RADIO	P-135	1936	6/DC	CONSOLE	
EMERSON RADIO	Q-157	1937	AC	MANTLE WALNUT, BLACK, IVORY	
EMERSON RADIO	Q-188	1937	AC	TABLE	
EMERSON RADIO	Q-236	1938	AC	MANTLE	
EMERSON RADIO	R-153	1937	AC	TABLE BC/SW	
EMERSON RADIO	R-156	1937	AC	TABLE BC/SW	

MANUFACTURER	MODEL	YEAR	POWER	STYLE	PRICE
EMERSON RADIO	R-158	1937	AC	TABLE BC/SW	
EMERSON RADIO	R-167	1937	AC	TABLE BC/SW	
EMERSON RADIO	R-189	1937	AC	TABLE BC/SW	
EMERSON RADIO	X-175	1937	AC	CONSOLE RADIO/PHONO BC/SW	
EMERSON RADIO	X-178	1937	AC	CONSOLE BC/SW	
EMERSON RADIO	X-183	1937	AC	CONSOLE BC/SW	
EMPIRE ELECTRICAL	10	1933	AC	TABLE	
EMPIRE ELECTRICAL	20	1933	AC	TABLE	
EMPIRE ELECTRICAL	30	1933	AC	TABLE	
EMPIRE ELECTRICAL	40	1933	AC	TABLE	
EMPIRE ELECTRICAL	74	1933	AC	COMPACT BC/SW	
EMPIRE ELECTRICAL	400AC	1933	AC	TABLE	
EMPIRE ELECTRICAL	400DC	1933	DC	TABLE	
EMPIRE ELECTRICAL	575	1933	AC	TABLE	
EMPIRE ELECTRICAL	600AC	1933	AC	COMPACT	
EMPIRE ELECTRICAL	600DC	1933	DC	COMPACT	
EMPIRE ELECTRICAL	700	1933	DC	COMPACT	
EQUITABLE RADIO	CLARATONE	1924	BAT	TABLE LP METER 5 TUBE 3 DIAL	
EQUITABLE RADIO	CLARATONE	1925	BAT	TABLE LP SLOPE PAN. 5T 3DL	
ESPEY	1-861	1938	AC	CONSOLE NIGHT STAND BC/SW	
ESPEY	2-861	1938	AC	CONSOLE CHAIRSIDE BC/SW	
ESPEY	3-861	1938	AC	CONSOLE END TABLE BC/SW	
ESPEY	7-861	1938	AC	CONSOLE END TABLE BC/SW	60
ESPEY	7A	1936	AC	CONSOLE RADIO/PHONO BC/SW	
ESPEY	7B	1936	AC	CONSOLE RADIO/PHONO BC/SW	
ESPEY	7C	1936	AC	CONSOLE RADIO/PHONO BC/SW	
ESPEY	7D	1936	AC	CONSOLE RADIO/PHONO BC/SW	
ESPEY	10-891	1938	AC	CONSOLE BC/SW	
ESPEY	11-7151	1938	AC	CONSOLE BC/SW	
ESPEY	13-861	1938	AC	CONSOLE END TABLE BC/SW	
ESPEY	45	1940	AC	CONSOLE RADIO/PHONO	
ESPEY	51	1939	AC	TABLE PB	
ESPEY	81	1939	AC	CONSOLE PB BC/SW	
ESPEY	101	1939	AC	CONSOLE PB BC/SW	
ESPEY	161	1939	AC	CONSOLE PB BC/SW	
ESPEY	423	1940	AC	CONSOLE RADIO/PHONO	
ESPEY	433	1940	AC	CONSOLE RADIO/PHONO	
ESPEY	458	1935	AC	TABLE BC/SW	
ESPEY	545	1935	2/DC	TABLE	
ESPEY	553	1935	AC	TABLE BC/SW	
ESPEY	555	1935	AC	TABLE BC/SW	
ESPEY	560	1935	AC	CHASSIS BC/SW	
ESPEY	564	1935	AC	TABLE BC/SW	

MANUFACTURER	MODEL	YEAR	POWER	STYLE	PRICE
ESPEY	565	1935	AC	TABLE BC/SW	
ESPEY	641	1936	2/DC	TABLE	
ESPEY	671	1936	AC	TABLE BC/SW	
ESPEY	771A THRU 771J	1937	AC	TABLE BC/SW	
ESPEY	861A	1938	AC	TABLE BC/SW	
ESPEY	861B,-C,-R	1938	AC	PORTABLE BC/SW	15
ESPEY	942A,-C,-F	1939	BAT	PORTABLE	
ESPEY	943	1939	BAT	PORTABLE	
ESPEY	958	1939	BAT	PORTABLE	
ESPEY	5191	1935	AC	CHASSIS BC/SW	
ESPEY	7111	1937	AC	TABLE BC/SW VERTICAL	
ESPEY	7151	1937	AC	TABLE BC/SW VERTICAL	
ESPEY	WQXR	1940	AC	CONSOLE RADIO/PHONO	
EXCELLO PRODUCTS	R- 71	1929	AC	CONSOLE LB RADIO/PHONO	145
EXCELLO PRODUCTS	R-154	1930	AC	CONSOLE HB RADIO/PHONO	135
EXCELLO PRODUCTS	R-171	1930	AC	CONSOLE LB RADIO/PHONO	140

F

MANUFACTURER	MODEL	YEAR	POWER	STYLE	PRICE
FADA RADIO & ELECTRIC				See also ANDREA F. A. D.	
FADA RADIO & ELECTRIC				See also ANDREA RADIO	
FADA RADIO & ELECTRIC	5-F-50	1939	AC	MANTLE PLASTIC WALNUT	
FADA RADIO & ELECTRIC	20-G	1938	AC	TABLE	
FADA RADIO & ELECTRIC	20-T	1938	AC	TABLE	20
FADA RADIO & ELECTRIC	20-V	1938	AC	TABLE	
FADA RADIO & ELECTRIC	20-W	1938	AC	TABLE	
FADA RADIO & ELECTRIC	31-L	1939	AC	TABLE	
FADA RADIO & ELECTRIC	32-L	1939	AC	TABLE	
FADA RADIO & ELECTRIC	33	1940	BAT	PORTABLE CAMERA TYPE	25
FADA RADIO & ELECTRIC	55	1932	AC	TABLE	
FADA RADIO & ELECTRIC	57	1939	AC	TABLE WALNUT	
FADA RADIO & ELECTRIC	59	1939	AC	MANTLE PLASTIC WALNUT	
FADA RADIO & ELECTRIC	61-PC	1939	AC	CONSOLE RADIO/PHONO	
FADA RADIO & ELECTRIC	64-AP	1940	AC	CONSOLE RADIO/PHONO	50
FADA RADIO & ELECTRIC	65-PC	1939	AC	CONSOLE RADIO/PHONO	
FADA RADIO & ELECTRIC	66	1932	AC	CONSOLE BC/SW	
FADA RADIO & ELECTRIC	69-PC	1939	AC	CONSOLE RADIO/PHONO	
FADA RADIO & ELECTRIC	73	1932	AC	TABLE 60/25 CYCLE	45
FADA RADIO & ELECTRIC	74	1932	AC	CONSOLE 60/25 CYCLE	
FADA RADIO & ELECTRIC	75-PC	1939	AC	CONSOLE RADIO/PHONO	
FADA RADIO & ELECTRIC	76	1932	AC	CONSOLE LB 60/25 CYCLE	
FADA RADIO & ELECTRIC	78	1932	AC	CONSOLE LB 60/25 CYCLE	
FADA RADIO & ELECTRIC	79	1932	AC	CONSOLE HB 60/25 CYCLE	
FADA RADIO & ELECTRIC	83	1932	AC	CONSOLE LB	
FADA RADIO & ELECTRIC	85	1932	AC	CONSOLE LB 60/25 CYCLE	
FADA RADIO & ELECTRIC	87	1932	AC	CONSOLE LB 60/25 CYCLE	
FADA RADIO & ELECTRIC	88	1932	AC	CONSOLE LB	
FADA RADIO & ELECTRIC	89	1932	AC	CONSOLE HB	
FADA RADIO & ELECTRIC	93	1932	AC	TABLE BC/LW	
FADA RADIO & ELECTRIC	95	1932	AC	CONSOLE LB BC/LW	
FADA RADIO & ELECTRIC	97	1932	AC	CONSOLE LB	

MANUFACTURER	MODEL	YEAR	POWER	STYLE	PRICE
FADA RADIO & ELECTRIC	98	1933	AC	CONSOLE LB	
FADA RADIO & ELECTRIC	103	1933	AC	COMPACT	
FADA RADIO & ELECTRIC	106	1933	AC	MANTLE BC/SW	
FADA RADIO & ELECTRIC	107	1933	AC	TABLE BC/SW	
FADA RADIO & ELECTRIC	112	1933	AC	PORTABLE RADIO/PHONO BC/SW	
FADA RADIO & ELECTRIC	115	1940	AC	TABLE CATALIN SERIES	70
FADA RADIO & ELECTRIC	115-BA	1940	AC	MANTLE TORPEDO STYLE	95
FADA RADIO & ELECTRIC	119-V,W	1940	AC	TABLE PLASTIC	25
FADA RADIO & ELECTRIC	126	1934	AC	TABLE	
FADA RADIO & ELECTRIC	127	1934	AC	CONSOLE	
FADA RADIO & ELECTRIC	128	1934	AC	CONSOLE	
FADA RADIO & ELECTRIC	131	1933	AC	CONSOLE	
FADA RADIO & ELECTRIC	132	1933	AC	CONSOLE	
FADA RADIO & ELECTRIC	133	1933	AC	CONSOLE BC/SW	
FADA RADIO & ELECTRIC	134	1933	AC	CONSOLE BC/SW	
FADA RADIO & ELECTRIC	135	1933	AC	CONSOLE BC/SW	
FADA RADIO & ELECTRIC	140-SW	1935	AC	COMPACT	
FADA RADIO & ELECTRIC	141	1933	AC	TABLE BC/SW	
FADA RADIO & ELECTRIC	145	1935	AC	COMPACT	
FADA RADIO & ELECTRIC	145-SW	1935	AC	COMPACT	
FADA RADIO & ELECTRIC	150-C	1935	AC	CONSOLE BC/SW	
FADA RADIO & ELECTRIC	150-CA	1935	AC	CONSOLE	
FADA RADIO & ELECTRIC	150-T	1936	AC	TABLE BC/SW	
FADA RADIO & ELECTRIC	155	1935	AC	TABLE	
FADA RADIO & ELECTRIC	156	1935	AC	TABLE	
FADA RADIO & ELECTRIC	157	1935	AC	TABLE BC/SW	
FADA RADIO & ELECTRIC	160-C	1935	AC	CONSOLE BC/SW	
FADA RADIO & ELECTRIC	160-T	1935	AC	TABLE BC/SW	
FADA RADIO & ELECTRIC	162-C	1935	2/DC	CONSOLE BC/SW	
FADA RADIO & ELECTRIC	162-T	1935	2/DC	TABLE BC/SW	
FADA RADIO & ELECTRIC	163-C	1936	6/DC	CONSOLE BC/SW	
FADA RADIO & ELECTRIC	163-T	1936	6/DC	TABLE BC/SW	
FADA RADIO & ELECTRIC	164-C	1936	32/DC	CONSOLE BC/SW	
FADA RADIO & ELECTRIC	164-T	1936	32/DC	TABLE BC/SW	
FADA RADIO & ELECTRIC	170-C	1935	AC	CONSOLE BC/SW	
FADA RADIO & ELECTRIC	170-CK	1935	AC	CONSOLE BC/SW	
FADA RADIO & ELECTRIC	170-T	1935	AC	TABLE BC/SW	
FADA RADIO & ELECTRIC	172	1936	AC	TABLE BC/SW	
FADA RADIO & ELECTRIC	182-G	1940	AC	TABLE MINATURE BABY GRAND	90
FADA RADIO & ELECTRIC	182-S	1940	AC	TABLE MINATURE SPINET	90
FADA RADIO & ELECTRIC	184-APC	1940	AC	CONSOLE RADIO/PHONO BC/SW	45
FADA RADIO & ELECTRIC	185-APC	1940	AC	CONSOLE RADIO/PHONO	
FADA RADIO & ELECTRIC	190-C	1935	AC	CONSOLE BC/SW	
FADA RADIO & ELECTRIC	190-CK	1935	AC	CONSOLE BC/SW	
FADA RADIO & ELECTRIC	190-T	1935	AC	TABLE BC/SW	
FADA RADIO & ELECTRIC	192-C	1935	AC	CONSOLE BC/SW	
FADA RADIO & ELECTRIC	192-CK	1935	AC	CONSOLE BC/SW	

MANUFACTURER	MODEL	YEAR	POWER	STYLE	PRICE
FADA RADIO & ELECTRIC	192-T	1935	AC	TABLE BC/SW	
FADA RADIO & ELECTRIC	194-APC	1940	AC	CONSOLE RADIO/PHONO	50
FADA RADIO & ELECTRIC	195-APC	1940	AC	CONSOLE RADIO/PHONO	
FADA RADIO & ELECTRIC	211-C	1936	AC	CONSOLE BC/SW	
FADA RADIO & ELECTRIC	211-T	1936	AC	TABLE BC/SW	
FADA RADIO & ELECTRIC	212-C	1936	AC	CONSOLE BC/SW	
FADA RADIO & ELECTRIC	212-T	1936	AC	TABLE BC/SW	
FADA RADIO & ELECTRIC	216-C	1936	AC	CONSOLE BC/SW	
FADA RADIO & ELECTRIC	216-T	1936	AC	TABLE BC/SW	
FADA RADIO & ELECTRIC	250-C	1936	AC	CONSOLE BC/SW	
FADA RADIO & ELECTRIC	250-T	1936	AC	TABLE BC/SW	
FADA RADIO & ELECTRIC	260-B	1936	AC	TABLE	
FADA RADIO & ELECTRIC	262-W	1936	AC	TABLE BC/SW	
FADA RADIO & ELECTRIC	270-C	1936	AC	CONSOLE BC/SW	
FADA RADIO & ELECTRIC	270-T	1936	AC	TABLE BC/SW	60
FADA RADIO & ELECTRIC	272-W	1936	AC	TABLE BC/SW	
FADA RADIO & ELECTRIC	280-C	1936	AC	CONSOLE BC/SW	
FADA RADIO & ELECTRIC	280-T	1936	AC	TABLE BC/SW	
FADA RADIO & ELECTRIC	290-C	1936	AC	CONSOLE BC/SW	
FADA RADIO & ELECTRIC	290-T	1936	AC	TABLE BC/SW	
FADA RADIO & ELECTRIC	350-V,-R	1937	AC	TABLE PLASTIC BC/SW	
FADA RADIO & ELECTRIC	351-J,-M	1937	AC	TABLE BC/SW	
FADA RADIO & ELECTRIC	351-K,-L	1937	AC	TABLE BC/SW	
FADA RADIO & ELECTRIC	352-J,-M	1937	AC	TABLE BC/SW	
FADA RADIO & ELECTRIC	352-K,-L	1937	AC	TABLE BC/SW	
FADA RADIO & ELECTRIC	353-J,-M	1937	AC	TABLE BC/SW	
FADA RADIO & ELECTRIC	353-K,-L	1937	AC	TABLE BC/SW	
FADA RADIO & ELECTRIC	354-B,-BG-,-D	1937	AC	TABLE PLASTIC BC/SW	
FADA RADIO & ELECTRIC	354-G,-R,-RG	1937	AC	TABLE PLASTIC BC/SW	
FADA RADIO & ELECTRIC	354-V,-W	1937	AC	TABLE PLASTIC BC/SW	
FADA RADIO & ELECTRIC	355-B,-BG,-D	1937	AC	TABLE PLASTIC BC/SW	
FADA RADIO & ELECTRIC	355-G,-R,-RG	1937	AC	TABLE PLASTIC BC/SW	
FADA RADIO & ELECTRIC	355-V,-W	1937	AC	TABLE PLASTIC BC/SW	
FADA RADIO & ELECTRIC	360-B,-BG,-D	1937	AC	TABLE PLASTIC BC/SW	
FADA RADIO & ELECTRIC	360-G,-R-,-RG	1937	AC	TABLE PLASTIC BC/SW	
FADA RADIO & ELECTRIC	360-V,-W	1937	AC	TABLE PLASTIC BC/SW	
FADA RADIO & ELECTRIC	361-J,-M	1937	AC	TABLE BC/SW	
FADA RADIO & ELECTRIC	361-K,-L	1937	AC	TABLE BC/SW	
FADA RADIO & ELECTRIC	362-B,-BG,-D	1937	AC	TABLE PLASTIC BC/SW	
FADA RADIO & ELECTRIC	362-G,-R,-RG	1937	AC	TABLE PLASTIC BC/SW	
FADA RADIO & ELECTRIC	362-V,-W	1937	AC	TABLE PLASTIC BC/SW	
FADA RADIO & ELECTRIC	363-K,-L	1937	AC	TABLE BC/SW	
FADA RADIO & ELECTRIC	365-B,-BG,-C	1937	AC	TABLE PLASTIC BC/SW	
FADA RADIO & ELECTRIC	365-C	1937	AC	CONSOLE BC/SW	
FADA RADIO & ELECTRIC	365-D,-R,-RG	1937	AC	TABLE PLASTIC BC/SW	
FADA RADIO & ELECTRIC	365-PF	1937	AC	PORTABLE RADIO/PHONO BC/SW	
FADA RADIO & ELECTRIC	365-PT	1937	AC	TABLERADIO/PHONO BC/SW	
FADA RADIO & ELECTRIC	365-T	1937	AC	TABLE BC/SW	
FADA RADIO & ELECTRIC	366-C	1937	AC	CONSOLE BC/SW	

MANUFACTURER	MODEL	YEAR	POWER	STYLE	PRICE
FADA RADIO & ELECTRIC	366-T	1937	AC	TABLE BC/SW	
FADA RADIO & ELECTRIC	370-B,-BG,-D	1937	AC	TABLE PLASTIC BC/SW	
FADA RADIO & ELECTRIC	370-C	1937	AC	CONSOLE BC/SW	
FADA RADIO & ELECTRIC	370-G,-R,-RG	1937	AC	TABLE PLASTIC BC/SW	
FADA RADIO & ELECTRIC	370-PF	1937	AC	PORTABLE RADIO/PHONO BC/SW	
FADA RADIO & ELECTRIC	370-PT	1937	AC	TABLE RADIO/PHONO BC/SW	
FADA RADIO & ELECTRIC	370-T	1937	AC	TABLE PLASTIC BC/SW	
FADA RADIO & ELECTRIC	370-V,-W	1937	AC	TABLE PLASTIC BC/SW	
FADA RADIO & ELECTRIC	372-C	1937	AC	TABLE BC/SW	
FADA RADIO & ELECTRIC	380-C	1937	AC	CONSOLE BC/SW	
FADA RADIO & ELECTRIC	380-T	1937	AC	TABLE BC/SW	
FADA RADIO & ELECTRIC	382-C	1937	AC	CONSOLE BC/SW	
FADA RADIO & ELECTRIC	390-C	1937	AC	CONSOLE BC/SW	
FADA RADIO & ELECTRIC	390-T	1937	AC	TABLE BC/SW	
FADA RADIO & ELECTRIC	392-C	1937	AC	CONSOLE BC/SW	
FADA RADIO & ELECTRIC	410-C	1937	AC	CONSOLE BC/SW	
FADA RADIO & ELECTRIC	410-CA	1937	AC	CONSOLE BC/SW	
FADA RADIO & ELECTRIC	410-PC	1937	AC	CONSOLE RADIO/PHONO BC/SW	
FADA RADIO & ELECTRIC	411-C	1937	AC	CONSOLE BC/SW	
FADA RADIO & ELECTRIC	411-CA	1937	AC	CONSOLE BC/SW	
FADA RADIO & ELECTRIC	411-PC	1937	AC	CONSOLE RADIO/PHONO BC/SW	
FADA RADIO & ELECTRIC	413-C	1937	AC	CONSOLE BC/SW	
FADA RADIO & ELECTRIC	413-PC	1937	AC	CONSOLE RADIO/PHONO BC/SW	
FADA RADIO & ELECTRIC	450-V	1938	AC	TABLE	
FADA RADIO & ELECTRIC	450-W	1938	AC	TABLE	
FADA RADIO & ELECTRIC	454-G	1938	AC	TABLE	
FADA RADIO & ELECTRIC	454-T	1938	AC	TABLE	
FADA RADIO & ELECTRIC	454-T	1938	AC	TABLE	
FADA RADIO & ELECTRIC	454-V	1938	AC	TABLE	
FADA RADIO & ELECTRIC	454-W	1938	AC	TABLE	
FADA RADIO & ELECTRIC	460-G	1938	AC	TABLE	
FADA RADIO & ELECTRIC	460-T	1938	AC	TABLE	
FADA RADIO & ELECTRIC	460-V	1938	AC	TABLE	
FADA RADIO & ELECTRIC	460-W	1938	AC	TABLE	
FADA RADIO & ELECTRIC	1216-T	1937	AC	TABLE BC/SW TO 60 MC.	
FADA RADIO & ELECTRIC	1242-B,-BG,-D	1937	2/DC	TABLE PLASTIC BC/SW	
FADA RADIO & ELECTRIC	1242-C	1937	2/DC	CONSOLE BC/SW	
FADA RADIO & ELECTRIC	1242-G,-R,-RG	1937	2/DC	TABLE PLASTIC BC/SW	
FADA RADIO & ELECTRIC	1242-T	1937	2/DC	TABLE BC/SW	
FADA RADIO & ELECTRIC	1242-V,-W	1937	2/DC	TABLE PLASTIC BC/SW	
FADA RADIO & ELECTRIC	1246-B,-BG	1937	6/DC	TABLE PLASTIC BC/SW	
FADA RADIO & ELECTRIC	1246-C	1937	6/DC	CONSOLE BC/SW	
FADA RADIO & ELECTRIC	1246-D,-W	1937	6/DC	TABLE PLASTIC BC/SW	
FADA RADIO & ELECTRIC	1246-G,-R	1937	6/DC	TABLE PLASTIC BC/SW	
FADA RADIO & ELECTRIC	1246-RG,-V	1937	6/DC	TABLE PLASTIC BC/SW	
FADA RADIO & ELECTRIC	1246-T	1937	6/DC	TABLE BC/SW	

MANUFACTURER	MODEL	YEAR	POWER	STYLE	PRICE
FADA RADIO & ELECTRIC	1450-A	1934	AC	TABLE	
FADA RADIO & ELECTRIC	1450-C	1934	AC	CONSOLE	
FADA RADIO & ELECTRIC	1452-A	1934	AC	TABLE	
FADA RADIO & ELECTRIC	1452-F	1934	AC	CONSOLE	
FADA RADIO & ELECTRIC	1460-B	1935	AC	MIDGET	
FADA RADIO & ELECTRIC	1462-D	1935	AC	TABLE BC/SW	
FADA RADIO & ELECTRIC	1470-C	1934	AC	TABLE	
FADA RADIO & ELECTRIC	1470-E	1934	AC	CONSOLE	
FADA RADIO & ELECTRIC	1470-EK	1935	AC	CONSOLE	
FADA RADIO & ELECTRIC	1480-C	1934	AC	TABLE	
FADA RADIO & ELECTRIC	1480-E	1935	AC	CONSOLE	
FADA RADIO & ELECTRIC	1480-EK	1935	AC	CONSOLE	
FADA RADIO & ELECTRIC	1582-H	1935	AC	TABLE	
FADA RADIO & ELECTRIC	1582-J	1935	AC	CONSOLE	
FADA RADIO & ELECTRIC	1582-JK	1935	AC	CONSOLE	
FADA RADIO & ELECTRIC	363-J,-M	1937	AC	TABLE BC/SW	
FADA RADIO & ELECTRIC	365-V,-W	1937	AC	TABLE PLASTIC BC/SW	
FADA RADIO & ELECTRIC	A-66-PC	1939	AC	CONSOLE R/P PB BC/SW	
FADA RADIO & ELECTRIC	A-66-SC	1939	AC	CONSOLE PB TUNE EYE BC/SW	120
FADA RADIO & ELECTRIC	A-66-T	1939	AC	TABLE PB BC/SW	
FADA RADIO & ELECTRIC	A-76-PC	1939	AC	CONSOLE R/P PB BC/SW	
FADA RADIO & ELECTRIC	A-76-R	1939	AC	TABLE PB BC/SW	
FADA RADIO & ELECTRIC	B-49	1939	BAT	CONSOLE	
FADA RADIO & ELECTRIC	B-49-T	1939	BAT	TABLE WALNUT	
FADA RADIO & ELECTRIC	B-49-TR	1939	BAT	TABLE WALNUT	
FADA RADIO & ELECTRIC	B-49-V	1939	BAT	MANTLE PLASTIC	
FADA RADIO & ELECTRIC	B-49-W	1939	BAT	MANTLE PLASTIC	
FADA RADIO & ELECTRIC	C-34	1941	AC	PORTABLE LEATHERETTE	25
FADA RADIO & ELECTRIC	F-45	1939	AC	MANTLE PLASTIC WALNUT	
FADA RADIO & ELECTRIC	F-55-W	1939	AC	MANTLE PLASTIC WALNUT	20
FADA RADIO & ELECTRIC	L-56	1939	AC	MANTLE PLASTIC WALNUT	95
FADA RADIO & ELECTRIC	L-67-T	1939	AC	TABLE	
FADA RADIO & ELECTRIC	L-96-W	1939	AC	TABLE PLASTIC	25
FADA RADIO & ELECTRIC	P-40	1939	BAT	PORTABLE CLOTH	
FADA RADIO & ELECTRIC	P-47	1939	BAT	PORTABLE CLOTH	
FADA RADIO & ELECTRIC	P-49-SW	1939	BAT	PORTABLE CLOTH BC/SW	
FADA RADIO & ELECTRIC	P-50	1939	AC	TABLE RADIO/PHONO	
FADA RADIO & ELECTRIC	P-58	1939	AC	PORTABLE CLOTH	
FADA RADIO & ELECTRIC	P-60	1939	AC	TABLE RADIO/PHONO	
FADA RADIO & ELECTRIC	PD-40	1939	BAT	PORTABLE LEATHERETTE	
FADA RADIO & ELECTRIC	PD-47	1939	BAT	PORTABLE LEATHERETTE	
FADA RADIO & ELECTRIC	PD-49-SW	1939	BAT	PORTABLE LEATHERETTE BC/SW	15
FADA RADIO & ELECTRIC	PL-24	1940	AC	PORTABLE LEATHERETTE	15
FADA RADIO & ELECTRIC	PL-40	1939	BAT	PORTABLE CLOTH	
FADA RADIO & ELECTRIC	PL-50	1939	AC	TABLE RADIO/PHONO	
FADA RADIO & ELECTRIC	PL-58	1939	AC	PORTABLE CLOTH	
FADA RADIO & ELECTRIC	PL-60	1939	AC	TABLE RADIO/PHONO	
FADA RADIO & ELECTRIC	PT-208	1941	AC	TABLE RADIO/PHONO	15

MANUFACTURER	MODEL	YEAR	POWER	STYLE	PRICE
FADA RADIO & ELECTRIC	PUL-60	1939	AC	TABLE RADIO/PHONO	
FADA RADIO & ELECTRIC	RE-197	1940	AC	CONSOLE	55
FADA RADIO & ELECTRIC	S-46-B,-G,-R,	1939	AC	TABLE PLASTIC	
FADA RADIO & ELECTRIC	S-46-T	1939	AC	TABLE WALNUT	
FADA RADIO & ELECTRIC	S-46-V,-W	1939	AC	TABLE PLASTIC	
FAIRVIEW ELECTRIC	J-400	1923	BAT	TABLE LP 5 TUBE 3 DIAL	
FARNSWORTH TV & RADIO	AC- 90	1939	AC	CONSOLE PB BC/SW	130
FARNSWORTH TV & RADIO	AK-59	1939	AC	CONSOLE R/P PB BC/SW	45
FARNSWORTH TV & RADIO	AT-50	1939	AC	TABLE PB BC/SW	45
FARNSWORTH TV & RADIO	BC-81	1940	AC	CONSOLE PB BC/SW	95
FARNSWORTH TV & RADIO	BC-82	1940	AC	CONSOLE BC/SW	115
FARNSWORTH TV & RADIO	BC-102	1940	AC	CONSOLE PB BC/SW	95
FARNSWORTH TV & RADIO	BC-1020	1940	AC	CONSOLE BC/SW	145
FARNSWORTH TV & RADIO	BK-69	1941	AC	TABLE RADIO/PHONO PB	15
FARNSWORTH TV & RADIO	BK-73	1940	AC	CONSOLE CHAIRSIDE R/P	60
FARNSWORTH TV & RADIO	BK-77	1940	AC	CONSOLE RADIO/PHONO	
FARNSWORTH TV & RADIO	BK-78	1940	AC	CONSOLE RADIO/PHONO	
FARNSWORTH TV & RADIO	BK-85	1940	AC	CONSOLE RADIO/PHONO	45
FARNSWORTH TV & RADIO	BK-87	1941	AC	CONSOLE R/P PB BC/SW	45
FARNSWORTH TV & RADIO	BK-88	1940	AC	CONSOLE RADIO/PHONO BC/SW	50
FARNSWORTH TV & RADIO	BK-89	1940	AC	CONSOLE RADIO/PHONO	45
FARNSWORTH TV & RADIO	BK-106	1940	AC	CONSOLE RADIO/PHONO	
FARNSWORTH TV & RADIO	BK-107	1940	AC	CONSOLE RADIO/PHONO	
FARNSWORTH TV & RADIO	BK-108	1940	AC	CONSOLE RADIO/PHONO	
FARNSWORTH TV & RADIO	BK-110	1940	AC	CONSOLE RADIO/PHONO	
FARNSWORTH TV & RADIO	BK-111	1940	AC	CONSOLE RADIO/PHONO	
FARNSWORTH TV & RADIO	BK-112	1940	AC	CONSOLE RADIO/PHONO	50
FARNSWORTH TV & RADIO	BK-1030	1940	AC	CONSOLE RADIO/PHONO BC/SW	50
FARNSWORTH TV & RADIO	BT-22	1940	AC	TABLE PB	30
FARNSWORTH TV & RADIO	BT-40	1940	2/DC	TABLE	30
FARNSWORTH TV & RADIO	BT-55	1940	AC	TABLE PB	40
FARNSWORTH TV & RADIO	BT-55	1940	AC	TABLE PB	35
FARNSWORTH TV & RADIO	BT-61	1940	AC	TABLE PB	40
FARNSWORTH TV & RADIO	BT-70	1940	AC	TABLE PB	45
FARNSWORTH TV & RADIO	BT-71	1940	AC	TABLE PB	45
FARNSWORTH TV & RADIO	BT-600	1940	AC	TABLE BC/SW	40
FARNSWORTH TV & RADIO	BT-601	1940	AC	CONSOLE BC/SW	90
FARNSWORTH TV & RADIO	CC-90	1941	AC	CONSOLE PB BC/SW	95
FARNSWORTH TV & RADIO	CK-58	1941	AC	TABLE RADIO/PHONO	20
FARNSWORTH TV & RADIO	CK-66	1941	AC	TABLE RADIO/PHONO	20
FARNSWORTH TV & RADIO	CK-70	1941	AC	CONSOLE PB BC/SW	95
FARNSWORTH TV & RADIO	CK-73	1941	AC	CONSOLE R/P CHAIRSIDE	55
FARNSWORTH TV & RADIO	CK-74	1941	AC	CONSOLE RADIO/PHONO	45
FARNSWORTH TV & RADIO	CK-75	1941	AC	CONSOLE RADIO/PHONO	50
FARNSWORTH TV & RADIO	CK-91	1941	AC	CONSOLE RADIO/PHONO	50
FARNSWORTH TV & RADIO	CK-92	1941	AC	CONSOLE RADIO/PHONO	50
FARNSWORTH TV & RADIO	CK-93	1941	AC	CONSOLE RADIO/PHONO	50
FARNSWORTH TV & RADIO	CK-111	1941	AC	CONSOLE R/P PB BC/SW	50

MANUFACTURER	MODEL	YEAR	POWER	STYLE	PRICE
FARNSWORTH TV & RADIO	CK-1112	1941	AC	CONSOLE R/P BC/FM	50
FARNSWORTH TV & RADIO	CT-43	1941	2/DC	TABLE	15
FARNSWORTH TV & RADIO	CT-50	1941	AC	TABLE PLASTIC MAHOGANY	25
FARNSWORTH TV & RADIO	CT-51	1941	AC	TABLE PLASTIC IVORY	25
FARNSWORTH TV & RADIO	CT-52	1941	AC	TABLE PLASTIC MAHOGANY	25
FARNSWORTH TV & RADIO	CT-53	1941	AC	TABLE PLASTIC IVORY	25
FARNSWORTH TV & RADIO	CT-54	1941	AC	TABLE	30
FARNSWORTH TV & RADIO	CT-59	1941	AC	PORTABLE CAMERAETTE	30
FARNSWORTH TV & RADIO	CT-60	1941	AC	PORTABLE LUGGAGE TYPE	15
FARNSWORTH TV & RADIO	CT-61	1941	AC	MANTLE PLASTIC MAHOGANY	20
FARNSWORTH TV & RADIO	CT-62	1941	AC	MANTLE PLASTIC IVORY	25
FARNSWORTH TV & RADIO	CT-63	1941	AC	TABLE BC/SW	15
FEDDERS	JRH RANGER	1922	BAT	TABLE LP LOOP ANT 6 T 1 DIAL	250
FEDERAL RADIO	8	1922	BAT	TABLE	325
FEDERAL RADIO	56	1922	BAT	TABLE METAL 2 TUBE 1 DIAL	350
FEDERAL RADIO	57	1922	BAT	TABLE METAL 4 TUBE 1 DIAL	425
FEDERAL RADIO	58-DX	1922	BAT	TABLE LP METAL 4 T 3 DIAL	370
FEDERAL RADIO	59	1923	BAT	TABLE METAL 4 TUBE 3 DIAL	600
FEDERAL RADIO	61	1924	BAT	TABLE HP 6 TUBE 3 DIAL	690
FEDERAL RADIO	102	1924	BAT	PORTABLE 4 TUBE 2 DIAL	330
FEDERAL RADIO	110	1924	BAT	TABLE METAL 3 TUBE 2 DIAL	450
FEDERAL RADIO	141	1925	BAT	TABLE LP 5 TUBE 2 DIAL	200
FEDERAL RADIO	142	1925	BAT	CONSOLE LB 5 TUBE 2 DIAL	175
FEDERAL RADIO	143	1925	BAT	CONSOLE LB 5 TUBE 2 DIAL	100
FEDERAL RADIO	144	1925	BAT	CONSOLE HB 5 TUBE 2 DIAL	
FEDERAL RADIO	161	1925	BAT	CONSOLE HB 6 TUBE 3 DIAL	
FEDERAL RADIO	200	1925	BAT	PORTABLE 3 TUBE 3 DIAL	270
FEDERAL RADIO	301	1940	AC	CONSOLE RADIO/PHONO	
FEDERAL RADIO	306	1940	AC	CONSOLE RADIO/PHONO	
FEDERAL RADIO	A-10 ORTHOSONIC	1925	BAT	TABLE LP 5 TUBE 3 DIAL	90
FEDERAL RADIO	B-20 ORTHOSONIC	1925	BAT	TABLE LP 5 TUBE 3 DIAL	85
FEDERAL RADIO	B-30 ORTHOSONIC	1925	BAT	TABLE LP 5 TUBE 3 DIAL	85
FEDERAL RADIO	B-35 ORTHOSONIC	1925	BAT	CONSOLE 5 TUBE 3 DIAL	
FEDERAL RADIO	B-36 ORTHOSONIC	1925	BAT	CONSOLE 5 TUBE 3 DIAL	
FEDERAL RADIO	C-20 ORTHOSONIC	1925	BAT	TABLE LP 7 TUBE 2 DIAL	
FEDERAL RADIO	C-30 ORTHOSONIC	1925	BAT	TABLE LP 7 TUBE 2 DIAL	85
FEDERAL RADIO	C-35 ORTHOSONIC	1925	BAT	CONSOLE 7 TUBE 2 DIAL	
FEDERAL RADIO	C-40 ORTHOSONIC	1925	BAT	CONSOLE 7 TUBE 2 DIAL	
FEDERAL RADIO	D-10 ORTHOSONIC	1926	BAT	TABLE LP 5 TUBE 1 DIAL	
FEDERAL RADIO	D-10/60	1927	AC	TABLE 6 TUBE 1 DIAL	
FEDERAL RADIO	D-40/60	1927	AC	CONSOLE 6 TUBE 1 DIAL	
FEDERAL RADIO	E-10 ORTHOSONIC	1926	BAT	TABLE LP 6 TUBE 1 DIAL	80
FEDERAL RADIO	E-10/60	1927	AC	TABLE 6 TUBE 1 DIAL	
FEDERAL RADIO	E-40/60	1928	AC	CONSOLE 6 TUBE 1 DIAL	
FEDERAL RADIO	E-41/60	1928	AC	CONSOLE 6 TUBE 1 DIAL	
FEDERAL RADIO	E-45/60	1927	AC	CONSOLE LB 6 TUBE 1 DIAL	145
FEDERAL RADIO	F-10	1928	AC	TABLE LP 8 TUBE 1 DIAL	
FEDERAL RADIO	F-11	1928	BAT	TABLE LP 7 TUBE 1 DIAL	
FEDERAL RADIO	F-11/60	1928	AC	TABLE 8 TUBE 1 DIAL	

MANUFACTURER	MODEL	YEAR	POWER	STYLE	PRICE
FEDERAL RADIO	F-40	1926	BAT	CONSOLE HB 7 TUBE 1 DIAL	160
FEDERAL RADIO	F-40/60	1928	AC	CONSOLE 8 TUBE 1 DIAL	
FEDERAL RADIO	F-41/60	1928	AC	CONSOLE 8 TUBE 1 DIAL	
FEDERAL RADIO	F-50	1928	BAT	CONSOLE 7 TUBE 1 DIAL	
FEDERAL RADIO	F-50/60	1928	AC	CONSOLE 8 TUBE 1 DIAL	
FEDERAL RADIO	F-51/60	1928	AC	CONSOLE 8 TUBE 1 DIAL	
FEDERAL RADIO	F-60	1928	BAT	CONSOLE 7 TUBE 1 DIAL	
FEDERAL RADIO	F-60/60	1928	AC	CONSOLE 8 TUBE 1 DIAL	
FEDERAL RADIO	F-61/60	1928	AC	CONSOLE 8 TUBE 1 DIAL	
FEDERAL RADIO	F-70	1928	BAT	CONSOLE 7 TUBE 1 DIAL	
FEDERAL RADIO	F-70/60	1928	AC	CONSOLE 8 TUBE 1 DIAL	
FEDERAL RADIO	F-71/60	1928	AC	CONSOLE 8 TUBE 1 DIAL	
FEDERAL RADIO	F-80	1928	BAT	CONSOLE 7 TUBE 1 DIAL	
FEDERAL RADIO	F-80/60	1928	AC	CONSOLE 8 TUBE 1 DIAL	
FEDERAL RADIO	F-81	1931	AC	COMPACT	
FEDERAL RADIO	F-81/60	1928	AC	CONSOLE 8 TUBE 1 DIAL	
FEDERAL RADIO	F-82	1931	AC	CONSOLE LB	
FEDERAL RADIO	F-83	1931	AC	CONSOLE HB	
FEDERAL RADIO	F-92	1931	AC	CONSOLE LB	
FEDERAL RADIO	F-93	1931	AC	CONSOLE HB	
FEDERAL RADIO	FE-94	1931	AC	TABLE	
FEDERAL RADIO	G-10/60	1928	AC	TABLE 8 TUBE 1 DIAL	
FEDERAL RADIO	G-40/60	1928	AC	CONSOLE 8 TUBE 1 DIAL	
FEDERAL RADIO	K-10	1929	AC	TABLE	
FEDERAL RADIO	K-40	1929	AC	CONSOLE	
FEDERAL RADIO	K-41	1929	AC	CONSOLE	
FEDERAL RADIO	L-36	1929	AC	CONSOLE LB	
FEDERAL RADIO	L-46	1929	AC	CONSOLE	
FEDERAL RADIO	M-10	1929	AC	TABLE	
FEDERAL RADIO	M-36	1929	AC	CONSOLE	
FEDERAL RADIO	M-41	1929	AC	CONSOLE	
FEDERAL RADIO	M-46	1929	AC	CONSOLE	
FEDERAL TELEGRAPH				See also KOLSTER RADIO	
FEDERAL TELEGRAPH				See also KOLSTER RADIO	
FEDERAL TELEGRAPH	6-A	1925	BAT	TABLE LP 6 TUBE 2 DIAL	
FEDERAL TELEGRAPH	6-B	1925	BAT	TABLE LP SPEAKER 6 T 2 DL	
FEDERAL TELEGRAPH	6-C	1925	BAT	CONSOLE LB 6 TUBE 2 DIAL	
FERGUSON	3	1923	BAT	TABLE LP 4 TUBE 3 DIAL	
FERGUSON	4	1924	BAT	TABLE LP 4 TUBE 2 DIAL	
FERGUSON	6	1925	BAT	TABLE LP 6 TUBE 2 DIAL	
FERGUSON	6	1925	BAT	CONSOLE 6 TUBE 2 DIAL	
FERGUSON	8	1925	BAT	TABLE LP 6 TUBE 1 DIAL	
FERGUSON	8	1925	BAT	CONSOLE 6 TUBE 1 DIAL	
FERGUSON	10	1926	BAT	TABLE LP 6 TUBE 1 DIAL	
FERGUSON	10-A	1927	BAT	TABLE LP 7 TUBE 1 DIAL	
FERGUSON	12	1926	BAT	TABLE LP 6 TUBE 2 DIAL	
FERGUSON	12	1927	BAT	CONSOLE LB 6 TUBE 2 DIAL	
FERGUSON	14	1927	BAT	TABLE LP LOOP ANT 10 T 1DL	
FERGUSON	18 HOMER	1927	AC	TABLE LP 7 TUBE 1 DIAL	130
FOREIGN & DOMESTIC ELECTRIC	ELECTRADYNE	1924	AC	TABLE HP 3 TUBE 2 DIAL	

MANUFACTURER	MODEL	YEAR	POWER	STYLE	PRICE
FREED EISEMANN	FE-15	1925	BAT	TABLE LP 5 TUBE 3 DIAL	95
FREED EISEMANN	FE-18	1925	BAT	TABLE LP 5 TUBE 3 DIAL	125
FREED EISEMANN	FE-30	1925	BAT	TABLE LP 6 TUBE 3 DIAL	
FREED EISEMANN	FM-40	1940	AC	TABLE BC/FM	45
FREED EISEMANN	FM-42	1941	AC	CONSOLE R/P BC/FM	50
FREED EISEMANN	FM-45	1940	AC	CONSOLE RADIO/PHONO BC/FM	50
FREED EISEMANN	NR-5	1923	BAT	TABLE LP 5 TUBE 3 DIAL	95
FREED EISEMANN	NR-6	1924	BAT	TABLE LP 5 TUBE 3 DIAL	85
FREED EISEMANN	NR-7	1925	BAT	TABLE LP 6 TUBE 3 DIAL	95
FREED EISEMANN	NR-8	1927	BAT	TABLE LP 6 TUBE 2 DIAL	
FREED EISEMANN	NR-9	1927	BAT	TABLE LP 6 TUBE 2 DIAL	80
FREED EISEMANN	NR-11	1927	AC	TABLE LP 6 TUBE 2 DIAL	
FREED EISEMANN	NR-12	1924	BAT	TABLE LP 4 TUBE 2 DIAL	195
FREED EISEMANN	NR-20	1924	BAT	TABLE LP 5 TUBE 3 DIAL	80
FREED EISEMANN	NR-45	1925	BAT	TABLE LP 6 TUBE 3 DIAL	125
FREED EISEMANN	NR-53	1929	BAT	TABLE LP METAL 7 T 1 DIAL	
FREED EISEMANN	NR-55	1929	DC	CONSOLE 7 TUBE 1 DIAL	
FREED EISEMANN	NR-55	1929	AC	CONSOLE 8 TUBE 1 DIAL	120
FREED EISEMANN	NR-56	1929	DC	TABLE LP METAL 7 T 1 DIAL	
FREED EISEMANN	NR-56	1929	AC	TABLE LP METAL 8 T 1 DIAL	
FREED EISEMANN	NR-57	1927	AC	TABLE LP 7 TUBE 1 DIAL	
FREED EISEMANN	NR-60 AC	1927	AC	TABLE LP 7 TUBE 1 DIAL	105
FREED EISEMANN	NR-60 DC	1927	DC	TABLE LP 7 TUBE 1 DIAL	90
FREED EISEMANN	NR-66	1927	BAT	TABLE LP 6 TUBE 1 DIAL	
FREED EISEMANN	NR-77	1927	BAT	TABLE LP 7 TUBE 1 DIAL	80
FREED EISEMANN	NR-78	1929	DC	CONSOLE 8 TUBE 1 DIAL	
FREED EISEMANN	NR-78	1929	AC	CONSOLE 9 TUBE 1 DIAL	120
FREED EISEMANN	NR-79	1929	DC	CONSOLE 8 TUBE 1 DIAL	
FREED EISEMANN	NR-79	1929	AC	CONSOLE 9 TUBE 1 DIAL	120
FREED EISEMANN	NR-80	1928	AC	TABLE LP METAL 8 T 1 DIAL	125
FREED EISEMANN	NR-85	1928	AC	TABLE LP METAL 8 T 1 DIAL	
FREED EISEMANN	NR-90	1929	AC	CONSOLE HB DOORS PB 8T 1 DIAL	250
FREED EISEMANN	NR-95	1929	DC	CONSOLE HB 8 TUBE 1 DIAL	
FREED EISEMANN	NR-95	1929	AC	CONSOLE HB 9 TUBE 1 DIAL	120
FREED EISEMANN	NR-215	1924	BAT	CONSOLE PANEL 4 TUBE 2 DL	
FREED EISEMANN	NR-400	1924	BAT	PORTABLE 4 TUBE 2 DIAL	
FREED EISEMANN	NR-800	1926	BAT	TABLE LP 8 TUBE 1 DIAL	
FREED EISEMANN	NR-800-C	1926	BAT	CONSOLE HB 8 TUBE 1 DIAL	
FREED EISEMANN	NR-850	1927	BAT	CONSOLE HB 8 TUBE 1 DIAL	
FREED RADIO & TELEVISION	24	1937	AC	TABLE	
FREED RADIO & TELEVISION	25	1937	AC	TABLE	
FREED RADIO & TELEVISION	26	1937	AC	TABLE TRAVELING CASE	
FREED RADIO & TELEVISION	27	1937	AC	TABLE BC/SW	
FREED RADIO & TELEVISION	28	1937	AC	TABLE BC/SW	
FREED RADIO & TELEVISION	33	1937	AC	MANTLE TRAVELING CASE B/SW	
FREED RADIO & TELEVISION	33W	1937	AC	MANTLE	
FREED RADIO & TELEVISION	51	1932	DC	TABLE	

MANUFACTURER	MODEL	YEAR	POWER	STYLE	PRICE
FREED RADIO & TELEVISION	52	1932	DC	TABLE CHEST	
FREED RADIO & TELEVISION	53	1932	DC	TABLE	
FREED RADIO & TELEVISION	54	1932	AC	TABLE	
FREED RADIO & TELEVISION	56	1932	AC	TABLE	
FREED RADIO & TELEVISION	58	1932	AC	TABLE	
FREED RADIO & TELEVISION	59	1932	AC	TABLE CHEST	
FREED RADIO & TELEVISION	97	1937	AC	TABLE BC/SW	
FREED RADIO & TELEVISION	98	1937	AC	TABLE BC/SW	
FREED RADIO & TELEVISION	310	1933	AC	CHASSIS BC/SW	
FREED RADIO & TELEVISION	346	1933	AC	TABLE	
FREED RADIO & TELEVISION	350	1933	AC	TABLE BC/SW	
FREED RADIO & TELEVISION	353	1934	AC	TABLE BC/LW	
FREED RADIO & TELEVISION	354	1934	AC	TABLE	
FREED RADIO & TELEVISION	355	1933	AC	TABLE	
FREED RADIO & TELEVISION	356	1933	AC	TABLE	
FREED RADIO & TELEVISION	358	1933	AC	PORTABLE RADIO/PHONO BC/SW	
FREED RADIO & TELEVISION	360	1933	AC	TABLE BC/SW	
FREED RADIO & TELEVISION	366	1934	AC	TABLE BC/SW	
FREED RADIO & TELEVISION	366-LW	1934	AC	TABLE BC/LW	
FREED RADIO & TELEVISION	367	1934	AC	TABLE BC/SW	
FREED RADIO & TELEVISION	368	1934	AC	TABLE BC/LW	
FREED RADIO & TELEVISION	406	1934	6/DC	TABLE	
FREED RADIO & TELEVISION	432	1934	32/DC	TABLE	
FREED RADIO & TELEVISION	450	1934	AC	TABLE BC/SW	
FREED RADIO & TELEVISION	466	1934	AC	TABLE BC/SW/LW	
FREED RADIO & TELEVISION	467	1934	AC	TABLE BC/SW/LW	
FREED RADIO & TELEVISION	469	1934	AC	TABLE BC/SW	
FREED RADIO & TELEVISION	475-X	1934	AC	TABLE BC/SW/LW	
FREED RADIO & TELEVISION	482	1934	AC	TABLE BC/SW	125
FREED RADIO & TELEVISION	A-342	1935	AC	TABLE	
FREED RADIO & TELEVISION	E-341	1935	AC	TABLE	
FREED RADIO & TELEVISION	FE-50	1936	AC	TABLE	
FREED RADIO & TELEVISION	FE-55	1936	2/DC	PORTABLE	
FREED RADIO & TELEVISION	FE-56	1936	AC	TABLE	
FREED RADIO & TELEVISION	FE-56-S	1936	AC	TABLE BC/SW	
FREED RADIO & TELEVISION	FE-58	1936	AC	MANTLE	
FREED RADIO & TELEVISION	FE-60	1936	AC	TABLE	
FREED RADIO & TELEVISION	FE-62-P	1936	AC	TABLE BC/SW	
FREED RADIO & TELEVISION	FE-64-S	1936	AC	TABLE BC/SW	
FREED RADIO & TELEVISION	FE-65	1936	AC	TABLE BC/SW	
FREED RADIO & TELEVISION	FE-96	1931	AC	COMPACT	
FREED RADIO & TELEVISION	FE-98	1931	AC	COMPACT	
FREED RADIO & TELEVISION	G-351-P	1935	AC	TABLE	
FREED RADIO & TELEVISION	H-357-L	1935	AC	TABLE	
FREED RADIO & TELEVISION	H-357-P	1935	AC	TABLE	
FREED RADIO & TELEVISION	Q-358S	1935	AC	TABLE BC/SW	
FREED RADIO & TELEVISION	R-369S	1935	AC	TABLE BC/SW	
FREED RADIO & TELEVISION	T-367S	1935	AC	TABLE BC/SW	
FREED RADIO & TELEVISION	T-368P	1935	AC	TABLE BC/SW	

MANUFACTURER	MODEL	YEAR	POWER	STYLE	PRICE
FREED RADIO & TELEVISION	U-363	1935	AC	TABLE BC/SW	
FREED RADIO & TELEVISION	V-360	1935	AC	TABLE	
FREED RADIO & TELEVISION	W-380	1935	AC	TABLE BC/SW	
FREED RADIO & TELEVISION	W-382	1935	AC	TABLE BC/SW	
FRENCH PIANO	JUNIOR	1930	AC	CATHEDRAL	150
FRESHMAN	3-Q	1928	AC	TABLE LP METAL 6 TUBE 1 DIAL	70
FRESHMAN	5-F-2	1925	BAT	TABLE LP 5 TUBE 3 DIAL	90
FRESHMAN	5-F-4	1925	BAT	TABLE LP 5 TUBE 3 DIAL	
FRESHMAN	5-F-5	1925	BAT	TABLE LP SPEAKER 5 T 3 DL	145
FRESHMAN	5-F-6	1925	BAT	TABLE LP 5 TUBE 3 DIAL	
FRESHMAN	5-F-7	1925	BAT	CONSOLE 5 TUBE 3 DIAL	195
FRESHMAN	6-F-2	1926	BAT	CONSOLE 5 TUBE 3 DIAL	
FRESHMAN	6-F-3	1926	BAT	TABLE LP 5 TUBE 3 DIAL	
FRESHMAN	6-F-4	1926	BAT	CONSOLE 5 TUBE 3 DIAL	160
FRESHMAN	6-F-5	1926	BAT	TABLE LP 5 TUBE 3 DIAL	
FRESHMAN	6-F-6	1926	BAT	TABLE LP 5 TUBE 3 DIAL	
FRESHMAN	6-F-7	1926	BAT	TABLE LP 5 TUBE 3 DIAL	75
FRESHMAN	6-F-9	1926	BAT	CONSOLE 5 TUBE 3 DIAL	
FRESHMAN	6-F-10	1926	BAT	CONSOLE 5 TUBE 3 DIAL	115
FRESHMAN	6-F-11	1926	BAT	CONSOLE 5 TUBE 3 DIAL	
FRESHMAN	6-F-16	1926	BAT	CONSOLE 5 TUBE 3 DIAL	
FRESHMAN	7-F-1	1927	BAT	TABLE LP 6 TUBE 1 DIAL	
FRESHMAN	7-F-2	1927	BAT	TABLE LP 6 TUBE 1 DIAL	
FRESHMAN	7-F-3	1927	BAT	CONSOLE 6 TUBE 1 DIAL	
FRESHMAN	7-F-4	1927	BAT	CONSOLE 6 TUBE 1 DIAL	
FRESHMAN	7-F-5	1927	BAT	CONSOLE 6 TUBE 1 DIAL	
FRESHMAN	7F-AC-2	1927	AC	TABLE 1 DIAL	
FRESHMAN	7F-AC-3	1927	AC	CONSOLE 1 DIAL	
FRESHMAN	7F-AC-4	1927	AC	CONSOLE 1 DIAL	
FRESHMAN	7F-AC-5	1927	AC	CONSOLE 1 DIAL	
FRESHMAN	21 EARL	1929	DC	TABLE LP METAL 7T 1 DIAL	
FRESHMAN	21 EARL	1929	AC	TABLE LP METAL 8T 1 DIAL	
FRESHMAN	22 EARL	1929	DC	CONSOLE 7 TUBE 1 DIAL	
FRESHMAN	22 EARL	1929	AC	CONSOLE 8 TUBE 1 DIAL	
FRESHMAN	31 EARL	1929	DC	CONSOLE 8 TUBE 1 DIAL	
FRESHMAN	31 EARL	1929	AC	CONSOLE 9 TUBE 1 DIAL	
FRESHMAN	32 EARL	1929	DC	CONSOLE 7 TUBE 1 DIAL	
FRESHMAN	32 EARL	1929	AC	CONSOLE 8 TUBE 1 DIAL	
FRESHMAN	41 EARL	1929	AC	CONSOLE HB 9 TUBE 1 DIAL	
FRESHMAN	121	1929	BAT	TABLE LP METAL	
FRESHMAN	F-1 EQUAPHASE	1927	BAT	TABLE LP 6 TUBE 1 DIAL	85
FRESHMAN	F-2 EQUAPHASE	1927	BAT	CONSOLE 6 TUBE 1 DIAL	
FRESHMAN	F-3 EQUAPHASE	1927	BAT	CONSOLE 6 TUBE 1 DIAL	
FRESHMAN	F-4 EQUAPHASE	1927	BAT	CONSOLE 6 TUBE 1 DIAL	
FRESHMAN	F-5 EQUAPHASE	1927	BAT	CONSOLE 6 TUBE 1 DIAL	
FRESHMAN	G-1 EQUAPHASE	1927	AC	TABLE LP 7 TUBE 1 DIAL	
FRESHMAN	G-2 EQUAPHASE	1927	AC	CONSOLE 7 TUBE 1 DIAL	
FRESHMAN	G-3 EQUAPHASE	1927	AC	CONSOLE LB 7 TUBE 1 DIAL	150
FRESHMAN	G-4 EQUAPHASE	1927	AC	CONSOLE HB 7 TUBE 1 DIAL	145

MANUFACTURER	MODEL	YEAR	POWER	STYLE	PRICE
FRESHMAN	G-5 EQUAPHASE	1927	AC	CONSOLE 7 TUBE 1 DIAL	
FRESHMAN	G-6 EQUAPHASE	1927	AC	CONSOLE RADIO/PHONO 7 T 1 DIAL	
FRESHMAN	G-7 EQUAPHASE	1927	AC	CONSOLE 7 TUBE 1 DIAL	
FRESHMAN	G-10 EQUAPHASE	1927	AC	CONSOLE 7 TUBE 1 DT	
FRESHMAN	MASTERPIECE	1924	BAT	TABLE LP 5 TUBE 3 DIAL	105
FRESHMAN	N-11	1928	AC	TABLE LP 7 TUBE 1 DIAL	
FRESHMAN	N-12	1928	AC	CONSOLE 7 TUBE 1 DIAL	
FRESHMAN	N-14	1928	AC	CONSOLE 7 TUBE 1 DIAL	
FRESHMAN	N-17	1928	AC	CONSOLE 7 TUBE 1 DIAL	
FRESHMAN	Q	1928	AC	TABLE LP METAL 5 TUBE 1 DIAL	65
FRESHMAN	Q-15	1928	AC	TABLE LP 5 TUBE 1 DIAL	
FRESHMAN	Q-16	1928	AC	CONSOLE 5 TUBE 1 DIAL	
FRESHMAN	QD-16	1928	AC	CONSOLE 5 TUBE 1 DIAL	140
G					
GALVIN MFG.	3-A-5	1941	AC	PORTABLE METAL MINATURE	25
GALVIN MFG.	5-T-1	1937	AC	TABLE BC/SW	
GALVIN MFG.	5-T-2	1937	AC	TABLE BC/SW	
GALVIN MFG.	5-Y	1937	AC	CONSOLE BC/SW	
GALVIN MFG.	6-A	1937	AC	CONSOLE BC/SW END TABLE	
GALVIN MFG.	6-T	1937	AC	TABLE BC/SW	
GALVIN MFG.	6-X	1937	AC	TABLE BC/SW	
GALVIN MFG.	6-Y	1937	AC	CONSOLE	
GALVIN MFG.	9-A	1937	AC	CONSOLE BC/SW END TABLE	
GALVIN MFG.	9-R	1937	AC	CONSOLE RADIO/PHONO BC/SW	
GALVIN MFG.	9-Y	1937	AC	CONSOLE BC/SW	
GALVIN MFG.	10-Y-1	1937	AC	CONSOLE BC/SW	
GALVIN MFG.	12-Y-1	1937	AC	CONSOLE BC/SW	
GALVIN MFG.	17-FM-41	1941	AC	CONSOLE PB BC/SW/FM	100
GALVIN MFG.	40-60	1940	AC	TABLE BC/SW	25
GALVIN MFG.	41-A	1939	BAT	MANTLE PLASTIC	
GALVIN MFG.	41-B	1939	BAT	TABLE PLASTIC	
GALVIN MFG.	41-D-1,-2	1939	BAT	PORTABLE	
GALVIN MFG.	41-E	1939	BAT	TABLE	
GALVIN MFG.	41-F	1939	BAT	TABLE BC/SW	
GALVIN MFG.	41-S	1939	BAT	PORTABLE	
GALVIN MFG.	50-XC	1940	AC	TABLE PLASTIC	20
GALVIN MFG.	51-A,-C	1939	AC	MANTLE PLASTIC	
GALVIN MFG.	51-D	1939	BAT	PORTABLE CLOTH	20
GALVIN MFG.	51x17 SERIES	1941	AC	TABLE LEATHERETTE 2 TONE	25
GALVIN MFG.	51x18 SERIES	1941	AC	TABLE	25
GALVIN MFG.	52-BK	1940	2/DC	CONSOLE BC/SW	70
GALVIN MFG.	52-BW	1940	2/DC	TABLE BC/SW	30
GALVIN MFG.	52-C	1939	AC	TABLE	45
GALVIN MFG.	52-C-1,2,3,4	1939	AC	MANTLE PLASTIC	
GALVIN MFG.	52-T	1937	BAT	TABLE	
GALVIN MFG.	53-A	1939	AC	TABLE	

MANUFACTURER	MODEL	YEAR	POWER	STYLE	PRICE
GALVIN MFG.	53-C	1939	AC	MANTLE PLASTIC	
GALVIN MFG.	56-T	1937	6/DC	TABLE BC/SW	
GALVIN MFG.	56-Y	1937	6/DC	CONSOLE BC/SW	
GALVIN MFG.	58-F-1	1940	AC	TABLE RADIO/PHONO	15
GALVIN MFG.	58-FRC	1940	AC	CONSOLE RADIO/PHONO	40
GALVIN MFG.	58-FRC	1941	AC	TABLE RADIO/PHONO	15
GALVIN MFG.	59-T-1	1938	AC	TABLE	
GALVIN MFG.	59-T-2	1938	AC	TABLE	
GALVIN MFG.	59-T-3	1938	AC	TABLE PB	
GALVIN MFG.	59-T-4	1938	AC	TABLE PB	
GALVIN MFG.	59-T-5	1938	AC	TABLE PB BC/SW	
GALVIN MFG.	60-X-2	1940	AC	MANTLE PLASTIC IVORY	25
GALVIN MFG.	61-A	1939	AC	TABLE PLASTIC PB	
GALVIN MFG.	61-B	1939	AC	TABLE PB BC/SW	
GALVIN MFG.	61-C	1939	AC	TABLE PB BC/SW	
GALVIN MFG.	61-D	1939	AC	CONSOLE PB BC/SW	
GALVIN MFG.	61-E	1939	AC	TABLE PB	
GALVIN MFG.	61-F	1939	AC	TABLE RADIO/PHONO	
GALVIN MFG.	62-B	1939	AC	TABLE PB BC/SW	
GALVIN MFG.	62-E	1939	AC	TABLE PB	
GALVIN MFG.	62-FI	1940	AC	CONSOLE RADIO/PHONO	
GALVIN MFG.	63-E	1939	AC	TABLE PB	
GALVIN MFG.	81-C	1939	AC	CONSOLE PB BC/SW	
GALVIN MFG.	81-F-21	1941	AC	CONSOLE R/P PB BC/SW	45
GALVIN MFG.	82-A	1939	AC	CONSOLE PB BC/SW	100
GALVIN MFG.	83-FI	1940	AC	CONSOLE RADIO/PHONO	
GALVIN MFG.	93-FI	1940	AC	CONSOLE RADIO/PHONO	
GALVIN MFG.	101-C-31	1941	AC	CONSOLE W/SELF TUN CLOCK	90
GALVIN MFG.	101-R-21	1941	AC	CONSOLE RADIO/PHONO	55
GALVIN MFG.	103-FI	1940	AC	CONSOLE RADIO/PHONO	
GALVIN MFG.	103-K-1	1940	AC	CONSOLE PB BC/SW	120
GALVIN MFG.	B-150	1940	BAT	PORTABLE BIKE RADIO	75
GAROD RADIO	EA	1926	AC	CONSOLE LB 5 TUBE 3 DIAL	135
GAROD RADIO	GEORGIAN	1924	BAT	CONSOLE 5 TUBE 3 DIAL	
GAROD RADIO	RAF	1925	BAT	TABLE 5 TUBE 3 DIAL	125
GAROD RADIO	V	1924	BAT	TABLE 5 TUBE 3 DIAL	60
GATES RADIO	212	1923	BAT	TABLE HP 5 TUBE 3 DIAL	175
GENERAL AMERICAN RADIO	D	1923	BAT	PORTABLE 6 TUBE 2 DIAL	175
GENERAL AMERICAN RADIO	E	1923	BAT	PORTABLE 6 TUBE 2 DIAL	200
GENERAL AMERICAN RADIO	F	1923	BAT	PORTABLE 6 TUBE 2 DIAL	225
GENERAL ELECTRIC	51	1935	AC	TABLE	
GENERAL ELECTRIC	66	1935	AC	CONSOLE	
GENERAL ELECTRIC	85	1935	AC	CONSOLE	
GENERAL ELECTRIC	A-51	1936	AC	TABLE BC/SW	
GENERAL ELECTRIC	A-53	1936	AC	TABLE	70
GENERAL ELECTRIC	A-53	1935	AC	TABLE BC/SW	100
GENERAL ELECTRIC	A-63	1936	AC	TABLE	60
GENERAL ELECTRIC	A-63	1935	AC	TABLE BC/SW	
GENERAL ELECTRIC	A-65	1936	AC	CONSOLE	
GENERAL ELECTRIC	A-65	1935	AC	CONSOLE BC/SW	

MANUFACTURER	MODEL	YEAR	POWER	STYLE	PRICE
GENERAL ELECTRIC	A-70	1935	AC	TABLE BC/SW	45
GENERAL ELECTRIC	A-75	1935	AC	CONSOLE BC/SW	
GENERAL ELECTRIC	A-81	1932	BAT	PORTABLE	40
GENERAL ELECTRIC	A-82	1935	AC	TABLE BC/SW	60
GENERAL ELECTRIC	A-87	1935	AC	CONSOLE BC/SW	
GENERAL ELECTRIC	A-125	1935	AC	CONSOLE BC/SW	75
GENERAL ELECTRIC	B-81	1934	AC	TABLE	
GENERAL ELECTRIC	B-86	1934	AC	CONSOLE	
GENERAL ELECTRIC	BX	1933	AC	COMPACT	
GENERAL ELECTRIC	C-61	1934	AC	PORTABLE	
GENERAL ELECTRIC	C-67	1934	AC	CONSOLE	
GENERAL ELECTRIC	C-70	1934	BAT	TABLE BC/SW	
GENERAL ELECTRIC	C-75	1934	BAT	CONSOLE BC/SW	
GENERAL ELECTRIC	E-52	1937	AC	TABLE BC/SW	
GENERAL ELECTRIC	E-61	1936	AC	TABLE BC/SW	
GENERAL ELECTRIC	E-62	1936	AC	TABLE BC/SW	
GENERAL ELECTRIC	E-68	1936	AC	CONSOLE BC/SW	85
GENERAL ELECTRIC	E-71	1936	AC	TABLE BC/SW	75
GENERAL ELECTRIC	E-72	1936	AC	TABLE BC/SW	
GENERAL ELECTRIC	E-76	1936	AC	CONSOLE BC/SW	
GENERAL ELECTRIC	E-79	1936	AC	CONSOLE RADIO/PHONO BC/SW	
GENERAL ELECTRIC	E-81	1936	AC	TABLE BC/SW	
GENERAL ELECTRIC	E-86	1936	AC	CONSOLE BC/SW	
GENERAL ELECTRIC	E-91	1936	AC	TABLE BC/SW	
GENERAL ELECTRIC	E-95	1936	AC	CONSOLE BC/SW	
GENERAL ELECTRIC	E-101	1936	AC	TABLE BC/SW	
GENERAL ELECTRIC	E-105	1936	AC	CONSOLE BC/SW	120
GENERAL ELECTRIC	E-106	1936	AC	CONSOLE BC/SW	
GENERAL ELECTRIC	E-126	1936	AC	CONSOLE BC/SW	
GENERAL ELECTRIC	E-129	1936	AC	CONSOLE RADIO/PHONO BC/SW	
GENERAL ELECTRIC	E-155	1936	AC	CONSOLE BC/SW	
GENERAL ELECTRIC	F-53	1937	AC	MANTLE BC/SW	45
GENERAL ELECTRIC	F-63	1937	AC	TABLE BC/SW	
GENERAL ELECTRIC	F-65	1937	AC	CONSOLE BC/SW	
GENERAL ELECTRIC	F-66	1937	AC	CONSOLE BC/SW	
GENERAL ELECTRIC	F-70	1937	AC	TABLE BC/SW	55
GENERAL ELECTRIC	F-74	1938	AC	TABLE TUNING EYE BC/SW	60
GENERAL ELECTRIC	F-75	1937	AC	CONSOLE BC/SW	
GENERAL ELECTRIC	F-78	1937	AC	CONSOLE RADIO/PHONO BC/SW	
GENERAL ELECTRIC	F-81	1937	AC	TABLE BC/SW	
GENERAL ELECTRIC	F-86	1937	AC	CONSOLE BC/SW	
GENERAL ELECTRIC	F-96	1937	AC	CONSOLE	135
GENERAL ELECTRIC	F-107	1937	AC	CONSOLE BC/SW	
GENERAL ELECTRIC	F-109	1937	AC	CONSOLE RADIO/PHONO BC/SW	
GENERAL ELECTRIC	F-135	1937	AC	CONSOLE BC/SW	
GENERAL ELECTRIC	F-665	1937	AC	CONSOLE BC/SW ARMCHAIR	

MANUFACTURER	MODEL	YEAR	POWER	STYLE	PRICE
GENERAL ELECTRIC	G-50	1938	AC	TABLE	
GENERAL ELECTRIC	G-53	1938	AC	TABLE BC/SW	
GENERAL ELECTRIC	G-55	1938	AC	CONSOLE	
GENERAL ELECTRIC	G-56	1938	AC	CONSOLE BC/SW	
GENERAL ELECTRIC	G-61	1938	AC	TABLE BC/SW	
GENERAL ELECTRIC	G-64	1939	AC	TABLE BC/SW	45
GENERAL ELECTRIC	G-66	1938	AC	CONSOLE BC/SW	
GENERAL ELECTRIC	G-68	1938	AC	CONSOLE RADIO/PHONO BC/SW	
GENERAL ELECTRIC	G-75	1938	AC	CONSOLE	150
GENERAL ELECTRIC	G-76	1938	AC	CONSOLE	140
GENERAL ELECTRIC	G-85	1938	AC	CONSOLE BC/SW	
GENERAL ELECTRIC	G-95	1938	AC	CONSOLE	
GENERAL ELECTRIC	G-97	1938	AC	CONSOLE BC/SW	
GENERAL ELECTRIC	G-105	1938	AC	CONSOLE BC/SW	
GENERAL ELECTRIC	G-106	1938	AC	CONSOLE BC/SW	
GENERAL ELECTRIC	GD-41	1938	AC	TABLE	
GENERAL ELECTRIC	GD-52	1938	AC	TABLE	50
GENERAL ELECTRIC	GD-62	1938	AC	TABLE	
GENERAL ELECTRIC	GD-67	1938	AC	CONSOLE	
GENERAL ELECTRIC	H-32	1931	AC	CONSOLE LB	
GENERAL ELECTRIC	H-72	1931	AC	CONSOLE RADIO/PHONO	
GENERAL ELECTRIC	H-73	1939	AC	TABLE R/P PB BC/SW	
GENERAL ELECTRIC	H-77	1939	AC	CONSOLE R/P PB BC/SW	265
GENERAL ELECTRIC	H-78	1939	AC	CONSOLE R/P PB BC/SW	
GENERAL ELECTRIC	H-79	1939	AC	CONSOLE R/P PB BC/SW	
GENERAL ELECTRIC	H-87	1939	AC	CONSOLE R/P PB BC/SW	135
GENERAL ELECTRIC	H-91	1931	AC	CLOCK GRANDFATHER	280
GENERAL ELECTRIC	H-116	1939	AC	CONSOLE R/P PB BC/SW	220
GENERAL ELECTRIC	H-118	1939	AC	CONSOLE R/P PB BC/SW	55
GENERAL ELECTRIC	H-400	1939	AC	MANTLE PLASTIC	
GENERAL ELECTRIC	H-500	1939	AC	MANTLE PLASTIC	
GENERAL ELECTRIC	H-510	1939	AC	MANTLE PLASTIC PB	
GENERAL ELECTRIC	H-520	1939	AC	MANTLE PLASTIC PB	20
GENERAL ELECTRIC	H-600	1939	AC	MANTLE PLASTIC	
GENERAL ELECTRIC	H-610	1939	AC	MANTLE PLASTIC PB	
GENERAL ELECTRIC	H-620	1939	AC	TABLE PB BC/SW	20
GENERAL ELECTRIC	H-630	1939	AC	TABLE PB BC/SW	
GENERAL ELECTRIC	H-632	1939	AC	TABLE PB BC/SW	35
GENERAL ELECTRIC	H-638	1939	AC	TABLE R/P PB BC/SW	
GENERAL ELECTRIC	H-639AC	1939	A	TABLE RADIO/PHONO	70
GENERAL ELECTRIC	H-640	1939	AC	TABLE R/P PB BC/SW	
GENERAL ELECTRIC	HB-403	1939	BAT	PORTABLE	
GENERAL ELECTRIC	HB-408	1939	BAT	PORTABLE RADIO/PHONO	
GENERAL ELECTRIC	HB-412	1939	AC	PORTABLE	15
GENERAL ELECTRIC	HJ-628	1939	AC	CONSOLE R/P PB BC/SW	45
GENERAL ELECTRIC	J-51	1940	AC	TABLE	35
GENERAL ELECTRIC	J-62	1940	AC	TABLE	35
GENERAL ELECTRIC	J-70	1932	AC	TABLE	
GENERAL ELECTRIC	J-72	1933	AC	TABLE	

MANUFACTURER	MODEL	YEAR	POWER	STYLE	PRICE
GENERAL ELECTRIC	J-75	1932	AC	CONSOLE LB	
GENERAL ELECTRIC	J-80	1932	AC	TABLE	
GENERAL ELECTRIC	J-82	1932	AC	TABLE	
GENERAL ELECTRIC	J-83	1933	AC	TABLE	
GENERAL ELECTRIC	J-85	1932	AC	CONSOLE LB	
GENERAL ELECTRIC	J-86	1932	AC	CONSOLE LB	260
GENERAL ELECTRIC	J-87-A	1933	AC	CONSOLE	
GENERAL ELECTRIC	J-88	1932	AC	CONSOLE LB RADIO/PHONO	
GENERAL ELECTRIC	J-100	1932	AC	CATHEDRAL	200
GENERAL ELECTRIC	J-105	1932	AC	CONSOLE LB	
GENERAL ELECTRIC	J-107	1932	AC	CONSOLE LB	
GENERAL ELECTRIC	J-109	1932	AC	CONSOLE HB RADIO/PHONO	120
GENERAL ELECTRIC	J-125	1932	AC	CONSOLE LB	95
GENERAL ELECTRIC	J-501W	1941	AC	TABLE PLASTIC IVORY	25
GENERAL ELECTRIC	J-620	1941	AC	TABLE	30
GENERAL ELECTRIC	J-629	1940	AC	TABLE R/P W/RECORDER	15
GENERAL ELECTRIC	J-644	1941	AC	TABLE PLASTIC	25
GENERAL ELECTRIC	J-709	1940	AC	CONSOLE RADIO/PHONO	
GENERAL ELECTRIC	J-718	1940	AC	CONSOLE RADIO/PHONO	
GENERAL ELECTRIC	J-728	1940	AC	CONSOLE RADIO/PHONO	
GENERAL ELECTRIC	J-805	1940	AC	CONSOLE PB BC/SW	120
GENERAL ELECTRIC	J-808	1940	AC	CONSOLE RADIO/PHONO	
GENERAL ELECTRIC	J-809	1940	AC	CONSOLE R/P PB BC/SW	50
GENERAL ELECTRIC	J-818	1940	AC	CONSOLE RADIO/PHONO	
GENERAL ELECTRIC	J-828	1940	AC	CONSOLE RADIO/PHONO	
GENERAL ELECTRIC	JB-410	1940	BAT	PORTABLE CAMERA-TYPE	30
GENERAL ELECTRIC	JFM-165	1940	AC	CONSOLE PB BC/SW/FM	95
GENERAL ELECTRIC	K-40-A	1933	AC	MANTLE	40
GENERAL ELECTRIC	K-41	1933	AC	COMPACT	
GENERAL ELECTRIC	K-43	1933	AC	TABLE BC/SW	
GENERAL ELECTRIC	K-48	1933	AC	TABLE RADIO/PHONO BC/SW	
GENERAL ELECTRIC	K-50	1933	AC	CATHEDRAL	230
GENERAL ELECTRIC	K-51	1933	AC	COMPACT	
GENERAL ELECTRIC	K-52	1933	AC	CATHEDRAL BC/SW	235
GENERAL ELECTRIC	K-53	1933	AC	TABLE CHEST BC/SW	
GENERAL ELECTRIC	K-53-M	1933	AC	TABLE	
GENERAL ELECTRIC	K-54-P	1933	AC	TABLE RADIO/PHONO CHEST	
GENERAL ELECTRIC	K-58	1933	AC	CONSOLE LB RADIO/PHONO	
GENERAL ELECTRIC	K-60	1933	AC	MANTLE WALNUT CLOCK DESIGN	260
GENERAL ELECTRIC	K-62	1931	AC	CONSOLE LB	
GENERAL ELECTRIC	K-63	1933	AC	CATHEDRAL BC/SW	250
GENERAL ELECTRIC	K-64	1933	AC	CATHEDRAL BC/SW	255
GENERAL ELECTRIC	K-65	1933	AC	CONSOLE	
GENERAL ELECTRIC	K-66	1933	AC	CONSOLE	
GENERAL ELECTRIC	K-78	1933	AC	CONSOLE RADIO/PHONO BC/SW	
GENERAL ELECTRIC	K-79	1933	AC	CONSOLE RADIO/PHONO BC/SW	
GENERAL ELECTRIC	K-80	1933	AC	TABLE	90

MANUFACTURER	MODEL	YEAR	POWER	STYLE	PRICE
GENERAL ELECTRIC	K-82	1931	AC	CLOCK GRANDFATHER	
GENERAL ELECTRIC	K-85	1933	AC	CONSOLE BC/SW	165
GENERAL ELECTRIC	K-105	1934	AC	CONSOLE	
GENERAL ELECTRIC	K-106	1933	AC	CONSOLE	
GENERAL ELECTRIC	K-107	1933	AC	CONSOLE BC/SW	
GENERAL ELECTRIC	K-126	1933	AC	CONSOLE BC/SW	150
GENERAL ELECTRIC	L-50	1933	AC	MANTLE PLASTIC W/HANDLE	35
GENERAL ELECTRIC	L-51	1933	AC	COMPACT	
GENERAL ELECTRIC	L-53	1934	AC	COMPACT	
GENERAL ELECTRIC	L-604	1941	AC	TABLE	30
GENERAL ELECTRIC	L-610	1941	AC	TABLE PLASTIC IVORY	25
GENERAL ELECTRIC	L-624	1941	AC	TABLE PLASTIC IVORY	25
GENERAL ELECTRIC	L-633	1941	AC	TABLE	35
GENERAL ELECTRIC	L-640	1941	AC	TABLE PB	35
GENERAL ELECTRIC	L-678	1941	AC	TABLE RADIO/PHONO	15
GENERAL ELECTRIC	L-740	1941	AC	TABLE PB BC/SW	45
GENERAL ELECTRIC	L-915	1941	AC	CONSOLE PB BC/SW	105
GENERAL ELECTRIC	LFC-1228	1941	AC	CONSOLE R/P PB BC/SW/FM	50
GENERAL ELECTRIC	M-40	1934	AC	COMPACT	
GENERAL ELECTRIC	M-42	1935	AC	TABLE	
GENERAL ELECTRIC	M-49	1934	AC	TABLE RADIO/PHONO	
GENERAL ELECTRIC	M-50	1935	AC	TABLE	
GENERAL ELECTRIC	M-51	1934	AC	TABLE BC/SW	75
GENERAL ELECTRIC	M-52	1935	AC	TABLE	
GENERAL ELECTRIC	M-55	1935	AC	CONSOLE	
GENERAL ELECTRIC	M-56	1934	AC	CONSOLE BC/SW	
GENERAL ELECTRIC	M-61	1934	AC	TABLE BC/SW	
GENERAL ELECTRIC	M-62	1935	AC	TABLE	70
GENERAL ELECTRIC	M-65	1934	AC	CONSOLE	
GENERAL ELECTRIC	M-67	1934	AC	CONSOLE BC/SW/LW	
GENERAL ELECTRIC	M-68	1934	AC	CONSOLE	
GENERAL ELECTRIC	M-81	1934	AC	TABLE BC/SW	
GENERAL ELECTRIC	M-86	1934	AC	CONSOLE BC/SW/LW	
GENERAL ELECTRIC	M-89	1934	AC	CONSOLE BC/SW/LW	
GENERAL ELECTRIC	M-106	1934	AC	CONSOLE BC/SW	
GENERAL ELECTRIC	M-125	1934	AC	CONSOLE BC/SW/LW	
GENERAL ELECTRIC	M-128	1934	AC	CONSOLE	
GENERAL ELECTRIC	M-129	1934	AC	CONSOLE R/P RC BC/SW/LW	
GENERAL ELECTRIC	M-655	1935	AC	CONSOLE	
GENERAL ELECTRIC	S-22	1931	AC	TOMBSTONE	85
GENERAL ELECTRIC	S-22D	1931	DC	TABLE	70
GENERAL ELECTRIC	S-22X	1931	AC	COMPACT RADIO/PHONO	
GENERAL ELECTRIC	S-42	1931	AC	CONSOLE LB	
GENERAL ELECTRIC	S-42B	1931	BAT	CONSOLE LB	
GENERAL ELECTRIC	S-42D	1931	DC	CONSOLE LB	
GENERAL ELECTRIC	T-12	1931	AC	CATHEDRAL	75
GENERAL ELECTRIC	T-12D	1931	DC	TABLE	
GENERAL ELECTRIC	U-50	1936	2/DC	TABLE	
GENERAL ELECTRIC	U-51	1936	6/DC	TABLE BC/SW	
GENERAL ELECTRIC	U-55	1936	6/DC	CONSOLE BC/SW	

MANUFACTURER	MODEL	YEAR	POWER	STYLE	PRICE
GENERAL ELECTRIC	U-70	1936	6/DC	TABLE BC/SW	
GENERAL ELECTRIC	U-75	1936	6/DC	CONSOLE BC/SW	
GENERAL HOUSEHOLD UTILITIES	410	1936	6/DC	TABLE	
GENERAL HOUSEHOLD UTILITIES	411	1936	6/DC	CONSOLE	
GENERAL HOUSEHOLD UTILITIES	450	1934	AC	TABLE BC/SW	
GENERAL HOUSEHOLD UTILITIES	470	1936	AC	TABLE	
GENERAL HOUSEHOLD UTILITIES	470	1935	AC	TABLE	
GENERAL HOUSEHOLD UTILITIES	500	1933	AC	TOMBSTONE	90
GENERAL HOUSEHOLD UTILITIES	501	1933	AC	TABLE	
GENERAL HOUSEHOLD UTILITIES	502	1933	AC	TABLE	
GENERAL HOUSEHOLD UTILITIES	503	1932	AC	CONSOLE	
GENERAL HOUSEHOLD UTILITIES	510	1936	2/DC	TABLE	
GENERAL HOUSEHOLD UTILITIES	520	1936	AC	TABLE	
GENERAL HOUSEHOLD UTILITIES	520	1935	AC	TABLE	
GENERAL HOUSEHOLD UTILITIES	532	1936	AC	TABLE BC/SW	
GENERAL HOUSEHOLD UTILITIES	542	1936	AC	TABLE BC/SW	
GENERAL HOUSEHOLD UTILITIES	550	1934	AC	TABLE BC/SW	
GENERAL HOUSEHOLD UTILITIES	551	1936	AC	CONSOLE BC/SW	
GENERAL HOUSEHOLD UTILITIES	553	1936	AC	CONSOLE BC/SW	
GENERAL HOUSEHOLD UTILITIES	555	1936	AC	CONSOLE BC/SW	
GENERAL HOUSEHOLD UTILITIES	564	1936	AC	TABLE	
GENERAL HOUSEHOLD UTILITIES	572	1936	AC	TABLE BC/SW	
GENERAL HOUSEHOLD UTILITIES	573	1936	AC	CONSOLE BC/SW	
GENERAL HOUSEHOLD UTILITIES	576	1937	AC	TABLE BC/SW	
GENERAL HOUSEHOLD UTILITIES	580	1935	AC	TABLE	
GENERAL HOUSEHOLD UTILITIES	581	1935	AC	TABLE	

MANUFACTURER	MODEL	YEAR	POWER	STYLE	PRICE
GENERAL HOUSEHOLD UTILITIES	585	1937	AC	CONSOLE BC/SW ARMCHAIR	
GENERAL HOUSEHOLD UTILITIES	588	1937	AC	TABLE BC/SW	
GENERAL HOUSEHOLD UTILITIES	589	1937	AC	CONSOLE BC/SW	
GENERAL HOUSEHOLD UTILITIES	592	1937	AC	COMPACT	
GENERAL HOUSEHOLD UTILITIES	599	1937	AC	CONSOLE BC/SW	
GENERAL HOUSEHOLD UTILITIES	621	1935	AC	CONSOLE	
GENERAL HOUSEHOLD UTILITIES	621	1935	AC	TABLE	
GENERAL HOUSEHOLD UTILITIES	622	1937	AC	COMPACT BC/SW	
GENERAL HOUSEHOLD UTILITIES	623	1937	AC	CONSOLE BC/SW	
GENERAL HOUSEHOLD UTILITIES	624	1937	AC	TABLE BC/SW	
GENERAL HOUSEHOLD UTILITIES	631	1936	AC	CONSOLE BC/SW	
GENERAL HOUSEHOLD UTILITIES	632	1937	AC	TABLE BC/SW	
GENERAL HOUSEHOLD UTILITIES	640	1935	AC	TABLE	
GENERAL HOUSEHOLD UTILITIES	643	1936	AC	CONSOLE BC/SW	
GENERAL HOUSEHOLD UTILITIES	650	1934	AC	TABLE BC/SW	
GENERAL HOUSEHOLD UTILITIES	651	1934	AC	CONSOLE BC/SW	
GENERAL HOUSEHOLD UTILITIES	654	1937	AC	TABLE BC/SW	
GENERAL HOUSEHOLD UTILITIES	660	1934	AC	TABLE BC/SW	
GENERAL HOUSEHOLD UTILITIES	661	1934	AC	CONSOLE BC/SW	
GENERAL HOUSEHOLD UTILITIES	662	1934	AC	CONSOLE BC/SW	
GENERAL HOUSEHOLD UTILITIES	663	1937	AC	CONSOLE BC/SW	
GENERAL HOUSEHOLD UTILITIES	670	1934	AC	TABLE BC/SW	
GENERAL HOUSEHOLD UTILITIES	671	1934	AC	CONSOLE BC/SW	
GENERAL HOUSEHOLD UTILITIES	681	1935	AC	CONSOLE	
GENERAL HOUSEHOLD UTILITIES	700	1933	AC	TOMBSTONE	

MANUFACTURER	MODEL	YEAR	POWER	STYLE	PRICE
GENERAL HOUSEHOLD UTILITIES	701	1932	AC	CONSOLE	
GENERAL HOUSEHOLD UTILITIES	711	1936	6/DC	CONSOLE BC/SW	
GENERAL HOUSEHOLD UTILITIES	720	1935	AC	TABLE	
GENERAL HOUSEHOLD UTILITIES	721	1935	AC	CONSOLE	
GENERAL HOUSEHOLD UTILITIES	723	1936	AC	CONSOLE BC/SW	
GENERAL HOUSEHOLD UTILITIES	731	1936	AC	CONSOLE BC/SW	
GENERAL HOUSEHOLD UTILITIES	733	1936	AC	CONSOLE BC/SW	
GENERAL HOUSEHOLD UTILITIES	735	1936	AC	CONSOLE BC/SW	
GENERAL HOUSEHOLD UTILITIES	750	1934	AC	TABLE BC/SW	
GENERAL HOUSEHOLD UTILITIES	751	1934	AC	CONSOLE BC/SW	
GENERAL HOUSEHOLD UTILITIES	752	1934	AC	CONSOLE BC/SW	
GENERAL HOUSEHOLD UTILITIES	753	1934	AC	CONSOLE BC/SW	
GENERAL HOUSEHOLD UTILITIES	755	1937	AC	CONSOLE BC/SW	
GENERAL HOUSEHOLD UTILITIES	761	1935	AC	CONSOLE	
GENERAL HOUSEHOLD UTILITIES	801	1933	AC	CONSOLE	
GENERAL HOUSEHOLD UTILITIES	823	1936	AC	CONSOLE BC/SW/LW	
GENERAL HOUSEHOLD UTILITIES	831	1936	AC	CONSOLE BC/SW/LW	
GENERAL HOUSEHOLD UTILITIES	833	1936	AC	CONSOLE BC/SW/LW	
GENERAL HOUSEHOLD UTILITIES	835	1936	AC	CONSOLE BC/SW/LW	
GENERAL HOUSEHOLD UTILITIES	871	1935	AC	CONSOLE	
GENERAL HOUSEHOLD UTILITIES	901	1933	AC	CONSOLE	
GENERAL HOUSEHOLD UTILITIES	902	1933	AC	CONSOLE	
GENERAL HOUSEHOLD UTILITIES	941	1936	AC	CONSOLE BC/SW/LW	
GENERAL HOUSEHOLD UTILITIES	1067	1937	AC	CONSOLE BC/SW	
GENERAL HOUSEHOLD UTILITIES	1081	1937	AC	CONSOLE BC/SW	

MANUFACTURER	MODEL	YEAR	POWER	STYLE	PRICE
GENERAL HOUSEHOLD UTILITIES	1091	1937	AC	CONSOLE BC/SW	
GENERAL HOUSEHOLD UTILITIES	1101	1933	AC	CONSOLE	
GENERAL HOUSEHOLD UTILITIES	1101	1934	AC	CONSOLE REMOTE CONTROL	75
GENERAL HOUSEHOLD UTILITIES	1151	1934	AC	CONSOLE BC/SW	
GENERAL HOUSEHOLD UTILITIES	1152	1934	AC	CONSOLE BC/SW	
GENERAL HOUSEHOLD UTILITIES	1171	1935	AC	CONSOLE	
GENERAL HOUSEHOLD UTILITIES	1181	1937	AC	CONSOLE BC/SW	
GENERAL HOUSEHOLD UTILITIES	1183	1937	AC	CONSOLE BC/SW	
GENERAL HOUSEHOLD UTILITIES	1185	1937	AC	CONSOLE BC/SW	
GENERAL HOUSEHOLD UTILITIES	1191	1937	AC	CONSOLE BC/SW	
GENERAL HOUSEHOLD UTILITIES	1191	1936	AC	CONSOLE BC/SW	
GENERAL HOUSEHOLD UTILITIES	1201	1933	AC	CONSOLE	
GENERAL HOUSEHOLD UTILITIES	1241	1935	AC	CONSOLE	
GENERAL HOUSEHOLD UTILITIES	1291	1937	AC	CONSOLE BC/SW	75
GENERAL HOUSEHOLD UTILITIES	1291	1936	AC	CONSOLE BC/SW	
GENERAL HOUSEHOLD UTILITIES	1293	1937	AC	CONSOLE BC/SW	
GENERAL HOUSEHOLD UTILITIES	1541	1936	AC	CONSOLE BC/SW–TO 70 MC.	
GENERAL HOUSEHOLD UTILITIES	1561	1937	AC	CONSOLE	
GENERAL MOTORS	6 DAYCRAFT	1927	BAT	CONSOLE 1 DIAL	
GENERAL MOTORS	6 DAYFAN	1927	BAT	CONSOLE 1 DIAL	
GENERAL MOTORS	7 DAYFAN	1927	BAT	CONSOLE 1 DIAL	
GENERAL MOTORS	25	1928	AC	TABLE LP 9 TUBE 1 DIAL	
GENERAL MOTORS	25-8-AC	1928	AC	TABLE LP DAYFAN 9 T 1 DIAL	
GENERAL MOTORS	26	1928	AC	CONSOLE 9 TUBE 1 DIAL	
GENERAL MOTORS	26-8-AC	1928	AC	CONSOLE DAYFAN 9 TUBE 1 DIAL	
GENERAL MOTORS	27	1928	AC	TABLE LP 9 TUBE 1 DIAL	
GENERAL MOTORS	27-8-AC	1928	AC	TABLE DAYFAN 9 TUBE 1 DIAL	
GENERAL MOTORS	28	1928	AC	CONSOLE 9 TUBE 1 DIAL	
GENERAL MOTORS	43	1929	AC	CONSOLE 9 TUBE 1 DIAL	
GENERAL MOTORS	48	1929	AC	TABLE LP 9 TUBE 1 DIAL	

MANUFACTURER	MODEL	YEAR	POWER	STYLE	PRICE
GENERAL MOTORS	54	1929	AC	CONSOLE 9 TUBE 1 DIAL	
GENERAL MOTORS	56	1929	AC	TABLE LP 9 TUBE 1 DIAL	
GENERAL MOTORS	66	1929	AC	TABLE LP 9 TUBE 1 DIAL	
GENERAL MOTORS	67	1929	BAT	TABLE LP 7 TUBE 1 DIAL	
GENERAL MOTORS	68	1929	AC	CONSOLE 9 TUBE 1 DIAL	
GENERAL MOTORS	69	1929	AC	CONSOLE 9 TUBE 1 DIAL	
GENERAL MOTORS	72	1929	AC	CONSOLE 9 TUBE 1 DIAL	
GENERAL MOTORS	73	1929	BAT	TABLE LP 7 TUBE 1 DIAL	
GENERAL MOTORS	74	1929	BAT	CONSOLE 7 TUBE 1 DIAL	
GENERAL MOTORS	80	1929	AC	CONSOLE 9 TUBE 1 DIAL	
GENERAL MOTORS	81	1929	AC	CONSOLE 9 TUBE 1 DIAL	
GENERAL MOTORS	82	1929	AC	CONSOLE 9 TUBE 1 DIAL	
GENERAL MOTORS	83	1929	AC	TABLE LP 9 TUBE 1 DIAL	
GENERAL MOTORS	90	1929	AC	CONSOLE 7 TUBE 1 DIAL	
GENERAL MOTORS	91	1929	AC	CONSOLE 7 TUBE 1 DIAL	
GENERAL MOTORS	93	1929	AC	CONSOLE 7 TUBE 1 DIAL	
GENERAL MOTORS	94	1929	AC	CONSOLE 7 TUBE 1 DIAL	
GENERAL MOTORS	120	1930	AC	CONSOLE LB	
GENERAL MOTORS	130	1930	AC	CONSOLE SHERATON	120
GENERAL MOTORS	140	1930	AC	CONSOLE ITALIAN	
GENERAL MOTORS	150	1930	AC	CONSOLE QUEEN ANNE	
GENERAL MOTORS	160	1929	AC	CONSOLE GEORGIA	
GENERAL MOTORS	201	1931	BAT	CONSOLE LB	
GENERAL MOTORS	216	1931	AC	CONSOLE END TABLE	
GENERAL MOTORS	217	1931	AC	CONSOLE END TABLE	
GENERAL MOTORS	219	1931	AC	COMPACT	
GENERAL MOTORS	250	1931	AC	TABLE	
GENERAL MOTORS	251	1931	AC	CONSOLE LB	
GENERAL MOTORS	252	1931	AC	CONSOLE LB	
GENERAL MOTORS	253	1931	AC	CONSOLE HB	
GENERAL MOTORS	254	1931	AC	CONSOLE LB	
GENERAL MOTORS	255	1931	AC	CONSOLE LB	
GENERAL MOTORS	256	1931	AC	CONSOLE HUTCH TABLE	
GENERAL MOTORS	257	1931	AC	CONSOLE HB	
GENERAL MOTORS	258	1931	AC	CONSOLE LB	
GENERAL MOTORS	291	1931	AC	CONSOLE RADIO/PHONO AUTO.	
GENERAL MOTORS	DAY GRAND 7	1926	BAT	CONSOLE 7 TUBE 1 DIAL	
GENERAL MOTORS	DAY ROYAL 7	1926	BAT	CONSOLE 7 TUBE 1 DIAL	
GENERAL MOTORS	DAYCRAFT	1925	BAT	TABLE 4 TUBE 3 DIAL	
GENERAL MOTORS	DAYCRAFT 5	1925	BAT	TABLE LP 5 TUBE 1 DIAL	
GENERAL MOTORS	DAYFAN 5	1925	BAT	TABLE LP 5 TUBE 1 DIAL	
GENERAL MOTORS	DAYFAN 7	1926	BAT	TABLE LP 7 TUBE 1 DIAL	
GENERAL MOTORS	DAYFAN DELUXE	1927	AC	CONSOLE LB W/MOTOR GEN. 1 DIAL	
GENERAL MOTORS	DAYFAN JR.	1927	BAT	CONSOLE 1 DIAL	
GENERAL MOTORS	DAYMAR 7	1926	BAT	CONSOLE 7 TUBE 1 DIAL	
GENERAL MOTORS	DAYOLA	1925	BAT	TABLE 4 TUBE 4 DIAL	65
GENERAL MOTORS	OEM-7	1924	BAT	TABLE 4 TUBE 3 DIAL	75
GENERAL MOTORS	OEM-11	1924	BAT	TABLE 3 TUBE 3 DIAL	70

MANUFACTURER	MODEL	YEAR	POWER	STYLE	PRICE
GENERAL MOTORS	OEM-12	1925	BAT	TABLE 4 TUBE 3 DIAL	80
GENERAL RADIO & TELEVISION	588	1940	AC	TABLE PLASTIC TORPEDO	25
GIBSON-SEARS	STERLING-FIVE	1924	BAT	TABLE LP SLOPE PAN. 5T 3DL	85
GILBERT RADIO	69	1930	AC	CATHEDRAL	
GILFILLAN BROTHERS	4-T	1934	AC	TABLE BC/SW	
GILFILLAN BROTHERS	5	1932	AC	CATHEDRAL	85
GILFILLAN BROTHERS	5-C	1934	AC	CONSOLE BC/SW	
GILFILLAN BROTHERS	5-M	1934	AC	TABLE	
GILFILLAN BROTHERS	5-T	1934	AC	TABLE BC/SW	
GILFILLAN BROTHERS	5-X	1934	AC	MANTLE	
GILFILLAN BROTHERS	6-C	1934	AC	CONSOLE BC/SW	
GILFILLAN BROTHERS	6-T	1934	AC	TABLE BC/SW	
GILFILLAN BROTHERS	7-C	1932	AC	CONSOLE	
GILFILLAN BROTHERS	7-M	1932	AC	TABLE	
GILFILLAN BROTHERS	8-C	1934	AC	CONSOLE BC/SW	
GILFILLAN BROTHERS	8-T	1934	AC	TABLE BC/SW	
GILFILLAN BROTHERS	10	1926	BAT	TABLE LP 5 TUBE 2 DIAL	
GILFILLAN BROTHERS	12	1932	AC	CONSOLE BC/SW	
GILFILLAN BROTHERS	20	1926	BAT	TABLE LP 6 TUBE 1 DIAL	
GILFILLAN BROTHERS	30	1926	BAT	CONSOLE 6 TUBE 2 DIAL	
GILFILLAN BROTHERS	33	1928	AC	CONSOLE LB 9 TUBE 1 DIAL	190
GILFILLAN BROTHERS	40	1926	AC	TABLE LP 6 TUBE 3 DIAL	
GILFILLAN BROTHERS	42A	1935	AC	TABLE	
GILFILLAN BROTHERS	43A	1935	AC	TABLE	
GILFILLAN BROTHERS	44	1928	AC	CONSOLE HB DOORS 9T 1 DIAL	
GILFILLAN BROTHERS	52A	1935	AC	MANTLE	
GILFILLAN BROTHERS	53A	1935	AC	MANTLE	
GILFILLAN BROTHERS	54A	1935	AC	TABLE	
GILFILLAN BROTHERS	55	1927	AC	CONSOLE 6 TUBE 1 DIAL	
GILFILLAN BROTHERS	55A	1935	AC	TABLE	
GILFILLAN BROTHERS	60	1927	AC	TABLE LP 6 TUBE 1 DIAL	
GILFILLAN BROTHERS	62B	1935	AC	TABLE BC/SW	
GILFILLAN BROTHERS	62X	1935	AC	TABLE BC/SW	
GILFILLAN BROTHERS	63B	1935	AC	TABLE BC/SW	
GILFILLAN BROTHERS	63X	1935	AC	TOMBSTONE BC/SW	55
GILFILLAN BROTHERS	65	1927	AC	CONSOLE 6 TUBE 1 DIAL	
GILFILLAN BROTHERS	66	1928	AC	CONSOLE HB DOORS 9T 1 DIAL	185
GILFILLAN BROTHERS	66S	1939	AC	TABLE PB	45
GILFILLAN BROTHERS	70	1927	AC	CONSOLE 6 TUBE 1 DIAL	
GILFILLAN BROTHERS	76A	1935	BAT	TABLE BC/SW	
GILFILLAN BROTHERS	77	1928	AC	CONSOLE HB RADIO/PHONO	220
GILFILLAN BROTHERS	77A	1935	BAT	TABLE BC/SW	
GILFILLAN BROTHERS	78B	1935	AC	CONSOLE	
GILFILLAN BROTHERS	78X	1935	AC	CONSOLE	
GILFILLAN BROTHERS	80	1927	AC	TABLE LP 8 TUBE 1 DIAL	
GILFILLAN BROTHERS	90	1927	AC	CONSOLE 8 TUBE 1 DIAL	
GILFILLAN BROTHERS	96B	1935	AC	TABLE BC/SW	

MANUFACTURER	MODEL	YEAR	POWER	STYLE	PRICE
GILFILLAN BROTHERS	96X	1935	AC	TABLE BC/SW	
GILFILLAN BROTHERS	97B	1935	AC	CONSOLE BC/SW	
GILFILLAN BROTHERS	97X	1935	AC	CONSOLE BC/SW	
GILFILLAN BROTHERS	100	1929	AC	CONSOLE LB 8 TUBE 1 DIAL	
GILFILLAN BROTHERS	101	1929	AC	TABLE 9 TUBE 1 DIAL	
GILFILLAN BROTHERS	102	1929	AC	CONSOLE HB	
GILFILLAN BROTHERS	103	1929	AC	CONSOLE HB	
GILFILLAN BROTHERS	105	1930	AC	CONSOLE LB	
GILFILLAN BROTHERS	106	1930	AC	CONSOLE HB	
GILFILLAN BROTHERS	107	1930	AC	CONSOLE HB RADIO/PHONO	
GILFILLAN BROTHERS	116B	1935	AC	TABLE BC/SW	
GILFILLAN BROTHERS	116X	1935	AC	TABLE BC/SW	
GILFILLAN BROTHERS	117B	1935	AC	CONSOLE BC/SW	
GILFILLAN BROTHERS	117X	1936	AC	CONSOLE	
GILFILLAN BROTHERS	200	1931	AC	CONSOLE LB	
GILFILLAN BROTHERS	250	1931	AC	CONSOLE LB	
GILFILLAN BROTHERS	GN-1	1924	BAT	TABLE LP 5 TUBE 3 DIAL	140
GILFILLAN BROTHERS	GN-2	1924	BAT	TABLE LP 5 TUBE 3 DIAL	135
GILFILLAN BROTHERS	GN-3	1925	BAT	TABLE SLOPE PANEL 4 T 2 DL	150
GILFILLAN BROTHERS	GN-4	1926	BAT	TABLE LP 4 TUBE 2 DIAL	125
GILFILLAN BROTHERS	GN-5	1926	BAT	TABLE LP 5 TUBE 2 DIAL	125
GILFILLAN BROTHERS	GN-6	1926	BAT	CONSOLE 5 TUBE 1 DIAL	120
GILFILLAN BROTHERS	PORTABLE	1925	BAT	PORTABLE 6 TUBE 1 DIAL	140
GLOBE ELECTRIC	2-53	1925	BAT	TABLE LP SLOPE PAN. 5T 3DL	75
GLOBE ELECTRIC	3-53	1925	BAT	TABLE HP SPEAKER 5 T 3 DL	
GLOBE ELECTRIC	770	1923	BAT	PORTABLE 4 TUBE 2 DIAL	
GLOBE ELECTRIC	772	1923	BAT	CONSOLE 4 TUBE 2 DIAL	
GLOBE ELECTRIC	775	1923	BAT	TABLE LP 4 TUBE 2 DIAL	85
GLOBE ELECTRIC	815	1923	BAT	TABLE LP METER 5 TUBE 3 DIAL	125
GLOBE ELECTRIC	825	1923	BAT	CONSOLE 5 TUBE 3 DIAL	
GLOBE ELECTRIC	830	1924	BAT	TABLE 5 TUBE 3 DIAL	80
GLOBE ELECTRIC	880-H	1924	BAT	TABLE 5 TUBE 3 DIAL	75
GLOBE ELECTRIC	900	1923	BAT	TABLE LP METER 5 TUBE 3 DIAL	145
GLOBE ELECTRIC	902	1923	BAT	CONSOLE 5 TUBE 3 DIAL	
GLOBE TROTTER	GLOBE TROTTER	1936	AC	TABLE WORLD GLOBE BC/SW	
GOLD MEDAL	5-60	1925	BAT	TABLE LP 3 DIAL	80
GOLD MEDAL	DELUXE	1925	BAT	TABLE LP 3 DIAL	95
GOLDEN LEUTZ				See LEUTZ	
GRAY DANIELSON	14	1930	AC	CATHEDRAL	60
GRAYBAR ELECTRIC	310	1928	AC	TABLE	60
GRAYBAR ELECTRIC	311	1929	AC	TABLE 9 TUBE 1 DIAL	
GRAYBAR ELECTRIC	311	1929	AC	CONSOLE	
GRAYBAR ELECTRIC	330	1929	AC	TABLE 9 TUBE 1 DIAL	
GRAYBAR ELECTRIC	340	1929	AC	CONSOLE 9 TUBE 1 DIAL	
GRAYBAR ELECTRIC	500	1929	AC	TABLE 5 TUBE 1 DIAL	
GRAYBAR ELECTRIC	550	1929	AC	CONSOLE 5 TUBE 1 DIAL	
GRAYBAR ELECTRIC	700	1929	AC	CONSOLE LB	
GRAYBAR ELECTRIC	770	1929	AC	CONSOLE HB	

MANUFACTURER	MODEL	YEAR	POWER	STYLE	PRICE
GRAYBAR ELECTRIC	900	1929	AC	CONSOLE LB RADIO/PHONO	
GREBE	2B2-I	1937	2/DC	CONSOLE BC/SW	
GREBE	2B2-L	1937	2/DC	TABLE BC/SW	
GREBE	2B2-T	1937	2/DC	TABLE BC/SW	
GREBE	2B6-I	1937	2/DC	CONSOLE BC/SW	
GREBE	2B6-L	1937	2/DC	TABLE BC/SW	
GREBE	2B6-T	1937	2/DC	TABLE BC/SW	
GREBE	3B2-3	1937	2/DC	CONSOLE BC/SW	
GREBE	3B2-L	1937	2/DC	TABLE BC/SW	
GREBE	3B2-T	1937	2/DC	TABLE BC/SW	
GREBE	3B6-3	1937	2/DC	CONSOLE BC/SW	
GREBE	3B6-L	1937	2/DC	TABLE BC/SW	
GREBE	3B6-T	1937	2/DC	TABLE BC/SW	
GREBE	5	1927	BAT	TABLE SYNCHROPHASE 1 DIAL	
GREBE	7	1927	BAT	TABLE LP SYNCHROPHASE 1 DIAL	120
GREBE	160	1930	AC	CONSOLE LB	
GREBE	206-JI	1937	AC	CONSOLE BC/SW	
GREBE	206-L	1937	AC	TABLE BC/SW	
GREBE	219-50-A	1929	AC	CONSOLE LB	100
GREBE	225-M	1930	AC	CONSOLE LB	100
GREBE	225-W	1930	AC	CONSOLE LB	
GREBE	265	1930	AC	CONSOLE LB	100
GREBE	270-A	1929	AC	CONSOLE HB	100
GREBE	270-B	1929	AC	CONSOLE HB	100
GREBE	270-C	1929	AC	CONSOLE HB	100
GREBE	285	1930	AC	CONSOLE HB	
GREBE	285-A	1929	AC	CONSOLE HB	100
GREBE	307-I	1937	AC	CONSOLE BC/SW	
GREBE	307-L	1937	AC	TABLE BC/SW	
GREBE	307-P5	1937	AC	CONSOLE BC/SW	
GREBE	307-T	1937	AC	TABLE BC/SW	
GREBE	309-2	1937	AC	CONSOLE BC/SW	
GREBE	309-E2	1937	AC	CONSOLE BC/SW/LW	
GREBE	309-EL	1937	AC	TABLE BC/SW/LW	
GREBE	309-ET	1937	AC	TABLE BC/SW/LW	
GREBE	309-L	1937	AC	TABLE BC/SW	
GREBE	309-P5	1937	AC	CONSOLE BC/SW	
GREBE	309-T	1937	AC	TABLE BC/SW	
GREBE	370-C	1936	AC	TABLE	
GREBE	411-OE	1936	AC	CONSOLE TUNING EYE BC/SW	
GREBE	415	1939	AC	TABLE PLASTIC	
GREBE	450-A	1929	AC	CONSOLE RADIO/PHONO	150
GREBE	451	1939	AC	TABLE PLASTIC	
GREBE	451-X	1939	AC	TABLE PLASTIC BC/SW	
GREBE	453	1939	AC	TABLE PLASTIC BC/SW	
GREBE	453-W	1939	AC	TABLE BC/SW	
GREBE	473	1939	AC	CONSOLE RADIO/PHONO BC/SW	

MANUFACTURER	MODEL	YEAR	POWER	STYLE	PRICE
GREBE	473	1939	AC	TABLE RADIO/PHONO BC/SW	
GREBE	493	1939	AC	CONSOLE RADIO/PHONO BC/SW	
GREBE	493	1939	AC	TABLE R/P BC/SW	
GREBE	703-L	1937	AC	TABLE BC/SW/LW	
GREBE	803-I	1937	AC	CONSOLE BC/SW	
GREBE	803-L	1937	AC	TABLE BC/SW	
GREBE	803-P5	1937	AC	CONSOLE BC/SW	
GREBE	803-T	1937	AC	TABLE BC/SW	
GREBE	903-2	1937	AC	CONSOLE BC/SW	
GREBE	903-E2	1937	AC	CONSOLE BC/SW/LW	
GREBE	903-EL	1937	AC	TABLE BC/SW/LW	
GREBE	903-ET	1937	AC	TABLE BC/SW/LW	
GREBE	903-L	1937	AC	TABLE BC/SW	
GREBE	903-P5	1937	AC	CONSOLE BC/SW	
GREBE	903-T	1937	AC	TABLE BC/SW	
GREBE	1203-3	1937	AC	CONSOLE BC/SW	
GREBE	1203-L	1937	AC	TABLE BC/SW	
GREBE	1203-P6	1937	AC	CONSOLE BC/SW	
GREBE	1203-P6A	1937	AC	CONSOLE RADIO/PHONO BC/SW	
GREBE	1203-T	1937	AC	TABLE BC/SW	
GREBE	1204-E3	1937	AC	CONSOLE BC/SW/LW	
GREBE	1204-EL	1937	AC	TABLE BC/SW/LW	
GREBE	1204-ET	1937	AC	TABLE BC/SW/LW	
GREBE	1603-4	1937	AC	CONSOLE BC/SW	
GREBE	1603-P6	1937	AC	CONSOLE BC/SW	
GREBE	1603-P6A	1937	AC	CONSOLE RADIO/PHONO BC/SW	
GREBE	1604-E4	1937	AC	CONSOLE BC/SW/LW	
GREBE	3012-3	1937	AC	CONSOLE BC/SW	
GREBE	3012-L	1937	AC	TABLE BC/SW	
GREBE	3012-P6	1937	AC	CONSOLE BC/SW	
GREBE	3012-P6A	1937	AC	CONSOLE RADIO/PHONO BC/SW	
GREBE	3012-T	1937	AC	TABLE BC/SW	
GREBE	3016-4	1937	AC	CONSOLE TELPHONE DL BC/SW	
GREBE	3016-P6	1937	AC	CONSOLE BC/SW	
GREBE	3016-P6A	1937	AC	CONSOLE RADIO/PHONO BC/SW	
GREBE	4012-E3	1937	AC	CONSOLE BC/SW/LW	
GREBE	4012-EL	1937	AC	TABLE BC/SW/LW	
GREBE	4012-ET	1937	AC	TABLE BC/SW/LW	
GREBE	4016-E4	1937	AC	CONSOLE BC/SW/LW	
GREBE	4123	1939	AC	CONSOLE R/P PB BC/SW	
GREBE	4123	1939	AC	TABLE R/P PB BC/SW	
GREBE	4124	1939	AC	CONSOLE R/P BC/SW/LW	
GREBE	4124	1939	AC	TABLE R/P BC/SW/LW	
GREBE	4310	1939	AC	CONSOLE R/P PB BC/SW	

MANUFACTURER	MODEL	YEAR	POWER	STYLE	PRICE
GREBE	4310	1939	AC	TABLE R/P PB BC/SW	
GREBE	43100	1939	AC	CONSOLE R/P PB BC/SW/LW	
GREBE	43100	1939	AC	TABLE R/P PB BC/SW/LW	
GREBE	4370	1939	AC	CONSOLE R/P BC/SW/LW	
GREBE	4370	1939	AC	TABLE RADIO/PHONO BC/SW/LW	
GREBE	4730	1939	AC	CONSOLE R/PBC/SW/LW	
GREBE	4730	1939	AC	TABLE RADIO/PHONO BC/SW/LW	
GREBE	5240-4	1937	AC	CONSOLE BC/SW TO 60 MC.	
GREBE	A	1929	AC	CONSOLE LB	
GREBE	AC-6	1928	AC	TABLE LP 7 TUBE 1 DIAL	110
GREBE	AC-6	1928	AC	CONSOLE MANY STYLES	150
GREBE	AC-7	1928	AC	TABLE LP 8 TUBE 1 DIAL	190
GREBE	ANDALUSIA	1926	BAT	CONSOLE LB W/MU-1	200
GREBE	B	1929	AC	CONSOLE LB	
GREBE	BP-7	1939	AC	PORTABLE	
GREBE	CHALLENGER 1	1937	AC	TABLE	
GREBE	CHALLENGER 2	1937	AC	TABLE BC/SW	
GREBE	CR-2	1920	BAT	TABLE LP 1 TUBE 2 DIAL	1200
GREBE	CR-3	1920	BAT	TABLE LP TUNER 1 TUBE 2 DIAL	850
GREBE	CR-4	1920	BAT	TABLE LP 1 TUBE 2 DIAL	1350
GREBE	CR-5	1921	BAT	TABLE LP TUNER/DECTECTOR	390
GREBE	CR-6	1921	BAT	TABLE LP 5 TUBE 3 DIAL	800
GREBE	CR-7	1922	BAT	TABLE LP LW 1 TUBE 2 DIAL	950
GREBE	CR-8	1922	BAT	TABLE LP SW 1 TUBE 3 DIAL	440
GREBE	CR-9	1922	BAT	TABLE LP 3 TUBE 2 DIAL	410
GREBE	CR-10	1922	BAT	CONSOLE 4 TUBE 2 DIAL	400
GREBE	CR-12	1923	BAT	TABLE LP 4 TUBE 2 DIAL	420
GREBE	CR-13	1923	BAT	TABLE LP 2 TUBE 3 DIAL	600
GREBE	CR-14	1924	BAT	TABLE LP 3 TUBE 2 DIAL	445
GREBE	CR-18	1926	BAT	TABLE LP PLUG-IN COILS 2DL	490
GREBE	CR-19	1928	BAT	TABLE LP SW 2 DIAL	500
GREBE	DELUXE	1925	BAT	CONSOLE W/MU-1	400
GREBE	KT-1	1921	BAT	PORTABLE	450
GREBE	LANCASTER	1926	BAT	CONSOLE LB W/MU-1	200
GREBE	M-1	1931	AC	TABLE SCOUT	
GREBE	M-2	1931	AC	CONSOLE LB	
GREBE	M-3	1931	AC	CONSOLE HB	
GREBE	M-4	1931	AC	CONSOLE DELUXE	
GREBE	MU-1	1924	BAT	TABLE LP W/OUT DIAL CHAIN	180
GREBE	MU-1	1925	BAT	TABLE LP W/DIAL CHAIN	140
GREBE	MU-2	1925	BAT	TABLE LP 5 TUBE 3 DIAL	200
GREBE	PRINCESS	1926	BAT	CONSOLE LB W/MU-1	200
GREBE	PURITAN	1926	BAT	CONSOLE LB W/MU-1	400
GREBE	RENAISSANCE	1926	BAT	CONSOLE LB GRAND W/MU-1	300
GREBE	SK-4	1930	AC	CONSOLE	
GREBE	SYNCHRONETTE	1933	AC	TABLE	
GRIFFIN SMITH	ROYALE	1930	AC	CATHEDRAL	95

MANUFACTURER	MODEL	YEAR	POWER	STYLE	PRICE
GRIGSBY-GRUNOW	15	1932	AC	CATHEDRAL MODERN	150
GRIGSBY-GRUNOW	15	1932	AC	CLOCK GRANDFATHER	425
GRIGSBY-GRUNOW	20	1932	AC	CLOCK GRANDFATHER	325
GRIGSBY-GRUNOW	44	1933	AC	TABLE BC/SW	
GRIGSBY-GRUNOW	49	1933	AC	TABLE BC/SW	
GRIGSBY-GRUNOW	50	1931	AC	CATHEDRAL MODERN	165
GRIGSBY-GRUNOW	52	1938	AC	TABLE PLASTIC	40
GRIGSBY-GRUNOW	55	1933	AC	TABLE BC/SW	140
GRIGSBY-GRUNOW	59	1933	AC	TABLE BC/SW	
GRIGSBY-GRUNOW	61	1931	AC	CONSOLE HB	140
GRIGSBY-GRUNOW	61	1928	AC	TABLE LP 8 TUBE 1 DIAL	85
GRIGSBY-GRUNOW	62	1928	AC	TABLE LP 8 TUBE 1 DIAL	85
GRIGSBY-GRUNOW	67	1933	AC	CONSOLE	
GRIGSBY-GRUNOW	68	1933	AC	CONSOLE	
GRIGSBY-GRUNOW	69	1933	AC	CONSOLE	
GRIGSBY-GRUNOW	70-B	1929	AC	CONSOLE LB	
GRIGSBY-GRUNOW	71	1928	AC	CONSOLE LB 8 TUBE 1 DIAL	85
GRIGSBY-GRUNOW	72	1928	AC	CONSOLE HB 8 TUBE 1 DIAL	140
GRIGSBY-GRUNOW	77	1933	AC	CONSOLETTE QUEEN ANNE	
GRIGSBY-GRUNOW	85	1933	AC	CONSOLE	
GRIGSBY-GRUNOW	86	1933	AC	CONSOLE	
GRIGSBY-GRUNOW	90	1930	AC	CONSOLE LB	185
GRIGSBY-GRUNOW	91	1929	AC	CONSOLE LB EARLY ENGLISH	110
GRIGSBY-GRUNOW	92	1929	AC	CONSOLE HB JACOBEAN	120
GRIGSBY-GRUNOW	93	1930	AC	CONSOLE LB	130
GRIGSBY-GRUNOW	101	1939	AC	CONSOLE LB RADIO/PHONO	
GRIGSBY-GRUNOW	102	1930	AC	CONSOLE LB RADIO/PHONO	140
GRIGSBY-GRUNOW	103	1930	AC	CONSOLE LB RADIO/PHONO	150
GRIGSBY-GRUNOW	105	1933	AC	TABLE ESTATE	
GRIGSBY-GRUNOW	130	1930	AC	CONSOLE LB	
GRIGSBY-GRUNOW	131	1930	AC	CONSOLE LB	
GRIGSBY-GRUNOW	132	1930	AC	CONSOLE HB	
GRIGSBY-GRUNOW	151	1931	AC	TABLE	
GRIGSBY-GRUNOW	153	1931	AC	CONSOLE LB	
GRIGSBY-GRUNOW	154	1931	AC	CONSOLE END TABLE	
GRIGSBY-GRUNOW	155	1931	AC	CONSOLE RADIO/PHONO	
GRIGSBY-GRUNOW	181	1929	AC	CONSOLE HB RADIO/PHONO	100
GRIGSBY-GRUNOW	194	1933	AC	TABLE BC/SW	
GRIGSBY-GRUNOW	195	1933	AC	TABLE BC/SW	
GRIGSBY-GRUNOW	196	1933	AC	COMPACT	125
GRIGSBY-GRUNOW	233	1930	AC	CONSOLE LB RADIO/PHONO	
GRIGSBY-GRUNOW	251	1931	AC	CONSOLE HB	
GRIGSBY-GRUNOW	253	1931	AC	CONSOLE LB	
GRIGSBY-GRUNOW	291	1932	AC	CATHEDRAL	170
GRIGSBY-GRUNOW	293	1932	AC	CONSOLE	175
GRIGSBY-GRUNOW	294	1932	AC	CONSOLE LB BC/SW	
GRIGSBY-GRUNOW	304	1932	AC	CONSOLE	175
GRIGSBY-GRUNOW	307	1932	AC	CONSOLE LB	
GRIGSBY-GRUNOW	311	1932	AC	TABLE	
GRIGSBY-GRUNOW	314	1932	AC	CONSOLE LB	

MANUFACTURER	MODEL	YEAR	POWER	STYLE	PRICE
GRIGSBY-GRUNOW	324	1932	AC	CONSOLE	
GRIGSBY-GRUNOW	331	1933	AC	TABLE	
GRIGSBY-GRUNOW	336	1933	AC	CONSOLE	
GRIGSBY- GRUNOW	337	1933	AC	CONSOLE	
GRIGSBY-GRUNOW	344	1933	AC	CONSOLE SHERATON	
GRIGSBY-GRUNOW	351	1931	AC	CONSOLE LB	
GRIGSBY-GRUNOW	353	1931	AC	CONSOLE RADIO/PHONO	
GRIGSBY-GRUNOW	363	1933	AC	CONSOLE CATHEDRAL	
GRIGSBY-GRUNOW	370	1933	AC	CATHEDRAL	95
GRIGSBY-GRUNOW	371	1933	AC	TABLE	
GRIGSBY-GRUNOW	373	1933	AC	TABLE	
GRIGSBY-GRUNOW	381	1933	AC	TABLE TREASURE CHEST	95
GRIGSBY-GRUNOW	393	1933	AC	CONSOLE	
GRIGSBY-GRUNOW	411	1933	AC	TABLE	
GRIGSBY-GRUNOW	461	1933	AC	COMPACT	70
GRIGSBY-GRUNOW	463	1933	AC	TABLE	90
GRIGSBY-GRUNOW	491	1933	AC	COMPACT	
GRIGSBY-GRUNOW	493	1933	32/DC	CONSOLE	
GRIGSBY-GRUNOW	511	1938	AC	TABLE PLASTIC IVORY	45
GRIMES	3-XP	1925	BAT	TABLE LP 3 TUBE 3 DIAL	150
GRIMES	4-DL	1925	BAT	TABLE LP 4 TUBE 3 DIAL	220
GRIMES	5-B	1925	BAT	TABLE 5 TUBE 3 DIAL	150
GRIMES	5-D	1925	BAT	TABLE LP 5 TUBE 3 DIAL	200
GRIMES	BABY GRAND	1925	BAT	TABLE LP 3 TUBE 2 DIAL	190
GRIMES	EMPIRE	1925	BAT	TABLE LP 6 TUBE 3 DIAL	150
GRIMES	MONOTUBE	1925	BAT	TABLE 1 TUBE 2 DIAL	125
GRIMES	RENAISSANCE	1925	BAT	TABLE LP 4 TUBE 3 DIAL	140
GRIMES	TRITUBE	1925	BAT	TABLE LP 3 TUBE 2 DIAL	100
GULBRANSEN	53	1930	AC	CONSOLE HB	
GULBRANSEN	130	1931	AC	MANTLE	
GULBRANSEN	135	1931	AC	CONSOLE	
GULBRANSEN	161	1930	AC	CONSOLE HB	
GULBRANSEN	200	1929	AC	CONSOLE RADIO/PHONO	
GULBRANSEN	235	1931	AC	CONSOLE	
GULBRANSEN	291	1929	AC	CONSOLE LB	130
GULBRANSEN	292	1929	AC	CONSOLE HB DOORS	145
GULBRANSEN	295	1929	AC	CONSOLE HB	
GULBRANSEN	330	1931	AC	MANTLE	
GULBRANSEN	3225	1931	AC	CONSOLE HB	
GULBRANSEN	3226	1931	AC	CONSOLE	
GULBRANSEN	3521	1931	AC	MANTLE BC/SW	
GULBRANSEN	3525	1931	AC	CONSOLE BC/SW	
GULBRANSEN	3925	1931	BAT	CONSOLE	
GULBRANSEN	8716	1931	AC	CONSOLE LB	
GULBRANSEN	9950	1930	AC	CONSOLE HB	175
GUTHRIE	GOLDFINCH	1924	BAT	TABLE LP 5 TUBE 2 DIAL	75
GUTHRIE	NIGHTINGALE	1924	BAT	TABLE LP 5 TUBE 3 DIAL	
H					
HALLDORSON	RD-400	1923	BAT	TABLE LP 4 TUBE 1 DIAL	85

MANUFACTURER	MODEL	YEAR	POWER	STYLE	PRICE
HALLDORSON	RF-500	1923	BAT	TABLE LP 4 TUBE 3 DIAL	70
HALSON RADIO & TELEVISION	4M	1935	AC	COMPACT	
HALSON RADIO & TELEVISION	5	1936	AC	TABLE BC/SW	
HALSON RADIO & TELEVISION	5-LE	1936	AC	TABLE BC/SW	
HALSON RADIO & TELEVISION	6L6	1936	AC	TABLE BC/SW	
HALSON RADIO & TELEVISION	15	1937	AC	TABLE	
HALSON RADIO & TELEVISION	18M	1935	AC	TABLE	
HALSON RADIO & TELEVISION	20	1933	AC	COMPACT	
HALSON RADIO & TELEVISION	20A	1933	AC	COMPACT	
HALSON RADIO & TELEVISION	25	1937	AC	TABLE BC/SW	
HALSON RADIO & TELEVISION	35	1937	AC	TABLE BC/SW	
HALSON RADIO & TELEVISION	40	1933	AC	MANTLE	
HALSON RADIO & TELEVISION	43B	1933	AC	COMPACT	
HALSON RADIO & TELEVISION	45	1934	AC	COMPACT	
HALSON RADIO & TELEVISION	50	1933	AC	TABLE BC/SW	
HALSON RADIO & TELEVISION	50M	1935	AC	COMPACT	
HALSON RADIO & TELEVISION	50RL	1936	AC	TABLE BC/SW	
HALSON RADIO & TELEVISION	50X	1936	AC	TABLE BC/SW	
HALSON RADIO & TELEVISION	51A	1933	AC	COMPACT	
HALSON RADIO & TELEVISION	55SW	1934	AC	COMPACT	
HALSON RADIO & TELEVISION	60	1933	AC	TABLE BC/SW	
HALSON RADIO & TELEVISION	60L	1936	AC	TABLE BC/SW	
HALSON RADIO & TELEVISION	60M	1936	AC	TABLE BC/SW	
HALSON RADIO & TELEVISION	66AW	1934	AC	TABLE	
HALSON RADIO & TELEVISION	72	1934	AC	CONSOLE BC/SW	
HALSON RADIO & TELEVISION	75	1934	AC	CONSOLE BC/SW	
HALSON RADIO & TELEVISION	78	1934	AC	CONSOLE BC/SW	
HALSON RADIO & TELEVISION	100D	1933	AC	COMPACT	
HALSON RADIO & TELEVISION	100M	1936	AC	TABLE BC/SW	
HALSON RADIO & TELEVISION	102	1937	AC	TABLE BC/SW	
HALSON RADIO & TELEVISION	103	1937	AC	TABLE BC/SW	
HALSON RADIO & TELEVISION	104	1937	AC	TABLE BC/SW	
HALSON RADIO & TELEVISION	106	1937	AC	TABLE BC/SW	
HALSON RADIO & TELEVISION	200	1932	AC	TABLE TREASURE CHEST	
HALSON RADIO & TELEVISION	410	1934	AC	TABLE	
HALSON RADIO & TELEVISION	412	1937	AC	TABLE BC/SW	
HALSON RADIO & TELEVISION	414	1932	AC	TABLE	
HALSON RADIO & TELEVISION	414-S	1932	AC	TABLE	
HALSON RADIO & TELEVISION	420	1934	AC	TABLE	
HALSON RADIO & TELEVISION	510	1932	AC	TABLE	
HALSON RADIO & TELEVISION	520	1934	AC	TABLE	
HALSON RADIO & TELEVISION	530	1934	AC	TABLE	
HALSON RADIO & TELEVISION	536	1936	AC	TABLE BC/SW	
HALSON RADIO & TELEVISION	540	1934	AC	TABLE BC/SW	
HALSON RADIO & TELEVISION	550-W	1934	AC	TABLE	
HALSON RADIO & TELEVISION	560	1934	AC	TABLE BC/SW	
HALSON RADIO & TELEVISION	606	1936	AC	TABLE BC/SW	
HALSON RADIO & TELEVISION	610	1934	AC	TABLE BC/SW	
HALSON RADIO & TELEVISION	612	1937	AC	TABLE BC/SW	
HALSON RADIO & TELEVISION	615	1932	AC	TABLE	

MANUFACTURER	MODEL	YEAR	POWER	STYLE	PRICE
HALSON RADIO & TELEVISION	620	1934	AC	TABLE BC/SW	
HALSON RADIO & TELEVISION	630	1934	AC	TABLE BC/SW	
HALSON RADIO & TELEVISION	770	1934	AC	TABLE BC/SW	
HALSON RADIO & TELEVISION	1100	1934	AC	CONSOLE DELUXE 11 TUBES	
HALSON RADIO & TELEVISION	1500	1936	AC	CONSOLE DELUXE 15 TUBES	
HALSON RADIO & TELEVISION	1900	1936	AC	CONSOLE DELUXE 19 TUBES	
HALSON RADIO & TELEVISION	AM- 6C	1936	AC	CONSOLE BC/SW	
HALSON RADIO & TELEVISION	AW- 6	1935	AC	TABLE	
HALSON RADIO & TELEVISION	CA-5	1935	AC	CONSOLE	
HALSON RADIO & TELEVISION	CA-8	1936	AC	CONSOLE	
HALSON RADIO & TELEVISION	CM-5	1935	AC	CONSOLE	
HALSON RADIO & TELEVISION	CM-8	1935	AC	CONSOLE	
HALSON RADIO & TELEVISION	CW-7	1935	AC	CONSOLE BC/SW	
HALSON RADIO & TELEVISION	DELUXE	1931	AC	COMPACT	
HALSON RADIO & TELEVISION	JR.	1931	AC	COMPACT	
HALSON RADIO & TELEVISION	MA-8	1935	AC	TABLE	
HALSON RADIO & TELEVISION	MA-53	1935	AC	TABLE	
HALSON RADIO & TELEVISION	MA-63	1935	AC	TABLE	
HALSON RADIO & TELEVISION	MG-5	1935	AC	COMPACT	
HALSON RADIO & TELEVISION	T 5	1937	AC	TABLE	
HALSON RADIO & TELEVISION	T 10	1937	AC	TABLE	
HALSON RADIO & TELEVISION	T 11	1937	AC	TABLE	
HALSON RADIO & TELEVISION	T1200	1937	AC	TABLE BC/SW/LW	
HANSEN STORAGE	BLUEBIRD	1923	BAT	TABLE LP SLOPE PAN. 4T 2DL	
HANSEN STORAGE	NIGHTINGALE	1923	BAT	MANTLE 4 TUBE 1 DIAL	
HARMON	HARMONSON	1924	BAT	TABLE HP SPEAKER 4 T 2 DL	
HARMONY	5	1923	BAT	TABLE LP 5 TUBE 1 DIAL	80
HARTMAN ELECTRIC	JR. UPRIGHT	1926	BAT	CONSOLE LP 6 TUBE 1 DIAL	120
HARTMAN ELECTRIC	SR. UPRIGHT	1926	BAT	CONSOLE LP 6 TUBE 1 DIAL	130
HARTMAN ELECTRIC	10-B	1924	BAT	TABLE LP 4 TUBE 2 DIAL	
HARTMAN ELECTRIC	10-CS	1924	BAT	TABLE LP 4 TUBE 2 DIAL	80
HARTMAN ELECTRIC	12-A	1923	BAT	CONSOLE ADAM 5 TUBE 3 DIAL	
HARTMAN ELECTRIC	12-A	1924	BAT	CONSOLE ITALIAN 5 T 3 DIAL	
HARTMAN ELECTRIC	12-A	1924	BAT	CONSOLE QUEEN ANNE 4T 3 DIAL	
HARTMAN ELECTRIC	12-B	1923	BAT	TABLE LP SPEAKER 5 T 3 DL	
HARTMAN ELECTRIC	12-C	1923	BAT	TABLE LP METER 5 TUBE 3 DIAL	
HIGH FREQUENCY LABORATORIES	ISOTONE	1928	AC	CONSOLE R/P DOORS 10T 2 DIAL	150
HOMER PRODUCTS	HOMER	1927	BAT	TABLE LP 7 TUBE 1 DIAL	
HORN RADIO	TIFFANY TONE	1930	AC	CATHEDRAL	85
HOWARD	0	1931	AC	CONSOLE LB	
HOWARD	5BT	1937	2/DC	TABLE	
HOWARD	6BT	1937	6/DC	TABLE BC/SW	
HOWARD	10	1931	AC	TABLE	
HOWARD	10-B	1939	BAT	PORTABLE	15
HOWARD	20	1931	AC	TABLE	

MANUFACTURER	MODEL	YEAR	POWER	STYLE	PRICE
HOWARD	25	1931	AC	TABLE W/TABLE	
HOWARD	30	1931	AC	CONSOLE	
HOWARD	35	1931	AC	CONSOLE LB	
HOWARD	45	1931	AC	CONSOLE LB	
HOWARD	45	1941	AC	CONSOLE R/P PB BC/SW/FM	45
HOWARD	47-U	1935	AC	TABLE	
HOWARD	57UA-S.W.	1935	AC	TABLE BC/SW	
HOWARD	60	1931	AC	CONSOLE RADIO/PHONO	
HOWARD	67-C	1935	AC	CONSOLE BC/SW	
HOWARD	67-T	1935	AC	TABLE BC/SW	
HOWARD	68	1936	AC	CONSOLE	
HOWARD	68	1936	AC	TABLE BC/SW	
HOWARD	77-C	1935	AC	CONSOLE BC/SW	
HOWARD	77-T	1935	AC	TABLE BC/SW	
HOWARD	99-C	1935	AC	CONSOLE BC/SW	
HOWARD	135	1927	AC	CONSOLE	
HOWARD	150	1925	BAT	TABLE LP 5 TUBE 3 DIAL	75
HOWARD	200	1925	BAT	TABLE LP 6 TUBE 3 DIAL	85
HOWARD	200	1937	AC	MANTLE	
HOWARD	220	1938	AC	MANTLE AUTOMATIC TUNING	20
HOWARD	225	1937	AC	TABLE BC/SW	
HOWARD	250	1925	BAT	CONSOLE 5 TUBE 3 DIAL	
HOWARD	250	1937	AC	TABLE BC/SW	
HOWARD	270	1938	AC	TABLE	
HOWARD	275	1937	AC	TABLE BC/SW	
HOWARD	300	1937	AC	TABLE BC/SW	45
HOWARD	300	1939	AC	TABLE PB	
HOWARD	301	1939	AC	CONSOLE RADIO/PHONO PB	
HOWARD	302-RA	1940	AC	CONSOLE RADIO/PHONO	
HOWARD	302-RAS	1940	AC	CONSOLE R/P W/RECORDER	
HOWARD	303	1939	AC	CONSOLE PB	
HOWARD	305	1939	AC	TABLE PB BC/SW	40
HOWARD	307	1940	AC	TABLE BC/SW	45
HOWARD	307-TP	1941	AC	TABLE RADIO/PHONO	15
HOWARD	308-APC	1940	AC	CONSOLE R/P PB BC/SW	45
HOWARD	308-C	1939	AC	CONSOLE PB BC/SW	95
HOWARD	308-TT	1939	AC	CONSOLE TILT TOP PB BC/SW	135
HOWARD	318	1937	AC	CONSOLE PUSH BUTTON BC/SW	100
HOWARD	325	1925	BAT	CONSOLE 6 TUBE 3 DIAL	
HOWARD	325D	1938	AC	CONSOLE AUTO. TUNING BC/SW	
HOWARD	368	1937	AC	TABLE BC/SW	
HOWARD	375	1938	AC	TABLE AUTO. TUNING BC/SW	
HOWARD	400	1937	AC	CONSOLE BC/SW	
HOWARD	418	1938	AC	CONSOLE AUTO. TUNING BC/SW	110
HOWARD	425	1937	AC	CONSOLE BC/SW	
HOWARD	430	1938	AC	CONSOLE BC/SW	
HOWARD	440	1938	AC	TABLE BC/SW	

MANUFACTURER	MODEL	YEAR	POWER	STYLE	PRICE
HOWARD	450A	1938	AC	TABLE BC/SW	
HOWARD	468	1938	AC	TABLE AUTO. TUNING BC/SW	
HOWARD	480	1938	AC	CONSOLE BC/SW	
HOWARD	500	1932	AC	CONSOLE	
HOWARD	501	1931	AC	CONSOLE BC/SW	
HOWARD	518	1939	AC	CONSOLE PB BC/SW	
HOWARD	525	1938	AC	CONSOLE AUTO. TUNING BC/SW	
HOWARD	568-C	1939	AC	CONSOLE PB BC/SW	
HOWARD	568-RA	1940	AC	CONSOLE RADIO/PHONO	
HOWARD	568-T	1939	AC	CONSOLE PB BC/SW	
HOWARD	575	1939	AC	TABLE PB BC/SW	45
HOWARD	580	1939	AC	CONSOLE PB BC/SW	50
HOWARD	700	1940	AC	MANTLE PLASTIC	20
HOWARD	718	1940	AC	CONSOLE PB BC/SW	145
HOWARD	718-APC	1940	AC	CONSOLE RADIO/PHONO	45
HOWARD	718-XFMA	1941	AC	CONSOLE BC/SW/FM	45
HOWARD	780	1941	AC	TABLE BC/SW	40
HOWARD	A	1923	BAT	TABLE LP 5 TUBE 3 DIAL	125
HOWARD	A-5	1925	BAT	TABLE LP 5 TUBE 3 DIAL	100
HOWARD	A-12	1934	AC	TABLE	
HOWARD	A-65	1926	BAT	TABLE LP 6 TUBE 2 DIAL	
HOWARD	A-65	1926	BAT	CONSOLE 6 TUBE 2 DIAL	
HOWARD	B-13	1934	AC	TABLE	
HOWARD	C-14	1934	AC	TABLE BC/SW	
HOWARD	CONSOLETTE	1928	AC	CONSOLE	
HOWARD	D	1923	BAT	TABLE LP 4 TUBE 3 DIAL	110
HOWARD	D-4	1925	BAT	TABLE LP 4 TUBE 3 DIAL	120
HOWARD	D-15	1934	AC	TABLE BC/SW	
HOWARD	D-16	1934	AC	CONSOLE BC/SW	
HOWARD	F-17	1934	AC	CONSOLE BC/SW	
HOWARD	F-18	1934	AC	CONSOLE RADIO/PHONO BC/SW	
HOWARD	FLORENTINE	1928	AC	CONSOLE	
HOWARD	GOTHIC	1928	AC	CONSOLE	
HOWARD	GRAND CONSOLE	1935	AC	CONSOLE BC/SW	
HOWARD	HEPPLEWHITE	1928	AC	CONSOLE	
HOWARD	HIGHBOY	1928	AC	CONSOLE HB	
HOWARD	K	1932	AC	CONSOLE	
HOWARD	L	1932	AC	CONSOLE HB	
HOWARD	LOUIS XVI	1929	AC	CONSOLE HB	
HOWARD	R-6	1933	AC	CONSOLE	
HOWARD	S-2	1933	AC	CONSOLE	
HOWARD	S-7	1926	BAT	TABLE LP 7 TUBE 2 DIAL	125
HOWARD	S-7	1926	BAT	CONSOLE 7 TUBE 2 DIAL	
HOWARD	S-7	1933	AC	CONSOLE	
HOWARD	SGA	1930	AC	CONSOLE LB PLYMOUTH	
HOWARD	SGA	1930	AC	CONSOLE HB CONSOLETTE	
HOWARD	SGA	1930	AC	CONSOLE HB PURITAN	
HOWARD	SGA	1930	AC	CONSOLE HB PATRICIAN	

MANUFACTURER	MODEL	YEAR	POWER	STYLE	PRICE
HOWARD	SGB	1930	AC	CATHEDRAL	
HOWARD	SGC	1930	AC	CONSOLE LB RADIO/PHONO	
HOWARD	SHERATON	1929	AC	CONSOLE HB	
HOWARD	V-11	1934	AC	TABLE BC/SW	
HOWARD	W-6	1933	AC	CONSOLE BC/SW/LW	
HOWARD	W-18	1934	AC	CONSOLE R/P BC/SW/LW	
HOWARD	X-2	1933	AC	TABLE	
HOWARD	Y-3	1933	AC	CONSOLE	
HOWARD	Z-4	1933	AC	CONSOLE	
HUNT	HEXADYNE	1926	BAT	PORTABLE 6 TUBE 1 DIAL	175
HYATT ELECTRIC	A	1927	BAT	PORTABLE BOX W/ANT 1 DIAL	120
HYATT ELECTRIC	A-6	1930	BAT	PORTABLE	
HYATT ELECTRIC	AC-7	1930	AC	CATHEDRAL W/CLOCK	
HYATT ELECTRIC	D	1930	BAT	PORTABLE	
HYMAN	V-60 BESTONE	1924	BAT	TABLE HP 4 TUBE 2 DIAL	200
I					
INCANDESCENT SUPPLY	DELUXE	1924	BAT	TABLE HP SPEAKER 5 T 3 DL	
INCANDESCENT SUPPLY	DELUXE	1924	BAT	CONSOLE HB 5 TUBE 3 DIAL	
INCANDESCENT SUPPLY	NEWLANDS	1924	BAT	TABLE LP 5 TUBE 3 DIAL	95
INDIANA ELECTRIC	60A CASE	1927	BAT	TABLE	
INDIANA ELECTRIC	61A CASE	1927	BAT	TABLE 1 DRUM DIAL	
INDIANA ELECTRIC	61C CASE	1927	BAT	CONSOLE HB 1 DRUM DAIL	
INDIANA ELECTRIC	62B CASE	1927	AC	TABLE 1 DRUM DIAL	
INDIANA ELECTRIC	62C CASE	1927	AC	CONSOLE HB 1 DRUM DIAL	
INDIANA ELECTRIC	90A CASE	1927	BAT	TABLE W/LOOP ANT 1 DRUM DIAL	
INDIANA ELECTRIC	90C CASE	1927	BAT	CONSOLE LB 9 TUBE 1 DRUM DIAL	135
INDIANA ELECTRIC	92A CASE	1927	AC	TABLE 1 DRUM DIAL	
INDIANA ELECTRIC	92C CASE	1927	AC	CONSOLE LP 9 TUBE 1 DRUM DIAL	145
INDIANA ELECTRIC	500 CASE	1925	BAT	TABLE LP 5 TUBE 3 DIAL	
INDIANA ELECTRIC	503 CASE	1926	BAT	TABLE 5 TUBE 3 DIAL	
INDIANA ELECTRIC	506 CASE	1926	BAT	TABLE LP 5 TUBE 3 DIAL	
INDIANA ELECTRIC	600 CASE	1926	BAT	TABLE 3 DIAL	
INDIANA ELECTRIC	603 CASE	1926	BAT	TABLE 3 DIAL	
INDIANA ELECTRIC	606 CASE	1926	BAT	TABLE 3 DIAL	
INDIANA ELECTRIC	701 CASE	1926	BAT	CONSOLE	
INDIANA ELECTRIC	702 CASE	1926	BAT	CONSOLE	
INDIANA ELECTRIC	703 CASE	1926	BAT	CONSOLE	
INDUSTRIAL RADIO	404	1923	BAT	TABLE LP ULTRAMARVEL 3T 2 DIAL	80
INDUSTRIAL RADIO	404-A	1923	BAT	TABLE LP ULTRAMARVEL 4T 2 DIAL	90
INSULINE	AIGLON	1933	AC	TABLE BC/SW	
INSULINE	AMERICUS	1933	AC	TABLE	
INSULINE	ATLANTIC	1933	AC	TABLE BC/LW	
INSULINE	BIJOU	1933	AC	TABLE BC/SW/LW	
INSULINE	ENVOY	1931	AC	CATHEDRAL	

MANUFACTURER	MODEL	YEAR	POWER	STYLE	PRICE
INSULINE	ENVOYETTE	1931	AC	TABLE	
INSULINE	GNOME	1933	AC	TABLE	
INSULINE	INSULINE	1931	AC	TABLE	
INSULINE	LATINIC	1933	AC	TABLE BC/SW	
INSULINE	MIGNON	1933	AC	TABLE BC/LW	
INSULINE	PACIFIC	1933	AC	TABLE BC/SW/LW	
INSULINE	SUPER CONQUEROR	1932	AC	CONSOLE RADIO/PHONO BC/SW	
INSULINE	SUPER CONQUEROR	1932	AC	TABLE BC/SW/LW	
INSULINE	SUPER-6	1932	AC	CONSOLE BC/LW	
INSULINE	SUPER-7	1932	AC	CONSOLE BC/LW	
INSULINE	UNA-9	1933	AC	TABLE BC/LW	
INSULINE	UNIV. COMPANION	1932	AC/DC	PORTABLE	
INTERNATIONAL				See also KADETTE RADIO	
INTERNATIONAL	26	1935	AC	TABLE KADETTE	
INTERNATIONAL	36	1935	AC	TABLE KADETTE	40
INTERNATIONAL	40	1935	AC	MANTLE KADETTE	
INTERNATIONAL	41	1935	AC	MANTLE KADETTE	
INTERNATIONAL	42	1935	AC	MANTLE KADETTE	
INTERNATIONAL	43	1935	AC	MANTLE KADETTE	
INTERNATIONAL	44	1935	AC	MANTLE KADETTE	
INTERNATIONAL	47	1935	AC	MANTLE KADETTE	
INTERNATIONAL	48	1935	AC	MANTLE KADETTE	
INTERNATIONAL	52	1935	AC	TABLE KADETTE BC/SW	
INTERNATIONAL	53	1935	AC	TABLE KADETTE BC/SW	
INTERNATIONAL	61	1935	AC	TABLE KADETTE BC/SW	
INTERNATIONAL	66	1935	AC	COMPACT KADETTE	30
INTERNATIONAL	66-X	1936	AC	COMPACT KADETTE BC/SW	35
INTERNATIONAL	70	1935	AC	TABLE KADETTE	
INTERNATIONAL	72	1935	2/DC	TABLE KADETTE BC/SW	
INTERNATIONAL	76	1936	AC	COMPACT KADETTE	
INTERNATIONAL	85	1935	AC	TABLE KADETTE	
INTERNATIONAL	86	1936	AC	TABLE KADETTE BC/SW	
INTERNATIONAL	87	1936	AC	TABLE KADETTE BC/SW	
INTERNATIONAL	90	1935	AC	COMPACT KADETTE	
INTERNATIONAL	120	1935	AC	TABLE KADETTE BC/SW	
INTERNATIONAL	400	1936	4/DC	TABLE KADETTE BC/SW	
INTERNATIONAL	500	1936	2/DC	TABLE KADETTE BC/SW	
INTERNATIONAL	550	1936	2/DC	CONSOLE BC/SW	
INTERNATIONAL	600	1936	6/DC	TABLE	
INTERNATIONAL	649-X	1937	AC	CONSOLE CHAIRSIDE BC/SW	70
INTERNATIONAL	660	1936	6/DC	CONSOLE	
INTERNATIONAL	1200	1935	AC	TABLE KADETTE BC/SW	
INTERNATIONAL	A-4	1933	AC	MANTLE KADETTE BC/SW	
INTERNATIONAL	A-5	1933	AC	MANTLE KADETTE BC/SW	
INTERNATIONAL	A-6	1933	AC	MANTLE KADETTE BC/SW	
INTERNATIONAL	B-1	1933	AC	MANTLE KADETTE BC/SW	
INTERNATIONAL	B-2	1933	AC	MANTLE KADETTE BC/SW	
INTERNATIONAL	B-3	1933	AC	MANTLE KADETTE BC/SW	
INTERNATIONAL	CB	1934	BAT	TABLE KADETTE	

MANUFACTURER	MODEL	YEAR	POWER	STYLE	PRICE
INTERNATIONAL	ES-19	1934	AC	TABLE KADETTE BC/SW	
INTERNATIONAL	ES-20	1934	AC	TABLE KADETTE BC/SW	
INTERNATIONAL	ES-25	1934	AC	TABLE KADETTE BC/SW	
INTERNATIONAL	JR	1933	AC	TABLE KADETTE	
INTERNATIONAL	K-10	1937	AC	MANTLE IVORY BLUE PLASTIC	
INTERNATIONAL	K-11	1937	AC	MANTLE IVORY YELLOW PLASTIC	
INTERNATIONAL	K-12	1937	AC	MANTLE IVORY GREEN PLASTIC	
INTERNATIONAL	K-13	1937	AC	MANTLE IVORY ROSE PLASTIC	
INTERNATIONAL	K-14	1937	AC	MANTLE BROWN TAN PLASTIC	
INTERNATIONAL	K-15	1937	AC	MANTLE DARK GREEN PLASTIC	
INTERNATIONAL	K-16	1937	AC	MANTLE BLACK RED PLASTIC	
INTERNATIONAL	K-25	1937	AC	MANTLE GREEN MARBLE	
INTERNATIONAL	K-26	1937	AC	MANTLE BLUE GLASS	
INTERNATIONAL	K-27	1937	AC	MANTLE RED MARBLE	
INTERNATIONAL	K-28	1937	AC	MANTLE BLUE MARBLE	
INTERNATIONAL	K-41	1937	AC	MANTLE PLASTIC	
INTERNATIONAL	K-43	1937	AC	MANTLE IVORY PLASTIC	
INTERNATIONAL	K-44	1937	AC	MANTLE RED PLASTIC	
INTERNATIONAL	K-85	1931	AC	TABLE	
INTERNATIONAL	K-90	1931	AC	CONSOLE LB	
INTERNATIONAL	K-95	1931	AC	CONSOLE LB	
INTERNATIONAL	K-150	1937	AC	MANTLE WALNUT PLASTIC	70
INTERNATIONAL	K-151	1937	AC	MANTLE IVORY PLASTIC	
INTERNATIONAL	K-617	1937	AC	TABLE BC/SW	
INTERNATIONAL	K-634	1937	AC	TABLE BC/SW	
INTERNATIONAL	K-739	1937	AC	TABLE BC/SW	
INTERNATIONAL	K-1019	1937	AC	TABLE BC/SW	
INTERNATIONAL	K-1129	1937	AC	TABLE BC/SW	
INTERNATIONAL	K-1149	1937	AC	CONSOLE BC/SW	
INTERNATIONAL	K-1159	1937	AC	CONSOLE BC/SW	
INTERNATIONAL	KADETTE	1932	AC/DC	MANTLE PLASTIC	95
INTERNATIONAL	LARK	1923	BAT	TABLE HP METER 1 T 1 DIAL	
INTERNATIONAL	PB	1932	AC	TABLE	
INTERNATIONAL	PD	1932	AC	TABLE	
INTERNATIONAL	PM	1932	AC	TABLE	
INTERNATIONAL	PW	1932	AC	TABLE	
IRVING RADIO	42	1925	BAT	TABLE LP 5 TUBE 3 DIAL	70
IRVING RADIO	44	1925	BAT	TABLE LP SLOPE PAN. 5T 3DL	
IRVING RADIO	46	1925	BAT	TABLE HP SPEAKER 5 T 3 DL	

J

MANUFACTURER	MODEL	YEAR	POWER	STYLE	PRICE
JACKSON-BELL	4	1935	AC	TABLE PETER PAN	200
JACKSON-BELL	5	1928	AC	TABLE	
JACKSON-BELL	8	1929	AC	TABLE	

MANUFACTURER	MODEL	YEAR	POWER	STYLE	PRICE
JACKSON-BELL	8	1929	AC	CONSOLE LB	
JACKSON-BELL	8	1929	AC	CONSOLE HB	
JACKSON-BELL	25-AV	1932	AC	TABLE	
JACKSON-BELL	26-AV	1932	AC	TABLE	
JACKSON-BELL	26-AV-SB	1932	AC	TABLE BC/SW	
JACKSON-BELL	27-AV	1932	AC	CONSOLE	
JACKSON-BELL	27-AV	1932	AC	TABLE	
JACKSON-BELL	28-AV-SB	1932	AC	CONSOLE BC/SW	
JACKSON-BELL	28-AV-SB	1932	AC	TABLE BC/SW	
JACKSON-BELL	59	1929	AC	TABLE	
JACKSON-BELL	60	1929	AC	TABLE	
JACKSON-BELL	62	1930	AC	CATHEDRAL MODERN	150
JACKSON-BELL	69	1931	AC	CATHEDRAL	
JACKSON-BELL	84	1931	AC	TABLE PETER PAN	
JACKSON-BELL	87	1931	AC	CONSOLE	
JACKSON-BELL	87	1931	AC	TABLE	
JACKSON-BELL	88	1931	AC	CONSOLE	
JACKSON-BELL	88	1931	AC	TABLE	
JACKSON-BELL	89	1931	AC	CONSOLE	260
JACKSON-BELL	89	1931	AC	TABLE	
JACKSON-BELL	212	1932	AC	CONSOLE	
JACKSON-BELL	401-A	1933	AC	TABLE	
JACKSON-BELL	503-C	1933	AC	TABLE	
JACKSON-BELL	504-B	1933	AC	TABLE	
JACKSON-BELL	505-D	1933	AC	TABLE	
JEWEL RADIO	M	1930	AC	CATHEDRAL	120
JEWETT RADIO	JEWETT	1925	BAT	TABLE LP 5 T 1 METER 2 DL	165
JONES RADIO	R HARMONIC	1926	BAT	TABLE LP 4 TUBE 3 DIAL	
JONES RADIO	S HARMONIC	1926	BAT	TABLE LP 5 TUBE 3 DIAL	
JONES, JOSEPH	J- 75-B	1925	BAT	TABLE LP SLOPE PAN. 5T 3DL	
JONES, JOSEPH	J-100-B	1925	BAT	TABLE LP 5 TUBE 3 DIAL	
JONES, JOSEPH	J-175	1925	BAT	TABLE LP 6 TUBE 3 DIAL	
JONES, JOSEPH	J-175-D	1925	BAT	CONSOLE 6 TUBE 3 DIAL	
JONES, JOSEPH	JONES	1924	BAT	TABLE SLOPE PAN. 4 T 2 DL	
JONES, JOSEPH	JONES	1924	BAT	TABLE SLOPE PAN. 5 T 3 DL	
K					
KADETTE RADIO				See also INTERNATIONAL	
KADETTE RADIO	L-20	1939	AC	TABLE	
KADETTE RADIO	L-21	1939	AC	MANTLE PLASTIC	
KADETTE RADIO	L-22	1939	AC	MANTLE PLASTIC	
KADETTE RADIO	L-23	1939	AC	MANTLE PLASTIC	
KADETTE RADIO	L-24	1939	AC	MANTLE PLASTIC	
KADETTE RADIO	L-25	1939	AC	TABLE PLASTIC	
KADETTE RADIO	L-26	1939	AC	TABLE PLASTIC	
KADETTE RADIO	L-27	1939	AC	TABLE PLASTIC	
KADETTE RADIO	L-28	1939	AC	TABLE PLASTIC	
KADETTE RADIO	L-29	1939	AC	TABLE ART DECO	
KADETTE RADIO	L-34	1939	BAT	PORTABLE	
KADETTE RADIO	L-36	1939	AC	TABLE PB	

MANUFACTURER	MODEL	YEAR	POWER	STYLE	PRICE
KADETTE RADIO	L-37	1939	AC	TABLE PB	
KADETTE RADIO	L-38	1939	AC	TABLE RADIO/PHONO	
KADETTE RADIO	L-39	1939	AC	TABLE RADIO/PHONO PB	
KADETTE RADIO	L-40	1939	AC	MANTLE W/CLOCK	
KARDON PRODUCTS	K-14	1925	BAT	TABLE LP 5 TUBE 3 DIAL	90
KELLER-FULLER	14	1930	AC	CATHEDRAL MODERN	75
KELLOGG	500 WAVEMASTER	1925	BAT	TABLE W/FREQ SWITCH 5T 1 DIAL	
KELLOGG	501 WAVEMASTER	1925	BAT	TABLE LP W/FREQ SWITCH 1 DIAL	250
KELLOGG	501-502	1925	BAT	CONSOLE 5 TUBE 1 DIAL	
KELLOGG	502	1925	BAT	TABLE SPEAKER 5 TUBE 1 DL	
KELLOGG	503 WAVEMASTER	1925	BAT	CONSOLE W/FREQ SWITCH 1 DIAL	320
KELLOGG	504 WAVEMASTER	1926	BAT	TABLE LP W/FREQ SWITCH 1 DIAL	270
KELLOGG	505 WAVEMASTER	1926	BAT	TABLE LP W/FREQ SWITCH 1 DIAL	320
KELLOGG	506 WAVEMASTER	1926	BAT	CONSOLE SPINET 1 DIAL	225
KELLOGG	507 WAVEMASTER	1926	BAT	TABLE LP W/FREQ SWITCH 1 DIAL	195
KELLOGG	508 WAVEMASTER	1926	BAT	CONSOLE W/FREQ SWITCH 1 DIAL	250
KELLOGG	510	1927	AC	CONSOLE W/FREQ SWITCH 1 DIAL	
KELLOGG	511	1927	AC	CONSOLE W/FREQ SWITCH 1 DIAL	
KELLOGG	512	1927	AC	CONSOLE W/FREQ SWITCH 1 DIAL	
KELLOGG	516	1928	AC	CONSOLE LB	145
KELLOGG	523	1929	AC	CONSOLE	
KELLOGG	524	1929	AC	CONSOLE	
KELLOGG	525	1929	AC	CONSOLE RADIO/PHONO	
KELLOGG	526	1929	AC	CONSOLE (25 CYCLE)	
KELLOGG	527	1929	AC	CONSOLE (25 CYCLE)	
KELLOGG	528	1929	AC	CONSOLE R/P (25 CYCLE)	
KELLOGG	532	1929	AC	CONSOLE	
KELLOGG	533	1930	AC	CONSOLE TUDOR	185
KELLOGG	534	1930	AC	CONSOLE HEPPELWHITE	
KELLOGG	535	1930	AC	CONSOLE	
KELLOGG	536	1930	AC	CONSOLE	
KELLOGG	KELLOGG	1922	BAT	TABLE 1 TUBE 1 DIAL	120
KEMPER RADIO	80	1930	BAT	TOMBSTONE KOMPAK	50
KEMPER RADIO	K-52	1926	BAT	PORTABLE 5 TUBE 2 DIAL	175
KEMPER RADIO	K-53	1927	BAT	PORTABLE 5 TUBE 2 DIAL	120
KEMPER RADIO	SG- 7	1930	BAT	TABLE	
KEMPER RADIO	SG-71	1930	BAT	CONSOLE LB	
KENMAN ELECTRIC	5 KENMAN	1925	BAT	TABLE LP 5 TUBE 3 DIAL	85
KENNEDY, COLIN B.	3 (III)	1923	BAT	PORTABLE 3 TUBE 1 DIAL	
KENNEDY, COLIN B.	7 CORONET	1927	DC	TABLE LP 7 TUBE 1 DIAL	85

MANUFACTURER	MODEL	YEAR	POWER	STYLE	PRICE
KENNEDY, COLIN B.	7 CORONET	1927	AC	TABLE LP 8 TUBE 1 DIAL	95
KENNEDY, COLIN B.	7 ROYAL	1927	BAT	CONSOLE LB 7 TUBE 1 DIAL	
KENNEDY, COLIN B.	10	1929	AC	TABLE LP	95
KENNEDY, COLIN B.	20	1925	BAT	TABLE 5 TUBE 1 DIAL	250
KENNEDY, COLIN B.	22	1924	BAT	TABLE 5 TUBE 1 DIAL	300
KENNEDY, COLIN B.	60 ROYAL	1928	AC	TABLE 7 TUBE 1 DIAL	
KENNEDY, COLIN B.	62-D	1932	AC	CONSOLE HB	
KENNEDY, COLIN B.	62-D	1932	AC	CONSOLE LB	
KENNEDY, COLIN B.	80 ROYAL	1928	AC	CONSOLE	
KENNEDY, COLIN B.	110	1922	BAT	TABLE HP 3 TUBE 4 DIAL	750
KENNEDY, COLIN B.	110/525	1922	BAT	TABLE HP 5 TUBE 4 DIAL	750
KENNEDY, COLIN B.	164-B	1932	AC	CONSOLE BC/SW	
KENNEDY, COLIN B.	210	1929	AC	CONSOLE	
KENNEDY, COLIN B.	220	1921	BAT	TABLE HP 2 DIAL 1 TUBE	440
KENNEDY, COLIN B.	220	1929	AC	CONSOLE	550
KENNEDY, COLIN B.	220/525	1922	BAT	TABLE HP 4 TUBE 2 DIAL	650
KENNEDY, COLIN B.	222	1930	DC	CONSOLE LB	
KENNEDY, COLIN B.	224	1930	BAT	CONSOLE LB	
KENNEDY, COLIN B.	266-B	1932	AC	CONSOLE BC/SW	
KENNEDY, COLIN B.	281	1921	BAT	TABLE LP 3 TUBE 3 DIAL	410
KENNEDY, COLIN B.	310	1929	AC	CONSOLE	
KENNEDY, COLIN B.	311/522	1923	BAT	PORTABLE CASE 5 TUBE 3 DL	380
KENNEDY, COLIN B.	320	1929	AC	CONSOLE	
KENNEDY, COLIN B.	366-B	1932	AC	CONSOLE HB BC/SW	
KENNEDY, COLIN B.	426	1930	AC	CONSOLE ASRMCHAIR	
KENNEDY, COLIN B.	455	1932	AC	TABLE BC/SW	
KENNEDY, COLIN B.	526	1930	AC	CONSOLE HB	
KENNEDY, COLIN B.	563	1932	AC	TABLE BC/SW	
KENNEDY, COLIN B.	632	1930	AC	CONSOLE LB	150
KENNEDY, COLIN B.	726	1930	AC	CONSOLE HB	
KENNEDY, COLIN B.	726-B	1930	AC	CONSOLE HB RADIO/PHONO	
KENNEDY, COLIN B.	826	1930	AC	CONSOLE	
KENNEDY, COLIN B.	826-A	1930	AC	CONSOLE RADIO/PHONO	
KENNEDY, COLIN B.	826-B	1930	AC	CONSOLE BC [SW UNITS	
KENNEDY, COLIN B.	1030	1930	AC	CONSOLE LB	
KENNEDY, COLIN B.	1656	1931	AC	CONSOLE LB	
KENNEDY, COLIN B.	1852	1931	AC	TABLE	
KENNEDY, COLIN B.	1872-B	1932	AC	TABLE BC/SW	
KENNEDY, COLIN B.	1952	1931	AC	CONSOLE LB	
KENNEDY, COLIN B.	JACOBEAN	1923	BAT	CONSOLE LP SLOPE PAN 3T 2D	1000
KENNEDY, COLIN B.	SPANISH DESK	1923	BAT	CONSOLE HP SLOPE PAN 3T 2D	1000
KENNEDY, COLIN B.	SPINET	1927	DC	CONSOLE DESK DOORS 7 T 1 DIAL	300
KENNEDY, COLIN B.	SPINET	1927	AC	CONSOLE DESK DOORS 8 T 1 DIAL	325
KENNEDY, COLIN B.	V	1923	BAT	TABLE HP SLOPE PAN 3T 2DL	300
KENNEDY, COLIN B.	VI	1925	BAT	TABLE HP SLOPE PAN 4 T 2 DIAL	300
KENNEDY, COLIN B.	X	1923	BAT	TABLE HP SLOPE PAN 3T 2DL	375
KENNEDY, COLIN B.	XI	1924	BAT	TABLE HP SLOPE PAN 4T 2 DL	350

MANUFACTURER	MODEL	YEAR	POWER	STYLE	PRICE
KENNEDY, COLIN B.	XV	1924	BAT	TABLE 5 TUBE 2 DIAL	320
KENNEDY, COLIN B.	XVI ROYAL	1925	BAT	CONSOLE 5 TUBE 2 DIAL	300
KENNEDY, COLIN B.	XXX	1925	BAT	TABLE SLOPE PANEL 5 T 2 DL	310
KILBOURNE & CLARK	AIR ROAMER	1924	BAT	TABLE LP 5 TUBE 3 DIAL	95
KILBOURNE & CLARK	AIR RULER	1924	BAT	TABLE LP 5 TUBE 3 DIAL	85
KING	10-K-1	1926	BAT	TABLE LP 5 TUBE 3 DIAL	
KING	25	1925	BAT	TABLE LP 5 TUBE 3 DIAL	65
KING	25-C	1925	BAT	CONSOLE 5 TUBE 3 DIAL	
KING	25-S	1925	BAT	TABLE LP SPEAKER 5 T 3 DL	
KING	30	1925	BAT	TABLE LP 5 TUBE 3 DIAL	90
KING	30-C	1925	BAT	CONSOLE 5 TUBE 3 DIAL	
KING	30-S	1925	BAT	TABLE LP SPEAKER 5 T 3 DL	
KING	35	1925	BAT	TABLE LP 5 TUBE 3 DIAL	
KING	35-C	1925	BAT	CONSOLE 5 TUBE 3 DIAL	
KING	35-S	1925	BAT	TABLE LP SPEAKER 5 T 3 DL	
KING	61	1926	BAT	TABLE LP 6 TUBE 3 DIAL	
KING	62	1926	BAT	TABLE LP 6 TUBE 1 DIAL	65
KING	63	1926	BAT	CONSOLE 6 TUBE 1 DIAL	
KING	71	1927	BAT	TABLE LP 6 TUBE 1 DIAL	
KING	71 COMMANDER	1927	BAT	TABLE LP LOOP ANT 6 T 1 DL	135
KING	80 BARONET	1927	BAT	TABLE LP 6 TUBE 1 DIAL	70
KING	80-H VIKING	1927	BAT	CONSOLE 6 TUBE 1 DIAL	
KING	81	1927	BAT	TABLE LP 6 TUBE 1 DIAL	90
KING	81 CHEVALIER	1927	BAT	CONSOLE 6 TUBE 1 DIAL	
KING	81 CRUSADER	1927	BAT	TABLE LP 6 TUBE 1 DIAL	
KING	81-H	1927	BAT	CONSOLE HB 6 TUBE 1 DIAL	
KING	82	1927	AC	CONSOLE 7 TUBE 1 DIAL	
KING	97 ROYAL	1929	AC	CONSOLE 8 TUBE 1 DIAL	
KING	98 IMPERIAL	1929	AC	CONSOLE 9 TUBE 1 DIAL	
KING	101 MONARCH	1929	AC	TABLE 7 TUBE 1 DIAL	
KING	F	1928	BAT	TABLE LP 5 TUBE 1 DIAL	
KING	G	1928	BAT	TABLE 6 TUBE 1 DIAL	
KING	H	1928	AC	TABLE LP 7 TUBE 1 DIAL	
KING	J	1928	AC	TABLE LP 8 TUBE 1 DIAL	
KING	VII KING COLE	1927	BAT	TABLE LP 7 TUBE 2 DIAL	
KING	VIII KING COLE	1927	BAT	CONSOLE 8 TUBE 1 DIAL	
KLITZEN RADIO	KENT	1924	BAT	TABLE LP 3 TUBE 3 DIAL	
KLITZEN RADIO	WOLVERINE	1924	BAT	TABLE LP 3 TUBE 3 DIAL	
KNIGHT	250-A	1930	AC	TABLE METAL	75
KODEL RADIO	1	1923	BAT	PORTABLE 1 TUBE 1 DIAL	150
KODEL RADIO	5 STANDARD	1925	BAT	TABLE LP 5 TUBE 3 DIAL	95
KODEL RADIO	5 STANDARD	1925	BAT	CONSOLE LB 5 TUBE 3 DIAL	150
KODEL RADIO	53 LOGODYNE	1925	BAT	CONSOLE 5 TUBE 3 DIAL	270
KODEL RADIO	BIG 5 LOGODYNE	1925	BAT	TABLE LP 5 TUBE 3 DIAL	110
KODEL RADIO	BIG 5 LOGODYNE	1925	BAT	CONSOLE LB ENCLOSED 5T 3 DIAL	150
KODEL RADIO	C-11 GOLD STAR	1924	BAT	TABLE LP 1 TUBE 1 DIAL	170
KODEL RADIO	C-12 GOLD STAR	1924	BAT	TABLE LP 2 TUBE 1 DIAL	160
KODEL RADIO	C-13 GOLD STAR	1924	BAT	TABLE LP 3 TUBE 2 DIAL	125
KODEL RADIO	C-14	1925	BAT	TABLE HP 4 TUBE 2 DIAL	155

MANUFACTURER	MODEL	YEAR	POWER	STYLE	PRICE
KODEL RADIO	C-L-5	1924	BAT	TABLE LP 5 TUBE 3 DIAL	
KODEL RADIO	P-11	1925	BAT	PORTABLE 1 TUBE 1 DIAL	125
KODEL RADIO	P-12	1925	BAT	PORTABLE 2 TUBE 1 DIAL	150
KOLSTER RADIO				See also FEDERAL TELEGRAPH	
KOLSTER RADIO	6-D	1926	BAT	TABLE 6 TUBE 1 DIAL	125
KOLSTER RADIO	6-D	1926	BAT	CONSOLE 6 TUBE 1 DIAL	
KOLSTER RADIO	6-E	1926	BAT	CONSOLE	
KOLSTER RADIO	6-F	1927	AC	TABLE LP 7 TUBE 1 DIAL	
KOLSTER RADIO	6-G	1926	BAT	CONSOLE 6 TUBE 1 DIAL	
KOLSTER RADIO	6-H	1926	BAT	CONSOLE 6 TUBE 1 DIAL	250
KOLSTER RADIO	6-J	1927	AC	TABLE LP 6 TUBE 1 DIAL	
KOLSTER RADIO	6-K	1927	AC	CONSOLE 7 TUBE 1 DIAL	
KOLSTER RADIO	6-R	1927	AC	CONSOLE 7 TUBE 1 DIAL	
KOLSTER RADIO	7-A	1926	BAT	TABLE LP 7 TUBE 1 DIAL	85
KOLSTER RADIO	7-B	1926	BAT	TABLE SPEAKER 7 TUBE 1 DL	
KOLSTER RADIO	8-A	1926	BAT	TABLE LP 8 TUBE 1 DIAL	
KOLSTER RADIO	8-B	1925	BAT	CONSOLE LOOP ANT 8 T 1 DL	300
KOLSTER RADIO	8-C	1925	BAT	CONSOLE LOOP ANT 8 T 1 DL	395
KOLSTER RADIO	B-10	1929	AC	TABLE LP 7 TUBE 1 DIAL	
KOLSTER RADIO	B-15	1929	AC	CONSOLE 8 TUBE 1 DIAL	
KOLSTER RADIO	B-16	1929	AC	CONSOLE 8 TUBE 1 DIAL	
KOLSTER RADIO	K-20	1928	AC	TABLE LP 7 TUBE 1 DIAL	75
KOLSTER RADIO	K-21	1928	AC	TABLE LP 8 TUBE 1 DIAL	90
KOLSTER RADIO	K-22	1928	AC	TABLE LP 7 TUBE 1 DIAL	
KOLSTER RADIO	K-23	1928	AC	CONSOLE 8 TUBE 1 DIAL	
KOLSTER RADIO	K-24	1928	AC	CONSOLE 9 TUBE 1 DIAL	240
KOLSTER RADIO	K-30	1928	DC	TABLE LP 6 TUBE 1 DIAL	
KOLSTER RADIO	K-32	1928	DC	CONSOLE 6 TUBE 1 DIAL	
KOLSTER RADIO	K-43	1929	AC	CONSOLE 8 TUBE 1 DIAL	70
KOLSTER RADIO	K-44	1929	AC	CONSOLE 9 TUBE 1 DIAL	
KOLSTER RADIO	K-45	1928	AC	CONSOLE 11 TUBE 1 DIAL	240
KOLSTER RADIO	K-60	1931	AC	TABLE	160
KOLSTER RADIO	K-70	1931	AC	CONSOLE LB	
KOLSTER RADIO	K-80	1931	AC	CONSOLE HB	
KOLSTER RADIO	K-110--K-112	1932	AC	MANTLE	
KOLSTER RADIO	K-113	1932	AC	MANTLE	
KOLSTER RADIO	K-114	1932	BAT	MANTLE	
KOLSTER RADIO	K-120--K-122	1932	BAT	CONSOLE	
KOLSTER RADIO	K-123	1932	AC	CONSOLE	
KOLSTER RADIO	K-130-K-132	1932	AC	CONSOLE	200
KOLSTER RADIO	K-133	1932	AC	CONSOLE	
KOLSTER RADIO	K-133-A	1932	AC	CONSOLE DELUXE	
KOLSTER RADIO	K-140--K-142	1932	AC	CONSOLE DELUXE	
L					
LACAULT RADIO	L-3	1925	BAT	TABLE ULTRADYNE "NO DIAL"	140
LAFAYETTE	BB-11	1940	AC	CONSOLE PB BC/SW	95
LAFAYETTE	BB-96	1940	AC	CONSOLE R/P RECORDER BC/SW	35

MANUFACTURER	MODEL	YEAR	POWER	STYLE	PRICE
LAFAYETTE	D-33	1940	AC	TABLE TUNING EYE BC/SW	30
LAMPOLIER	LAMP RADIO	1924	BAT	TABLE SPEAKER 5 TUBE 1 DL	
LANGBEIN-KAUFMANN	4S	1925	BAT	TABLE LP SUPER ELKAY 4T 2D	85
LANGBEIN-KAUFMANN	5S	1925	BAT	TABLE LP SUPER ELKAY 5T 2D	95
LANGBEIN-KAUFMANN	6 ELKAY ELEC.	1926	AC	TABLE LP 6 TUBE 2 DIAL	120
LANGBEIN-KAUFMANN	6 ELKAY JR.	1926	BAT	TABLE LP 6 TUBE 2 DIAL	90
LANGBEIN-KAUFMANN	6 ELKAY SR.	1926	BAT	TABLE LP 6 TUBE 2 DIAL	90
LANGBEIN-KAUFMANN	6 ELKAY SUPER	1926	BAT	TABLE LP 6 TUBE 2 DIAL	95
LANGBEIN-KAUFMANN	7 ELKAY JR.	1926	BAT	TABLE LP 6 TUBE 2 DIAL	110
LEUTZ	6 PLIO	1926	BAT	TABLE LP ADMIRALITY 6T 2 DIAL	300
LEUTZ	6 PLIO	1926	BAT	TABLE LP IMPERIAL 6 T 2 DL	350
LEUTZ	6 PLIO JR.	1926	BAT	TABLE LP ADMIRALTY 6T 2 DL	350
LEUTZ	6 PLIO JR.	1926	BAT	CONSOLE IMPERIAL 6 T 2DL	300
LEUTZ	6 PLIO SR.	1926	BAT	CONSOLE 6 TUBE 2 DIAL	350
LEUTZ	6 PLIODYNE	1925	BAT	TABLE LP 6 TUBE 2 DIAL	375
LEUTZ	7 TRANSOCEANIC	1927	BAT	TABLE LP MTL PHANTOM 7T 5 DIAL	400
LEUTZ	7-CS	1930	AC	CONSOLE RADIO/PHONO	
LEUTZ	8 SUPER	1926	BAT	TABLE LP ADMIRALITY 7T 2 DIAL	400
LEUTZ	8 SUPER	1926	BAT	TABLE LP IMPERIAL 7 T 2 DL	400
LEUTZ	8 SUPER	1926	BAT	TABLE LP UNIVERSAL 7T 2 DL	375
LEUTZ	9 PLIODYNE	1924	BAT	TABLE LP SUPER 9 TUBE 2 DL	400
LEUTZ	9 TRANSOCEANIC	1927	BAT	TABLE LP MTL PHANTOM 9T 5 DIAL	400
LEUTZ	9 TRANSOCEANIC	1928	AC	TABLE LP MTL UNIV. 9T 2 DL	450
LEUTZ	SEVEN SEAS	1929	AC	CONSOLE HB 9 TUBE 1 DIAL	375
LEUTZ	SG- 7	1929	AC	CONSOLE R/P SILVER GHOST	400
LEUTZ	UNIVERSAL PLIO.	1925	BAT	TABLE 6 TUBE 2 DIAL	
LIBERTY TRANSFORMER	LIBERTY	1925	BAT	TABLE LP 5 TUBE 3 DIAL	75
LIBERTY TRANSFORMER	LIBERTY	1925	BAT	CONSOLE 5 TUBE 3 DIAL	
LIBERTY TRANSFORMER	STANDARD	1925	BAT	TABLE LP 5 TUBE 3 DIAL	65
M					
MACHINE SPECIALITY	ARBORPHONE	1925	BAT	TABLE LP 5 TUBE 3 DIAL	80
MACK	SINCROFLEX	1924	BAT	TABLE LP 3 TUBE 2 DIAL	
MacLAREN	DOMINAIRE	1925	AC	TABLE LP 5 TUBE 3 DIAL	100
MAGNAVOX	5-F	1940	AC	CONSOLE RADIO/PHONO	45
MAGNAVOX	10	1925	BAT	TABLE LP 5 TUBE 1 DIAL	150
MAGNAVOX	25	1925	BAT	TABLE HP SPEAKER 5T 1 DIAL	175
MAGNAVOX	31F	1941	AC	TABLE RADIO/PHONO PB	15
MAGNAVOX	32D	1940	AC	CONSOLE CHAIRSIDE R/P	
MAGNAVOX	34G	1940	AC	CONSOLE RADIO/PHONO	
MAGNAVOX	38G	1940	AC	CONSOLE RADIO/PHONO	
MAGNAVOX	42A	1940	AC	CONSOLE RADIO/PHONO	
MAGNAVOX	47	1940	AC	CONSOLE RADIO/PHONO	
MAGNAVOX	48	1940	AC	CONSOLE RADIO/PHONO	
MAGNAVOX	50	1940	AC	CONSOLE RADIO/PHONO	

MANUFACTURER	MODEL	YEAR	POWER	STYLE	PRICE
MAGNAVOX	51A	1940	AC	CONSOLE RADIO/PHONO	
MAGNAVOX	62	1940	AC	CONSOLE RADIO/PHONO	
MAGNAVOX	72	1940	AC	CONSOLE RADIO/PHONO	
MAGNAVOX	75	1925	BAT	CONSOLE HP DESK 5 T 1 DIAL	250
MAGNAVOX	498	1940	AC	CONSOLE RADIO/PHONO	
MAGNAVOX	A	1925	BAT	TABLE 5 TUBE 1 DIAL	175
MAGNAVOX	D	1925	BAT	TABLE 5 TUBE 1 DIAL	125
MAGNAVOX	JUNIOR	1925	BAT	TABLE LP 5 TUBE 1 DIAL	95
MAGNAVOX	TRF- 5	1924	BAT	TABLE LP 5 TUBE 1 DIAL	125
MAGNAVOX	TRF-50	1924	BAT	TABLE HP DOORS SPKER 5T 1D	175
MAJESTIC RADIO & TELEVISION	1-A-50	1939	AC	TABLE WAL W/MAPLE TRIM	35
MAJESTIC RADIO & TELEVISION	1-A-59	1939	AC	TABLE	35
MAJESTIC RADIO & TELEVISION	1-BR-50-B	1939	BAT	PORTABLE CLOTH	
MAJESTIC RADIO & TELEVISION	1-BR-50-BP	1939	BAT	PORTABLE RADIO/PHONO	
MAJESTIC RADIO & TELEVISION	1-M-40	1939	AC	MANTL PLASTIC	
MAJESTIC RADIO & TELEVISION	2-A-50-F	1939	AC	TABLE LEATHERETTE PB	
MAJESTIC RADIO & TELEVISION	2-C-60	1939	AC	CONSOLE PB BC/SW	
MAJESTIC RADIO & TELEVISION	2-C-60-P	1939	AC	CONSOLE R/P PB BC/SW	45
MAJESTIC RADIO & TELEVISION	3-BC-90-B	1939	AC	CONSOLE PB TUN EYE BC/SW	150
MAJESTIC RADIO & TELEVISION	3-C-70	1939	AC	CONSOLE PB BC/SW	
MAJESTIC RADIO & TELEVISION	3-C-80-P	1939	AC	CONSOLE R/P PB BC/SW	
MAJESTIC RADIO & TELEVISION	3-C-90	1939	AC	CONSOLE R/P PB TE BC/SW	150
MAJESTIC RADIO & TELEVISION	5-BD	1940	AC	PORTABLE RAWHIDE	15
MAJESTIC RADIO & TELEVISION	5-T	1939	AC	MANTLE PLASTIC W/CLOCK	25
MAJESTIC RADIO & TELEVISION	76	1936	AC	TABLE LAYDOWN TE BC/SW	
MAJESTIC RADIO & TELEVISION	85	1936	AC	TOMBSTONE TE BC/SW	
MAJESTIC RADIO & TELEVISION	130	1939	BAT	PORTABLE LEATHERETTE	
MAJESTIC RADIO & TELEVISION	130-U	1939	BAT	PORTABLE CLOTH	
MAJESTIC RADIO & TELEVISION	167	1939	AC	TOMBSTONE BC/SW	70
MAJESTIC RADIO & TELEVISION	250	1939	AC	TABLE PLASTIC	
MAJESTIC RADIO & TELEVISION	259-EB	1939	AC	TABLE PB BC/SW	40
MAJESTIC RADIO & TELEVISION	260	1939	AC	TABLE WALNUT PB BC/SW	
MAJESTIC RADIO & TELEVISION	360	1939	AC	TABLE PB BC/SW	
MAJESTIC RADIO & TELEVISION	448-2	1940	2/DC	TABLE	30
MAJESTIC RADIO & TELEVISION	1250	1936	AC	CONSOLE TE BC/SW/LW	
MAJESTIC RADIO & TELEVISION	3-C-80	1939	AC	CONSOLE PB BC/SW	
MAJESTIC RADIO & TELEVISION	H-191-A	1940	AC	PORTABLE	15
MAJESTIC RADIO & TELEVISION	P-1-A-50	1939	AC	TABLE RADIO/PHONO	
MAJESTIC RADIO & TELEVISION	T-101-LA	1940	AC	TABLE	25
MAJESTIC RADIO & TELEVISION	T-202-A	1940	AC	TABLE	35
MAJESTIC RADIO & TELEVISION	TP-221-A	1941	AC	TABLE RADIO/PHONO	15
MARATHON RADIO	MP-5	1924	BAT	TABLE LP 5 TUBE 3 DIAL	70
MARATHON RADIO	MR-3	1924	BAT	TABLE SLOPE PAN. 3 T 2 DL	
MARATHON RADIO	MR-5	1924	BAT	TABLE LP 5 TUBE 3 DIAL	75
MARATHON RADIO	MRS-3	1924	BAT	TABLE SLOPE PAN. 3 T 2 DL	
MARATHON RADIO	MX-5	1924	BAT	TABLE LP 5 TUBE 3 DIAL	65
MARTI ELECTRIC	CS-2	1927	AC	CONSOLE LB W/SP. 9T 2 DIAL	
MARTI ELECTRIC	CS-10	1927	AC	CONSOLE	
MARTI ELECTRIC	DC-2	1926	AC	CONSOLE DESK 6 TUBE 2 DIAL	150

MANUFACTURER	MODEL	YEAR	POWER	STYLE	PRICE
MARTI ELECTRIC	TA-2	1927	AC	TABLE LP 6 TUBE 2 DIAL	95
MARTI ELECTRIC	TA-10	1927	AC	TABLE	110
MARWOL RADIO	A-1	1925	BAT	TABLE FLAT BOX 3 DIAL	200
MARWOL RADIO	BABY GRAND	1925	BAT	TABLE LP PANEL ON TOP 3 DL	225
MARWOL RADIO	CONSOLE GRAND	1925	BAT	CONSOLE LB ENCLOSED 3 DIAL	150
MARWOL RADIO	JEWEL	1925	BAT	TABLE LP 5 TUBE 3 DIAL	75
MARWOL RADIO	SIX	1925	BAT	TABLE LP 6 TUBE 3 DIAL	
MARWOOD RADIO	6	1927	AC	TABLE LP	85
MARWOOD RADIO	8	1927	AC	TABLE LP	95
MARWOOD RADIO	8	1927	AC	CONSOLE HB	140
MARWOOD RADIO	MARWOOD	1926	BAT	TABLE LP 6 TUBE 1 DIAL	70
MARWOOD RADIO	MARWOOD	1926	BAT	TABLE LP 6 TUBE 3 DIAL	70
MASTER RADIO	50	1930	AC	TABLE	
MASTER RADIO	70	1930	AC	CATHEDRAL	130
MASTER RADIO	CONSOLE	1930	AC	CONSOLE	
MAZDA RADIO	CONSOMELLO	1923	BAT	TABLE HP 6 TUBE 3 DIAL	
McMILLAN RADIO	6 OXFORD	1927	AC	TABLE LP 6 TUBE 1 DIAL	135
McMILLAN RADIO	925	1929	AC	CONSOLE	
McMILLAN RADIO	935	1929	AC	CONSOLE	
McMILLAN RADIO	937	1929	AC	CONSOLE	
McMILLAN RADIO	959	1929	AC	CONSOLE	
McMILLAN RADIO	965	1929	AC	TABLE	
McMILLAN RADIO	965	1929	AC	CONSOLE	
McMILLAN RADIO	975	1929	AC	CONSOLE	
McMILLAN RADIO	999	1929	AC	CONSOLE	
McMILLAN RADIO	RIDGEWAY	1927	BAT	CONSOLE LB DOORS 6T 1 DL	115
McMURDO SILVER	MASTERPIECE III	1935	AC	CHASSIS BC/SW/LW	
McMURDO SILVER	MASTERPIECE IVA	1935	AC	TABLE BC/SW	
McMURDO SILVER	MASTERPIECE VI	1935	AC	CHASSIS BC/SW/LW	
MEISSNER RADIO	MEISSNER	1924	AC	TABLE LP 6 TUBE 3 DIAL	
METRO ELECTRIC	6 SUPER	1927	AC	TABLE LP	110
METRO ELECTRIC	7 METRODYNE	1927	AC	TABLE LP 7 TUBE 1 DIAL	130
METRO ELECTRIC	7 METRODYNE	1927	AC	CONSOLE 7 TUBE 1 DIAL	75
METROPOLITAN ENGINEERING	VERIPHONE	1925	BAT	TABLE LP 4 TUBE 3 DIAL	
METROPOLITIAN ELECTRIC	5	1924	BAT	CONSOLE LB 5 TUBE 2 DIAL	
METROPOLITIAN ELECTRIC	10	1924	BAT	TABLE LP 5 TUBE 2 DIAL	80
MICHIGAN RADIO	JUNIOR	1923	BAT	TABLE LP 2 TUBE 2 DIAL	165
MICHIGAN RADIO	M-10	1924	BAT	TABLE HP 1 TUBE 2 DIAL	
MICHIGAN RADIO	M-12	1924	BAT	TABLE HP 3 TUBE 2 DIAL	
MICHIGAN RADIO	MIDGET	1923	BAT	TABLE 2 TUBE 2 DIAL	300
MICHIGAN RADIO	MRC-4	1923	BAT	TABLE LP 4 TUBE 2 DIAL	210
MICHIGAN RADIO	MRC-14	1925	BAT	TABLE 4 TUBE 2 DIAL	195
MICHIGAN RADIO	SENIOR	1922	BAT	TABLE HP 3 TUBE 2 DIAL	290
MIDWEST RADIO	5 UNITUNE	1926	BAT	TABLE LP 5 TUBE 1 DIAL	75
MIDWEST RADIO	6 SUPER	1927	AC	TABLE LP 6 TUBE 3 DIAL	85
MIDWEST RADIO	7	1927	AC	TABLE LP 7 TUBE 1 DIAL	75
MIDWEST RADIO	8	1927	AC	TABLE LP 1 DIAL	85
MIDWEST RADIO	8 AC	1929	AC	CONSOLE	130
MIDWEST RADIO	A	1930	AC	CONSOLE HB	

MANUFACTURER	MODEL	YEAR	POWER	STYLE	PRICE
MIDWEST RADIO	A-16	1933	AC	CONSOLE BC/SW/LW	
MIDWEST RADIO	AA-18	1935	AC	CONSOLE BC/SW/LW	
MIDWEST RADIO	B	1930	AC	CONSOLE	
MIDWEST RADIO	B-16	1933	AC	CONSOLE BC/SW/LW	
MIDWEST RADIO	BB-18	1935	AC	CONSOLE BC/SW/LW	
MIDWEST RADIO	CC-18	1935	AC	CONSOLE BC/SW/LW	
MIDWEST RADIO	D-16	1933	AC	CONSOLE BC/SW/LW	
MIDWEST RADIO	DD-18	1935	AC	CONSOLE BC/SW/LW	
MIDWEST RADIO	EE-18	1935	AC	CONSOLE BC/SW/LW	
MIDWEST RADIO	F-9	1933	AC	CONSOLE BC/SW/LW	
MIDWEST RADIO	G-9	1933	AC	TABLE BC/SW/LW	
MIDWEST RADIO	GG-11	1935	AC	TABLE BC/SW/LW	
MIDWEST RADIO	H	1930	AC	CONSOLE HB	
MIDWEST RADIO	H-6	1933	AC	TABLE	
MIDWEST RADIO	HE-6	1933	AC	TABLE BC/SW	
MIDWEST RADIO	HH-7	1935	AC	TABLE BC/SW/LW	
MIDWEST RADIO	J	1930	AC	CONSOLE EASY CHAIR	
MIDWEST RADIO	K	1922	BAT	TABLE LP 2 TUBE 2 DIAL	150
MIDWEST RADIO	K	1930	AC	CONSOLE HB	
MIDWEST RADIO	L	1930	AC	CONSOLE LB	
MIDWEST RADIO	M	1930	AC	CONSOLE HB	
MIDWEST RADIO	MIRACO	1922	BAT	MANTLE 1 TUBE 1 DIAL	130
MIDWEST RADIO	MM-11	1935	AC	CONSOLE BC/SW/LW	
MIDWEST RADIO	MW	1923	BAT	TABLE LP 4 TUBE 2 DIAL	160
MIDWEST RADIO	PR-16	1933	AC	CONSOLE RADIO/PHONO BC/SW	
MIDWEST RADIO	R	1925	BAT	TABLE LP 1 TUBE 1 DIAL	150
MIDWEST RADIO	R- 3	1925	BAT	TABLE LP 3 TUBE 1 DIAL	175
MIDWEST RADIO	RM-16	1933	AC	CONSOLE REMOTE CONTROL	
MIDWEST RADIO	SS- 7	1935	AC	CONSOLE BC/SW/LW	
MIDWEST RADIO	ULTRA 5	1924	BAT	TABLE LP 5 TUBE 3 DIAL	125
MINERVA RADIO	1	1929	AC	CONSOLE	
MINERVA RADIO	2	1929	AC	CONSOLE	
MINERVA RADIO	3	1929	AC	CONSOLE	
MINERVA RADIO	4	1929	AC	CONSOLE	
MISSION BELL	MANTLE	1930	AC	CATHEDRAL MODERN	140
MITCHELL	LULLABY BEDLAMP	1939	AC	MANTLE SIZE BEDLAMP	45
MOGUL ELECTRIC	DEXAPHONE	1922	BAT	MANTLE 1 TUBE 2 DIAL	225
MOHAWK	100	1923	BAT	TABLE HP 5 TUBE 1 DIAL	85
MOHAWK	105	1925	BAT	PORTABLE 6 TUBE 3 DIAL	
MOHAWK	A- 5	1925	BAT	TABLE 5 TUBE 1 DIAL	200
MOHAWK	CHEROKEE	1926	BAT	TABLE	
MOHAWK	CHIPPEWA	1926	BAT	CONSOLE	
MOHAWK	CORTEZ	1927	AC	CONSOLE	
MOHAWK	GENEVA	1926	BAT	CONSOLE	
MOHAWK	HIAWATHA	1927	AC	CONSOLE	
MOHAWK	IROQUOIS	1927	AC	CONSOLE HB DOORS 6 T 1 DIAL	140
MOHAWK	NAVAHO	1927	AC	TABLE HP 6 TUBE 1 DIAL	95
MOHAWK	PAWNEE	1927	AC	CONSOLE	

MANUFACTURER	MODEL	YEAR	POWER	STYLE	PRICE
MOHAWK	PONTIAC	1926	BAT	CONSOLE	
MOHAWK	PORTABLE	1926	BAT	PORTABLE CASE 6 TUBE 2 DL	195
MOHAWK	SEMINOLE	1927	AC	CONSOLE	
MOHAWK	WINONA	1926	BAT	TABLE	
MONTGOMERY WARD	13	1932	AC	CATHEDRAL	55
MONTGOMERY WARD	19	1932	AC	CONSOLE	50
MONTGOMERY WARD	21	1932	AC	CONSOLE	50
MONTGOMERY WARD	23	1932	BAT	CONSOLE	50
MONTGOMERY WARD	25	1932	AC	CATHEDRAL	80
MONTGOMERY WARD	39	1932	BAT	CONSOLE W/KROMATIC DIAL	65
MONTGOMERY WARD	41	1932	BAT	CONSOLE W/KROMATIC DIAL	70
MONTGOMERY WARD	43	1932	AC	CATHEDRAL BC/SW	130
MONTGOMERY WARD	45	1932	AC	CONSOLE LB W/KROMATIC DIAL	45
MONTGOMERY WARD	47	1932	AC	CATHEDRAL	90
MONTGOMERY WARD	49	1932	AC	CONSOLE LB W/KROMATIC DIAL	80
MONTGOMERY WARD	51	1932	AC	CONSOLE LB W/KROMATIC DIAL	85
MONTGOMERY WARD	53	1933	AC	CONSOLE	60
MONTGOMERY WARD	57	1933	AC	TABLE	35
MONTGOMERY WARD	67	1933	AC	CATHEDRAL	80
MONTGOMERY WARD	75	1933	BAT	CONSOLE	50
MONTGOMERY WARD	77	1933	BAT	CATHEDRAL	70
MONTGOMERY WARD	79	1933	AC	CONSOLE	60
MONTGOMERY WARD	81	1933	AC	CATHEDRAL	95
MONTGOMERY WARD	89	1933	AC	CONSOLE LOG-O-MATIC DL	90
MONTGOMERY WARD	91	1933	BAT	CONSOLE LOG-O-MATIC DL	75
MONTGOMERY WARD	93	1933	32/DC	CONSOLE LOG-O-MATIC DIAL	65
MONTGOMERY WARD	95	1933	BAT	CONSOLE	65
MONTGOMERY WARD	97	1933	AC	CATHEDRAL	60
MONTGOMERY WARD	98	1933	AC	TABLE W/BRONZE HANDLE	40
MONTGOMERY WARD	99	1933	AC	CONSOLE LB LOG-O-MATIC DL	60
MONTGOMERY WARD	101	1933	AC	CONSOLE LOG-O-MATIC DL	80
MONTGOMERY WARD	103	1934	AC	CATHEDRAL	50
MONTGOMERY WARD	105	1934	AC	CONSOLE	60
MONTGOMERY WARD	114	1934	6/DC	CATHEDRAL BC/SW	40
MONTGOMERY WARD	115	1934	AC	TABLE	40
MONTGOMERY WARD	116	1934	6/DC	CONSOLE BC/SW	55
MONTGOMERY WARD	117	1934	AC	TABLE	40
MONTGOMERY WARD	119	1934	BAT	CONSOLE	50
MONTGOMERY WARD	120	1934	2/DC	TOMBSTONE BC/SW	35
MONTGOMERY WARD	121	1934	AC	CONSOLE	50
MONTGOMERY WARD	122	1934	2/DC	CONSOLE BC/SW	45
MONTGOMERY WARD	123	1928	BAT	TABLE LOW PROFILE	
MONTGOMERY WARD	129	1934	BAT	CONSOLE BC/SW	55
MONTGOMERY WARD	131	1934	AC	CATHEDRAL BC/SW	40
MONTGOMERY WARD	133	1934	AC	CONSOLE BC/SW	50
MONTGOMERY WARD	135	1934	AC	TOMBSTONE BC/SW	40
MONTGOMERY WARD	136	1934	32/DC	CONSOLE BC/SW	55

MANUFACTURER	MODEL	YEAR	POWER	STYLE	PRICE
MONTGOMERY WARD	137	1934	AC	CONSOLE BC/SW	35
MONTGOMERY WARD	138	1934	32/DC	CATHEDRAL BC/SW	45
MONTGOMERY WARD	139	1934	AC	CONSOLE BC/SW	60
MONTGOMERY WARD	141	1934	2/DC	TOMBSTONE BC/SW	
MONTGOMERY WARD	143	1934	2/DC	CONSOLE BC/SW	
MONTGOMERY WARD	145	1934	BAT	CONSOLE BC/SW	
MONTGOMERY WARD	147	1935	AC	TOMBSTONE BC/SW	35
MONTGOMERY WARD	148	1934	AC	TOMBSTONE	40
MONTGOMERY WARD	149	1935	BAT	CATHEDRAL BC/SW	35
MONTGOMERY WARD	151	1935	32/DC	CATHEDRAL BC/SW	45
MONTGOMERY WARD	152	1935	AC	CONSOLE BC/SW	60
MONTGOMERY WARD	153	1935	BAT	CONSOLE BC/SW	55
MONTGOMERY WARD	154	1935	AC	CONSOLE BC/SW	50
MONTGOMERY WARD	155	1935	BAT	CONSOLE	50
MONTGOMERY WARD	159	1935	BAT	CATHEDRAL BC/SW	35
MONTGOMERY WARD	161	1935	BAT	CONSOLE	50
MONTGOMERY WARD	163	1935	BAT	CONSOLE BC/SW	55
MONTGOMERY WARD	165	1935	AC	CONSOLE BC/SW	50
MONTGOMERY WARD	167	1935	AC	CONSOLE BC/SW	60
MONTGOMERY WARD	169	1936	2/DC	TABLE TOMBSTONE	25
MONTGOMERY WARD	171	1936	BAT	TABLE TOMBSTONE	25
MONTGOMERY WARD	175	1936	AC	CONSOLE BC/SW	50
MONTGOMERY WARD	177	1936	AC	TOMBSTONE BC/SW	70
MONTGOMERY WARD	179	1936	AC	CONSOLE BC/SW	
MONTGOMERY WARD	181	1936	2/DC	CONSOLE BC/SW	
MONTGOMERY WARD	183	1936	BAT	CONSOLE BC/SW	
MONTGOMERY WARD	185	1936	AC	TOMBSTONE BC/SW	40
MONTGOMERY WARD	187	1936	AC	CONSOLE BC/SW	55
MONTGOMERY WARD	203	1936	BAT	CONSOLE BC/SW	45
MONTGOMERY WARD	205	1936	BAT	CONSOLE BC/SW	45
MONTGOMERY WARD	207	1936	BAT	CONSOLE BC/SW	45
MONTGOMERY WARD	209	1936	BAT	CONSOLE BC/SW	45
MONTGOMERY WARD	211	1936	2/DC	TOMBSTONE BC/SW	30
MONTGOMERY WARD	213	1936	BAT	TOMBSTONE BC/SW	30
MONTGOMERY WARD	217	1936	BAT	TABLE BC/SW	35
MONTGOMERY WARD	219	1936	BAT	TABLE BC/SW	35
MONTGOMERY WARD	222	1927	BAT	TABLE LOW PROFILE 2 DL 6	
MONTGOMERY WARD	226	1936	AC	TABLE MOVIE DIAL	60
MONTGOMERY WARD	229	1936	32/DC	TABLE CATHEDRAL MOVIE DL	35
MONTGOMERY WARD	230	1936	2/DC	TABLE BC/SW	25
MONTGOMERY WARD	233	1936	AC	TABLE	25
MONTGOMERY WARD	235	1936	AC	TABLE BC/SW	30
MONTGOMERY WARD	240	1936	BAT	TABLE BC/SW	25
MONTGOMERY WARD	249	1936	AC	TABLE BC/SW	30
MONTGOMERY WARD	251	1936	BAT	TABLE BC/SW	30
MONTGOMERY WARD	253	1936	BAT	TABLE BC/SW	30
MONTGOMERY WARD	254	1937	BAT	TABLE	20
MONTGOMERY WARD	256	1937	AC	TABLE	30
MONTGOMERY WARD	267	1937	AC	CONSOLE MOVIE DIAL	60
MONTGOMERY WARD	271	1937	AC	CONSOLE MOVIE DIAL	75

MANUFACTURER	MODEL	YEAR	POWER	STYLE	PRICE
MONTGOMERY WARD	277	1937	AC	TABLE MOVIE DIAL	50
MONTGOMERY WARD	288	1939	AC	TABLE BROWN PLASTIC PB	15
MONTGOMERY WARD	290	1939	AC	TABLE IVORY PLASTIC PB	20
MONTGOMERY WARD	292	1938	2/DC	CONSOLE	45
MONTGOMERY WARD	293	1937	AC	CONSOLE MOVIE DIAL	85
MONTGOMERY WARD	296	1938	AC	CONSOLE ARMCHAIR BC/SW	45
MONTGOMERY WARD	297	1937	AC	TABLE PB BC.SW	35
MONTGOMERY WARD	320	1939	AC	MANTLE IVORY PLASTIC	20
MONTGOMERY WARD	325	1939	AC	MANTLE BROWN PLASTIC	15
MONTGOMERY WARD	333	1927	BAT	CONSOLE 3 DL W/SPEAKER	
MONTGOMERY WARD	350	1938	AC	TABLE	15
MONTGOMERY WARD	351	1939	AC	TABLE BROWN PLASTIC PB	15
MONTGOMERY WARD	352	1939	AC	TABLE IVORY PLASTIC PB	20
MONTGOMERY WARD	354	1939	2/DC	TABLE	25
MONTGOMERY WARD	361	1939	AC	TABLE	30
MONTGOMERY WARD	362	1939	AC	TABLE RADIO/PHONO PB BC/SW	15
MONTGOMERY WARD	363	1927	BAT	TABLE LP ENCHANTER 5	25
MONTGOMERY WARD	369	1927	BAT	TABLE LOW PROFILE 2 DL	
MONTGOMERY WARD	369	1927	BAT	TABLE LP 3 DL W/SPEAKER	
MONTGOMERY WARD	370	1939	AC	TABLE	30
MONTGOMERY WARD	390	1939	AC	TABLE	30
MONTGOMERY WARD	401	1939	AC	CONSOLE	65
MONTGOMERY WARD	403	1939	AC	CONSOLE MOVIE DIAL	85
MONTGOMERY WARD	404	1937	2/DC	TABLE	20
MONTGOMERY WARD	405	1938	2/DC	TABLE BC/SW	20
MONTGOMERY WARD	406	1937	AC	TABLE PB BC/SW	30
MONTGOMERY WARD	407	1936	AC	CONSOLE BC/SW	50
MONTGOMERY WARD	408	1936	AC	CONSOLE MOVIE DIAL BC/SW	75
MONTGOMERY WARD	409	1937	AC	CONSOLE PB TUN EYE BC/SW	60
MONTGOMERY WARD	410	1936	2/DC	CONSOLE MOVIE DIAL BC/SW	
MONTGOMERY WARD	411	1936	AC	CONSOLE MOVIE DIAL BC/SW	
MONTGOMERY WARD	412	1936	2/DC	CONSOLE MOVIE DIAL BC/SW	60
MONTGOMERY WARD	413	1936	AC	CONSOLE MOVIE DIAL BC/SW	
MONTGOMERY WARD	414	1938	BAT	TABLE BC/SW	20
MONTGOMERY WARD	415	1936	AC	TABLE	25
MONTGOMERY WARD	416	1936	AC	TABLE	30
MONTGOMERY WARD	417	1937	AC	CONSOLE PB BC/SW	75
MONTGOMERY WARD	418	1936	AC	TABLE MOVIE DIAL BC/SW	75
MONTGOMERY WARD	419	1937	AC	MANTLE TUNING EYE BC/SW	30
MONTGOMERY WARD	420	1940	AC	MANTLE WALNUT PLASTIC PB	20
MONTGOMERY WARD	421	1940	AC	MANTLE PLASTIC IVORY	20
MONTGOMERY WARD	423	1940	AC	TABLE CORD RED	
MONTGOMERY WARD	424	1940	AC	TABLE TURQUOISE BLUE	
MONTGOMERY WARD	425	1936	AC	TABLE	25
MONTGOMERY WARD	427	1936	6/DC	CONSOLE MOVIE DIAL BC/SW	55
MONTGOMERY WARD	428	1936	BAT	CONSOLE BC/SW	45
MONTGOMERY WARD	429	1936	BAT	CONSOLE BC/SW	45
MONTGOMERY WARD	430	1936	BAT	CONSOLE BC/SW	45
MONTGOMERY WARD	431	1940	AC	TABLE JADE GREEN	

MANUFACTURER	MODEL	YEAR	POWER	STYLE	PRICE
MONTGOMERY WARD	432	1936	32/DC	CONSOLE MOVIE DIAL BC/SW	70
MONTGOMERY WARD	433	1937	AC	CONSOLE MOVIE DIAL	75
MONTGOMERY WARD	434	1939	BAT	TABLE PB	20
MONTGOMERY WARD	435	1939	BAT	TABLE PB	20
MONTGOMERY WARD	437	1936	6/DC	TABLE CATHEDRAL MOVIE DL	50
MONTGOMERY WARD	439	1936	BAT	TABLE BC/SW	25
MONTGOMERY WARD	441	1937	32/DC	CONSOLE MOVIE DIAL BC/SW	60
MONTGOMERY WARD	442	1937	32/DC	CONSOLE BC/SW	50
MONTGOMERY WARD	443	1938	AC	CONSOLE MOVIE DIAL BC/SW	80
MONTGOMERY WARD	445	1937	AC	MANTLE IVORY PLASTIC PB	
MONTGOMERY WARD	446	1938	2/DC	CONSOLE	45
MONTGOMERY WARD	448	1938	AC	CONSOLE BC/SW	50
MONTGOMERY WARD	449	1937	AC	CONSOLE MOVIE DIAL BC/SW	80
MONTGOMERY WARD	451	1937	AC	CONSOLE MOVIE DIAL BC/SW	85
MONTGOMERY WARD	452	1937	32/DC	TABLE TELEPHONE DIAL BC/SW	35
MONTGOMERY WARD	455	1937	AC	MANTLE WALNUT PLASTIC PB	20
MONTGOMERY WARD	458	1938	AC	TABLE BC/SW	30
MONTGOMERY WARD	465	1937	BAT	MANTLE TEL DIAL TE BC/SW	15
MONTGOMERY WARD	466	1937	BAT	CONSOLE TEL DIAL TE BC/SW	45
MONTGOMERY WARD	470	1939	AC	CONSOLE	50
MONTGOMERY WARD	471	1938	AC	CONSOLE TUNING EYE BC/SW	50
MONTGOMERY WARD	472	1938	AC	TABLE BC/SW	30
MONTGOMERY WARD	475	1937	BAT	MANTLE PB	20
MONTGOMERY WARD	476	1937	BAT	TABLE TUNING EYE PB	20
MONTGOMERY WARD	486	1937	BAT	TABLE MOVIE DIAL BC/SW	25
MONTGOMERY WARD	490	1939	AC	CONSOLE	60
MONTGOMERY WARD	497	1937	AC	CONSOLE PB BC/SW	55
MONTGOMERY WARD	500	1938	AC	TABLE RADIO/PHONO BC/SW	15
MONTGOMERY WARD	501	1939	AC	MNATLE BROWN PLASTIC PB	15
MONTGOMERY WARD	502	1939	AC	MANTLE IVORY PLASTIC PB	20
MONTGOMERY WARD	508	1940	AC	MANTLE BROWN PLASTIC PB	15
MONTGOMERY WARD	509	1940	AC	MANTLE IVORY PLASTIC PB	20
MONTGOMERY WARD	510	1940	AC	TABLE RADIO/PHONO	
MONTGOMERY WARD	511	1941	AC	MANTLE BROWN PLASTIC	20
MONTGOMERY WARD	512	1941	AC	MANTLE IVORY PLASTIC	20
MONTGOMERY WARD	513	1941	AC	MANTLE BROWN PLASTIC PB	15
MONTGOMERY WARD	514	1941	AC	MANTLE IVORY PLASTIC PB	15
MONTGOMERY WARD	515	1941	AC	TABLE RADIO/PHONO BC/SW	15
MONTGOMERY WARD	552	1939	BAT	TABLE PB	20
MONTGOMERY WARD	553	1939	BAT	TABLE	20
MONTGOMERY WARD	570	1941	6/DC	TABLE PB	30
MONTGOMERY WARD	600	1939	AC	TABLE PB BC/SW	30
MONTGOMERY WARD	601	1939	AC	TABLE PB	30
MONTGOMERY WARD	602	1940	AC	TABLE BROWN PLASTIC PB	20

MANUFACTURER	MODEL	YEAR	POWER	STYLE	PRICE
MONTGOMERY WARD	603	1940	AC	TABLE IVORY PLASTIC PB	20
MONTGOMERY WARD	604	1940	AC	TABLE PB	20
MONTGOMERY WARD	605	1940	AC	TABLE PB	20
MONTGOMERY WARD	606	1937	AC	TABLE WALNUT PLASTIC TE PB	20
MONTGOMERY WARD	607	1937	2/DC	CONSOLE PB BC/SW	45
MONTGOMERY WARD	610	1941	AC	MANTLE IVORY PLASTIC PB	20
MONTGOMERY WARD	611	1941	AC	MANTLE IVORY PLASTIC PB	20
MONTGOMERY WARD	612	1941	AC	TABLE PB BC/SW	30
MONTGOMERY WARD	614	1941	AC	TABLE RADIO/PHONO	15
MONTGOMERY WARD	615	1941	AC	TABLE R/P W/RECORDER	20
MONTGOMERY WARD	616	1937	AC	TABLE IVORY PLASTIC TE PB	
MONTGOMERY WARD	617	1937	2/DC	MANTLE PB BC/SW	25
MONTGOMERY WARD	619	1941	AC	CONSOLE BC/SW	55
MONTGOMERY WARD	620	1941	AC	CONSOLE RADIO/PHONO	40
MONTGOMERY WARD	651	1939	BAT	TABLE PB BC/SW	30
MONTGOMERY WARD	652	1939	BAT	CONSOLE	40
MONTGOMERY WARD	675	1941	6/DC	TABLE PB BC/SW	30
MONTGOMERY WARD	676	1941	6/DC	CONSOLE PB BC/SW	60
MONTGOMERY WARD	701	1939	AC	CONSOLE BC/SW	45
MONTGOMERY WARD	702	1939	AC	TABLE BC/SW	25
MONTGOMERY WARD	704	1939	AC	TABLE PB BC/SW	30
MONTGOMERY WARD	705	1939	AC	CONSOLE PB BC/SW	50
MONTGOMERY WARD	709	1939	AC	CONSOLE EARLY AMERICAN PB	50
MONTGOMERY WARD	711	1939	AC	CONSOLE SWEDISH MDRN CHEST	50
MONTGOMERY WARD	712	1939	AC	CONSOLE HB QUEEN ANNE PB	80
MONTGOMERY WARD	714	1939	AC	CONSOLE	50
MONTGOMERY WARD	715	1939	AC	TABLE	35
MONTGOMERY WARD	717	1940	AC	TABLE RADIO/PHONO PB BC/SW	15
MONTGOMERY WARD	720	1940	AC	TABLE PB BC/SW	35
MONTGOMERY WARD	722	1940	AC	TABLE PB AC/SW	
MONTGOMERY WARD	725	1940	AC	CONSOLE PB BC/SW	55
MONTGOMERY WARD	729	1941	AC	TABLE HIGH PRO PB BC/SW	35
MONTGOMERY WARD	730	1941	AC	CONSOLE HIGH PRO PB BC/SW	55
MONTGOMERY WARD	731	1941	AC	TABLE	25
MONTGOMERY WARD	732	1941	AC	CONSOLE RADIO/PHONO	40
MONTGOMERY WARD	750	1939	BAT	TABLE	20
MONTGOMERY WARD	751	1939	BAT	CONSOLE	40
MONTGOMERY WARD	752	1939	32/DC	TABLE	30
MONTGOMERY WARD	753	1939	32/DC	CONSOLE	45
MONTGOMERY WARD	754	1941	6/DC	TABLE PB	30
MONTGOMERY WARD	755	1941	32/DC	CONSOLE	50
MONTGOMERY WARD	800	1940	AC	TABLE PB BC/SW	30
MONTGOMERY WARD	801	1940	AC	CONSOLE PB BC/SW	115
MONTGOMERY WARD	803	1941	AC	TABLE PB BC/SW	40
MONTGOMERY WARD	805	1940	AC	CONSOLE PB BC/SW	40
MONTGOMERY WARD	902	1939	AC	CONSOLE R/P PB T EYE BC/SW	40
MONTGOMERY WARD	903	1927	BAT	CONSOLE CADENZA 7	70

MANUFACTURER	MODEL	YEAR	POWER	STYLE	PRICE
MONTGOMERY WARD	903	1941	AC	CONSOLE PB BC/SW	85
MONTGOMERY WARD	904	1941	AC	CONSOLE RADIO/PHONO BC/SW	40
MONTGOMERY WARD	905	1939	AC	CONSOLE PB TUN EYE BC/SW	60
MONTGOMERY WARD	906	1941	AC	CONSOLE RADIO/PHONO BC/SW	45
MONTGOMERY WARD	907	1941	AC	CONSOLE PB BC/SW	85
MONTGOMERY WARD	908	1941	AC	CONSOLE RADIO/PHONO BC/SW	40
MONTGOMERY WARD	910	1939	AC	CONSOLE 18TH CENT. CHEST	
MONTGOMERY WARD	923	1927	BAT	TABLE LOW PROFILE 2 DL	55
MONTGOMERY WARD	933	1927	BAT	TABLE HIGH PROFILE 7 TUBE	60
MONTGOMERY WARD	943	1927	BAT	CONSOLE DESK MODEL 7	70
MONTGOMERY WARD	983	1927	BAT	CONSOLE TUDOR 7	70
MONTGOMERY WARD	993	1927	BAT	CONSOLE HB 7 TUBE	60
MONTGOMERY WARD	1000	1940	AC	CONSOLE PB BC/SW	85
MONTGOMERY WARD	1001	1940	AC	CONSOLE PB BC/SW	90
MONTGOMERY WARD	1023	1928	BAT	TABLE LOW PROFILE (DC-6)	
MONTGOMERY WARD	1053	1928	BAT	CONSOLE HB (DC-6)	
MONTGOMERY WARD	1073	1928	BAT	CONSOLE SPANISH (DC-6)	
MONTGOMERY WARD	1083	1928	BAT	CONSOLE TUDOR (DC-6)	
MONTGOMERY WARD	1093	1928	BAT	CONSOLE PRESIDENT (DC-6)	
MONTGOMERY WARD	1102	1939	AC	CONSOLE PB TUN EYE BC/SW	55
MONTGOMERY WARD	1104	1940	AC	CONSOLE PB BC/SW	55
MONTGOMERY WARD	1105	1941	AC	CONSOLE	85
MONTGOMERY WARD	1106	1941	AC	CONSOLE RADIO/PHONO BC/SW	45
MONTGOMERY WARD	1107	1941	AC	CONSOLE RADIO/PHONO BC/SW	45
MONTGOMERY WARD	1108	1941	AC	CONSOLE	55
MONTGOMERY WARD	1110	1939	AC	CONSOLE	70
MONTGOMERY WARD	1115	1941	AC	CONSOLE PB BC/SW	
MONTGOMERY WARD	1123	1929	BAT	TABLE LOW PROFILE	
MONTGOMERY WARD	1123	1928	BAT	TABLE LOW PROFILE (DC-7)	
MONTGOMERY WARD	1153	1928	BAT	CONSOLE HB (DC-7)	
MONTGOMERY WARD	1173	1928	BAT	CONSOLE SPANISH (DC-7)	
MONTGOMERY WARD	1183	1928	BAT	CONSOLE TUDOR (DC-7)	
MONTGOMERY WARD	1193	1928	BAT	CONSOLE PRESIDENT (DC-7)	
MONTGOMERY WARD	1323	1929	BAT	TABLE METAL	
MONTGOMERY WARD	1373	1929	BAT	CONSOLE	
MONTGOMERY WARD	1400	1941	AC	CONSOLE	55
MONTGOMERY WARD	1420	1931	AC	CATHEDRAL	45
MONTGOMERY WARD	1422	1929	BAT	TABLE METAL	35
MONTGOMERY WARD	1453	1939	BAT	TABLE	15
MONTGOMERY WARD	1454	1939	BAT	TABLE PB	10
MONTGOMERY WARD	1461	1940	BAT	PORTABLE	15
MONTGOMERY WARD	1464	1941	BAT	PORTABLE	15
MONTGOMERY WARD	1468	1941	BAT	PORTABLE	15
MONTGOMERY WARD	1483	1929	BAT	CONSOLE HB	
MONTGOMERY WARD	1522	1930	BAT	TABLE METAL	35

MANUFACTURER	MODEL	YEAR	POWER	STYLE	PRICE
MONTGOMERY WARD	1562	1930	BAT	CONSOLE LB	50
MONTGOMERY WARD	1568	1941	BAT	TABLE PB	25
MONTGOMERY WARD	1611	1931	AC	CATHEDRAL	80
MONTGOMERY WARD	1623	1930	BAT	TABLE METAL	35
MONTGOMERY WARD	1663	1930	BAT	CONSOLE HB	55
MONTGOMERY WARD	1711	1931	AC	CONSOLE	55
MONTGOMERY WARD	1800	1931	AC	CATHEDRAL	85
MONTGOMERY WARD	1838	1931	AC	CONSOLE	
MONTGOMERY WARD	1955	1931	AC	CONSOLE	65
MONTGOMERY WARD	2023	1928	AC	TABLE LOW PROFILE (AC-7)	
MONTGOMERY WARD	2026	1931	AC	CONSOLE LB	45
MONTGOMERY WARD	2053	1928	AC	CONSOLE HB AC-7)	
MONTGOMERY WARD	2073	1928	AC	CONSOLE SPANISH (AC-7)	
MONTGOMERY WARD	2083	1928	AC	CONSOLE TUDOR (AC-7)	
MONTGOMERY WARD	2093	1928	AC	CONSOLE PRESIDENT (AC-7)	
MONTGOMERY WARD	2123	1928	AC	TABLE LOW PROFILE (AC-8)	
MONTGOMERY WARD	2126	1931	AC	TABLE COLLEGIAN	70
MONTGOMERY WARD	2153	1928	AC	CONSOLE HB (AC-8)	
MONTGOMERY WARD	2173	1928	AC	CONSOLE SPANISH (AC-8)	
MONTGOMERY WARD	2183	1928	AC	CONSOLE TUDOR AC-8)	
MONTGOMERY WARD	2193	1928	AC	CONSOLE PRESIDENT (AC-8)	
MONTGOMERY WARD	2243	1929	BAT	CONSOLE	
MONTGOMERY WARD	2253	1929	AC	CONSOLE	
MONTGOMERY WARD	2423	1929	AC	TABLE METAL	35
MONTGOMERY WARD	2443	1930	AC	CONSOLE W/REMOTE CONTROL	60
MONTGOMERY WARD	2466	1929	AC	CONSOLE HB	65
MONTGOMERY WARD	2493	1929	AC	CONSOLE HB	55
MONTGOMERY WARD	2555	1939	BAT	PORTABLE	15
MONTGOMERY WARD	2557	1939	BAT	TABLE	20
MONTGOMERY WARD	2565	1940	BAT	PORTABLE	15
MONTGOMERY WARD	2568	1941	AC	TABLE PB BC/SW	25
MONTGOMERY WARD	2569	1941	BAT	PORTABLE	15
MONTGOMERY WARD	2623	1929	AC	TABLE	
MONTGOMERY WARD	2643	1929	AC	CONSOLE HP RADIO/PHONO	65
MONTGOMERY WARD	2654	1939	BAT	TABLE PB BC/SW	25
MONTGOMERY WARD	2655	1939	BAT	CONSOLE PB BC/SW	45
MONTGOMERY WARD	2656	1929	AC	CONSOLE HB	
MONTGOMERY WARD	2660	1939	BAT	CONSOLE	40
MONTGOMERY WARD	2663	1940	BAT	PORTABLE	15
MONTGOMERY WARD	2668	1940	BAT	PORTABLE	15
MONTGOMERY WARD	2672	1941	BAT	PORTABLE	15
MONTGOMERY WARD	2673	1941	AC	TABLE PB BC/SW	30
MONTGOMERY WARD	2674	1941	AC	CONSOLE PB BC/SW	55
MONTGOMERY WARD	2776	1930	AC	CONSOLE HB	60
MONTGOMERY WARD	2823	1930	AC	TABLE METAL	35
MONTGOMERY WARD	2895	1930	AC	CONSOLE HB	55
MONTGOMERY WARD	2955	1930	AC	CONSOLE HB	
MONTGOMERY WARD	2956	1930	AC	CONSOLE HB W/CLOCK	70
MONTGOMERY WARD	3035	1930	AC	CONSOLE LB	45

MANUFACTURER	MODEL	YEAR	POWER	STYLE	PRICE
MONTGOMERY WARD	3065	1930	AC	CONSOLE HB	60
MONTGOMERY WARD	3296	1931	AC	CONSOLE	55
MONTGOMERY WARD	3335	1931	AC	CONSOLE	65
MONTGOMERY WARD	6552	1939	BAT	TABLE PB BC/SW	
MONTGOMERY WARD	A	1927	BAT	CONSOLE	65
MONTGOMERY WARD	A	1927	BAT	TABLE	40
MONTGOMERY WARD	AIRLINE	1925	BAT	MANTLE 1 TUBE ULTRA	
MONTGOMERY WARD	AIRLINE	1925	BAT	TABLE 3 DL SUPER	
MONTGOMERY WARD	AIRLINE	1925	BAT	TABLE 3 DL SUPER W/SPEAKER	
MONTGOMERY WARD	AIRLINE	1925	BAT	TABLE BOX 2 TUBE	
MONTGOMERY WARD	AIRLINE	1925	BAT	TABLE HP 3 DL GRAND W/SPKR	
MONTGOMERY WARD	AIRLINE	1925	BAT	TABLE LP 1 TUBE SPECIAL	
MONTGOMERY WARD	AIRLINE	1925	BAT	TABLE LP 3 DL GRAND	
MONTGOMERY WARD	AIRLINE	1925	BAT	CONSOLE 3 TUBE QUEEN ANNE	
MONTGOMERY WARD	B	1927	BAT	CONSOLE	65
MONTGOMERY WARD	B	1927	BAT	TABLE	50
MONTGOMERY WARD	C	1927	BAT	CONSOLE	65
MONTGOMERY WARD	C	1927	BAT	CONSOLE DESK	75
MONTGOMERY WARD	D	1927	BAT	CONSOLE	65
MONTGOMERY WARD	D	1927	BAT	CONSOLE DESK	75
MONTGOMERY WARD	E	1927	BAT	CONSOLE	65
MONTGOMERY WARD	F	1927	BAT	TABLE	50
MONTGOMERY WARD	G	1927	BAT	TABLE	40
MONTROSE RADIO	MONTROSET	1924	BAT	TABLE LP 5 TUBE 3 DIAL	65
MOON RADIO	C-2	1923	BAT	TABLE HP 4 TUBE 2 DIAL	225
MOON RADIO	C-2-A	1923	BAT	TABLE HP 4 TUBE 2 DIAL	
MOON RADIO	C-3-A	1923	BAT	CONSOLE HB 4 TUBE 2 DIAL	
MOON RADIO	TA	1923	BAT	TABLE LP 5 TUBE 1 DIAL	
MORRISON RADIO	III MORRADYNE	1924	BAT	TABLE LP 5 TUBE 3 DIAL	80
MU-RAD RADIO	A	1925	BAT	TABLE LP 5 TUBE 1 DIAL	165
MU-RAD RADIO	B	1925	BAT	TABLE LP 5 TUBE 1 DIAL	140
MU-RAD RADIO	MA-12	1922	BAT	TABLE LP 4 TUBE 2 DIAL	
MU-RAD RADIO	MA-13	1922	BAT	TABLE LP 6 TUBE 2 DIAL	100
MU-RAD RADIO	MA-15	1924	BAT	TABLE LP 6 TUBE 2 DIAL	80
MU-RAD RADIO	MA-17	1923	BAT	TABLE LP 6 TUBE 2 DIAL	
MU-RAD RADIO	MA-18	1924	BAT	TABLE LP 5 TUBE 2 DIAL	95
MU-RAD RADIO	MA-20	1924	AC	TABLE LP 5 TUBE 3 DIAL	160
MU-RAD RADIO	MA-20	1925	AC	CONSOLE 5 TUBE 3 DIAL	
MU-RAD RADIO	SC	1927	BAT	CONSOLE SEMI 6 TUBE 1 DIAL	
MU-RAD RADIO	SE	1927	AC	CONSOLE SEMI 6 TUBE 1 DIAL	
MU-RAD RADIO	SUPER SIX	1926	BAT	TABLE LP 6 TUBE 1 DIAL	95
MU-RAD RADIO	TRIPLEX	1925	BAT	TABLE LP 4 TUBE 2 DIAL	85
MURDOCK RADIO	5 MARVELO	1925	BAT	TABLE LP 5 TUBE 3 DIAL	130
MURDOCK RADIO	7	1927	BAT	TABLE LP METAL 7 TUBE 1 DIAL	
MURDOCK RADIO	60	1927	BAT	TABLE	
MURDOCK RADIO	65	1927	BAT	TABLE 7 TUBE 1 DIAL	
MURDOCK RADIO	70	1927	BAT	TABLE	

MANUFACTURER	MODEL	YEAR	POWER	STYLE	PRICE
MURDOCK RADIO	75	1927	BAT	TABLE 7 TUBE 1 DIAL	
MURDOCK RADIO	100	1924	BAT	TABLE LP 5 TUBE 3 DIAL	
MURDOCK RADIO	101	1925	BAT	TABLE LP 5 TUBE 3 DIAL	
MURDOCK RADIO	110	1925	BAT	TABLE LP 5 TUBE 3 DIAL	
MURDOCK RADIO	125	1927	BAT	CONSOLE	
MURDOCK RADIO	140	1927	BAT	CONSOLE	
MURDOCK RADIO	155	1927	BAT	CONSOLE	
MURDOCK RADIO	175	1927	BAT	CONSOLE	
MURDOCK RADIO	201	1925	BAT	TABLE LP 5 TUBE 3 DIAL	190
MURDOCK RADIO	203	1925	BAT	TABLE LP 6 TUBE 3 DIAL	175
MURDOCK RADIO	204	1925	BAT	CONSOLE HB 6 TUBE 2 DIAL	200
MURDOCK RADIO	CS-32	1923	BAT	TABLE LP 5 TUBE 3 DIAL	75
MURDOCK RADIO	CS-33	1923	BAT	TABLE LP 5 TUBE 3 DIAL	85
MURDOCK RADIO	M-26	1926	BAT	TABLE LP 5 TUBE 3 DIAL	80
MUSIC MASTER	60	1925	BAT	TABLE LP 5 TUBE 3 DIAL	80
MUSIC MASTER	100	1925	BAT	TABLE LP 5 TUBE 3 DIAL	120
MUSIC MASTER	140	1925	BAT	TABLE HP 5 TUBE 1 DIAL	95
MUSIC MASTER	175	1925	BAT	TABLE LP SLOPE PAN. 6T 2DL	175
MUSIC MASTER	300	1925	BAT	CONSOLE LB 5 TUBE 1 DIAL	
MUSIC MASTER	400	1925	AC	CONSOLE LB 5 TUBE 1 DIAL	
MUSIC MASTER	460	1925	BAT	CONSOLE 7 TUBE 1 DIAL	
N					
NASSAU RADIO	LR-70	1923	BAT	TABLE LP 7 TUBE 2 DIAL	195
NASSAU RADIO	MR-60	1925	BAT	TABLE LP MAGNADYNE 6T 3 DIAL	95
NASSAU RADIO	VR-215	1925	BAT	TABLE LP MAGNADYNE 6T 3 DIAL	
NASSAU RADIO	VR-400	1925	BAT	TABLE LP MAGNADYNE 5T 3 DIAL	
NATIONAL AIRPHONE	MONODYNE	1923	BAT	MANTLE 1 TUBE 1 DIAL	
NATIONAL CARBON	1	1927	AC	TABLE 8 TUBE 1 DIAL	180
NATIONAL CARBON	2	1928	AC	TABLE 8 TUBE 1 DIAL	155
NATIONAL CARBON	3	1928	AC	TABLE 8 TUBE 1 DIAL	
NATIONAL CARBON	11	1928	AC	CONSOLE 8 TUBE 1 DIAL	
NATIONAL CARBON	20	1928	BAT	TABLE 6 TUBE 1 DIAL	65
NATIONAL CARBON	21	1928	BAT	TABLE 6 TUBE 1 DIAL	80
NATIONAL CARBON	31	1929	AC	TABLE	
NATIONAL CARBON	32	1929	AC	CONSOLE HB	145
NATIONAL CARBON	33	1929	AC	CONSOLE	
NATIONAL CARBON	34	1929	AC	CONSOLE	
NATIONAL CARBON	42	1929	AC	CONSOLE	
NATIONAL CARBON	43	1929	AC	CONSOLE	
NATIONAL CARBON	44	1929	AC	CONSOLE	
NATIONAL CARBON	52	1929	AC	CONSOLE	
NATIONAL CARBON	53	1929	AC	CONSOLE	
NATIONAL CARBON	54	1929	AC	CONSOLE	130
NATIONAL RADIO	SG-5	1929	AC	TABLE METAL	
NATIONAL RADIO	SW-3	1931	AC	TABLE METAL	
NATIONAL RADIO	SW-4	1929	AC	TABLE METAL	

MANUFACTURER	MODEL	YEAR	POWER	STYLE	PRICE
NATIONAL RADIO	SW-5	1929	AC	TABLE METAL	
NEUTROWOUND RADIO				See ADVANCE AUTO. ACC.	
NOBLITT-SPARKS	40-A	1939	AC	MANTLE STEEL	35
NOBLITT-SPARKS	41	1935	AC	TABLE	
NOBLITT-SPARKS	51	1935	AC	TABLE BC/SW	
NOBLITT-SPARKS	58	1938	AC	MANTLE BC/SW	20
NOBLITT-SPARKS	58A	1939	AC	MANTLE PLASTIC	25
NOBLITT-SPARKS	61	1935	AC	TABLE BC/SW	
NOBLITT-SPARKS	62	1935	AC	CONSOLE BC/SW	
NOBLITT-SPARKS	68	1938	AC	TABLE PLASTIC PB	20
NOBLITT-SPARKS	71	1939	AC	TABLE PLASTIC PB	
NOBLITT-SPARKS	71A	1939	AC	TABLE PLASTIC PB	
NOBLITT-SPARKS	78	1938	AC	TABLE PB BC/SW	30
NOBLITT-SPARKS	81	1935	AC	CONSOLE BC/SW	
NOBLITT-SPARKS	88	1938	AC	TABLE RADIO/PHONO BC/SW	20
NOBLITT-SPARKS	89	1939	AC	TABLE PB BC/SW	
NOBLITT-SPARKS	91	1939	AC	TABLE PB BC/SW	
NOBLITT-SPARKS	92	1939	AC	TABLE PB BC/SW	
NOBLITT-SPARKS	302-A	1940	AC	TABLE PLASTIC RADIO/PHONO	20
NOBLITT-SPARKS	402-A	1939	AC	MANTLE	35
NOBLITT-SPARKS	407	1936	AC	TABLE BC/SW	
NOBLITT-SPARKS	417	1936	AC	TOMBSTONE BC/SW	45
NOBLITT-SPARKS	422	1941	AC	MANTLE PLASTIC BROWN	25
NOBLITT-SPARKS	422 -A	1941	AC	MANTLE PLASTIC	35
NOBLITT-SPARKS	467	1936	AC	TABLE BC/SW	45
NOBLITT-SPARKS	502	1939	AC	MANTLE STEEL	30
NOBLITT-SPARKS	507	1936	AC	TABLE BC/SW	
NOBLITT-SPARKS	508	1937	AC	TABLE BC/SW	
NOBLITT-SPARKS	517	1936	AC	TOMBSTONE BC/SW	45
NOBLITT-SPARKS	517-B	1936	6/DC	TOMBSTONE BC/SW	45
NOBLITT-SPARKS	518	1937	AC	TABLE BC/SW	30
NOBLITT-SPARKS	518B	1938	AC	CONSOLE BC/SW	
NOBLITT-SPARKS	522 -A	1941	AC	MANTLE PLASTIC	30
NOBLITT-SPARKS	527	1936	AC	CONSOLE BC/SW	125
NOBLITT-SPARKS	527-B	1936	6/DC	CONSOLE BC/SW	125
NOBLITT-SPARKS	528CS	1938	AC	CONSOLE BC/SW CHAIRSIDE	
NOBLITT-SPARKS	532 -A	1941	AC	MANTLE CATALIN	25
NOBLITT-SPARKS	568	1937	AC	TABLE BC/SW	30
NOBLITT-SPARKS	578-B	1937	2/DC	TABLE	
NOBLITT-SPARKS	602	1939	AC	MANTLE PLASTIC	30
NOBLITT-SPARKS	608	1937	AC	TABLE BC/SW	
NOBLITT-SPARKS	616-A	1941	AC	MANTLE PLASTIC	25
NOBLITT-SPARKS	617	1936	AC	TOMBSTONE BC/SW	50
NOBLITT-SPARKS	617-B	1936	6/DC	TOMBSTONE BC/SW	50
NOBLITT-SPARKS	618	1937	AC	TABLE BC/SW	45
NOBLITT-SPARKS	618-B	1937	2/DC	TABLE	30
NOBLITT-SPARKS	622-A	1941	AC	TABLE PLASTIC	25
NOBLITT-SPARKS	627	1936	AC	CONSOLE BC/SW	140
NOBLITT-SPARKS	627-B	1936	6/DC	CONSOLE BC/SW	140

MANUFACTURER	MODEL	YEAR	POWER	STYLE	PRICE
NOBLITT-SPARKS	628	1937	AC	CONSOLE BC/SW	
NOBLITT-SPARKS	628-B	1937	2/DC	CONSOLE	
NOBLITT-SPARKS	628-CS	1937	AC	CONSOLE CHAIRSIDE BC/SW	150
NOBLITT-SPARKS	632	1941	AC	TABLE	30
NOBLITT-SPARKS	638	1938	AC	CONSOLE BC/SW	
NOBLITT-SPARKS	638CS	1938	AC	CONSOLE BC/SW CHAIRSIDE	
NOBLITT-SPARKS	722, -A	1941	AC	TABLE PLASTIC	25
NOBLITT-SPARKS	732	1941	AC	TABLE	40
NOBLITT-SPARKS	802	1939	AC	PORTABLE CLOTH	15
NOBLITT-SPARKS	818	1937	AC	TABLE BC/SW	
NOBLITT-SPARKS	828	1937	AC	CONSOLE BC/SW	
NOBLITT-SPARKS	838AT	1938	AC	CONSOLE BC/SW	
NOBLITT-SPARKS	838CS	1938	AC	CONSOLE BC/SW CHAIRSIDE	
NOBLITT-SPARKS	848CS	1938	AC	CONSOLE RP BC/SW CHAIRSIDE	
NOBLITT-SPARKS	927	1936	AC	CONSOLE BC/SW	190
NOBLITT-SPARKS	1127	1936	AC	CONSOLE BC/SW	210
NOBLITT-SPARKS	1237	1937	AC	CONSOLE	135
NOBLITT-SPARKS	1247	1937	AC	CONSOLE BC/SW BOOK CASE	120
NOBLITT-SPARKS	1427	1937	AC	CONSOLE BC/SW	
NORBERT RADIO	2	1926	BAT	TABLE LP 2 T (1 MV) 2 DIAL	
NORBERT RADIO	MIDGET	1926	BAT	TABLE 1 MULTIVALVE 2 DIAL	
NORDEN-HAUCK	6 PLIO	1926	BAT	TABLE LP UNIVERSAL 6T 2 DL	195
NORDEN-HAUCK	6 PLIODYNE	1925	BAT	TABLE LP 6 TUBE 2 DIAL	
NORDEN-HAUCK	10 SUPER	1926	BAT	TABLE LP 10 TUBE 2 DIAL	390
NORDEN-HAUCK	10 SUPER	1928	AC	TABLE LP STANDARD 10T 2 DL	325
NORDEN-HAUCK	10 SUPER	1928	AC	TABLE LP ADMIRALTY 10T 2DL	400
NORDEN-HAUCK	C-7	1925	BAT	TABLE LP 7 TUBE 2 DIAL	
NORDEN-HAUCK	C-10 NAVY MODEL	1925	BAT	TABLE LP 10 T 2 DL BC/SW	1200
NORTHWESTERN	55 NORCO	1926	BAT	TABLE LP 5 TUBE 3 DIAL	
NORTHWESTERN	66 NORCO	1927	BAT	TABLE LP 6 TUBE 3 DIAL	
O					
OHIO STAMPING & ENGINEERING	5 NO-DIAL	1925	BAT	TABLE 5 T ROTATE LID DIAL	250
OPERADIO	2	1924	BAT	PORTABLE 6 TUBE 1 DIAL	200
OPERADIO	7	1927	BAT	PORTABLE 7 TUBE 1 DIAL	
OPERADIO	1925	1925	BAT	PORTABLE 6 TUBE 2 DIAL	
OPERADIO	1926	1925	BAT	PORTABLE 7 TUBE 2 DIAL	
OPERADIO	EMPIRE	1925	BAT	CONSOLE	295
OZARKA	S-1 SENIOR	1925	BAT	TABLE LP SLOPE PANEL 3 DL	95
OZARKA	V-6	1932	AC	TABLE	
OZARKA	V-8	1932	AC	CONSOLE	
OZARKA	V-10	1932	AC	CONSOLE	
OZARKA	V-16	1932	AC	CONSOLE	
P					
P & M RADIO	5 PENN-C	1926	BAT	TABLE LP 5 TUBE 3 DIAL	
P & M RADIO	6 PENN-C	1927	BAT	TABLE LP 6 TUBE 1 DIAL	
PACKARD RADIO	5	1925	BAT	TABLE LP 5 TUBE 3 DIAL	80
PACKARD RADIO	8 SUPER	1928	AC	CONSOLE HB DOORS 8 T 1 DL	115

MANUFACTURER	MODEL	YEAR	POWER	STYLE	PRICE
PARAGON ELECTRIC	LINCOLN	1927	AC	CONSOLE	
PARAGON ELECTRIC	MONROE	1927	AC	CONSOLE	
PARAMOUNT	GENERAL	1924	BAT	TABLE LP SLOPE PAN. 6T 3DL	85
PARAMOUNT	STANDARD	1924	BAT	TABLE LP 6 TUBE 3 DIAL	70
PARAMOUNT	STANDARD	1924	BAT	CONSOLE 6 TUBE 3 DIAL	
PATHE	GRAND HB SIX	1925	BAT	CONSOLE HB 6 TUBE 3 DIAL	
PATHE	MINUTE MAN	1923	BAT	TABLE LP 5 TUBE 3 DIAL	75
PATHE	UNIVERSAL FIVE	1925	BAT	TABLE LP SLOPE PAN. 5T 3DL	80
PATHE	UNIVERSAL FIVE	1925	BAT	CONSOLE LB 5 TUBE 3 DIAL	
PATHE	UNIVERSAL SIX	1925	BAT	TABLE LP 6 TUBE 3 DIAL	
PATTERSON RADIO	5-69	1930	AC	CATHEDRAL MODERN	95
PATTERSON RADIO	6-59	1930	AC	TABLE	
PATTERSON RADIO	6-69	1930	AC	CONSOLE LB	
PATTERSON RADIO	7-69	1930	AC	CATHEDRAL MODERN	95
PATTERSON RADIO	7-79	1930	AC	CONSOLE LB	
PEARSON	DELUXE	1924	BAT	TABLE LP 5 TUBE 3 DIAL	70
PEARSON	PARAMOUNT	1924	BAT	CONSOLE LB 5 TUBE 3 DIAL	
PEARSON	PS-5 DELUXE	1925	BAT	TABLE LP 5 TUBE 3 DIAL	
PEARSON	PS-5 STANDARD	1925	BAT	TABLE LP 5 TUBE 3 DIAL	80
PEARSON	PS-5 STANDARD	1925	BAT	CONSOLE 5 TUBE 3 DIAL	
PEARSON	SPECIAL	1925	BAT	TABLE LP 5 TUBE 3 DIAL	65
PEARSON	STANDARD	1924	BAT	TABLE LP 5 TUBE 3 DIAL	65
PENNSYLVANIA WIRELESS	DELUXE FIVE	1924	BAT	TABLE LP 5 TUBE 3 DIAL	70
PENNSYLVANIA WIRELESS	GT	1923	BAT	TABLE LP 1 TUBE 1 DIAL	
PENNSYLVANIA WIRELESS	GT-2	1923	BAT	TABLE LP 3 TUBE 2 DIAL	
PENNSYLVANIA WIRELESS	GT-3	1923	BAT	TABLE LP 4 TUBE 2 DIAL	
PENNSYLVANIA WIRELESS	GT-4	1923	BAT	CONSOLE 3 TUBE 2 DIAL	
PENNSYLVANIA WIRELESS	SPECIAL	1924	BAT	TABLE LP 3 TUBE 2 DIAL	65
PERRY RADIO	PRD-11	1923	BAT	TABLE LP 1 TUBE 2 DIAL	
PFANSTIEHL RADIO	7	1924	BAT	TABLE LP 5 TUBE 3 DIAL	65
PFANSTIEHL RADIO	8	1924	BAT	TABLE LP 5 TUBE 2 DIAL	
PFANSTIEHL RADIO	8-C	1924	BAT	CONSOLE 5 TUBE 2 DIAL	
PFANSTIEHL RADIO	10	1924	BAT	TABLE LP 6 TUBE 1 DIAL	
PFANSTIEHL RADIO	10-C	1924	BAT	CONSOLE 6 TUBE 1 DIAL	
PFANSTIEHL RADIO	10-S	1924	BAT	CONSOLE 6 TUBE 1 DIAL	
PFANSTIEHL RADIO	18	1926	BAT	TABLE LP 5 TUBE 2 DIAL	
PFANSTIEHL RADIO	20	1926	BAT	TABLE LP 6 TUBE 1 DIAL	80
PFANSTIEHL RADIO	28	1927	BAT	TABLE LP 6 TUBE 1 DIAL	
PFANSTIEHL RADIO	30	1927	BAT	TABLE LP 6 TUBE 1 DIAL	125
PFANSTIEHL RADIO	32	1927	BAT	TABLE LP 7 TUBE 1 DIAL	135
PFANSTIEHL RADIO	34	1927	AC	CONSOLE 7 TUBE 1 DIAL	
PFANSTIEHL RADIO	50	1927	AC	TABLE LP 7 TUBE 1 DIAL	95
PFANSTIEHL RADIO	181	1926	BAT	CONSOLE	
PFANSTIEHL RADIO	182	1926	BAT	CONSOLE 5 TUBE 2 DIAL	
PFANSTIEHL RADIO	201	1926	BAT	CONSOLE HB 6 TUBE 1 DIAL	160
PFANSTIEHL RADIO	202	1926	BAT	CONSOLE HB DOORS 6 T 1 DL	180
PFANSTIEHL RADIO	302	1927	BAT	CONSOLE 6 TUBE 1 DIAL	
PFANSTIEHL RADIO	322	1927	BAT	CONSOLE 7 TUBE 1 DIAL	
PHENIX RADIO	L-2 ULTRADYNE	1924	BAT	TABLE LP 6 TUBE 2 DIAL	95
PHENIX RADIO	L-3 ULTRADYNE	1925	BAT	TABLE LP 6 TUBE 2 DIAL	

MANUFACTURER	MODEL	YEAR	POWER	STYLE	PRICE
PHENIX RADIO	ULTRADYNE	1923	BAT	TABLE LP 2 DIAL	95
PHILCO	8-IT	1940	BAT	PORTABLE	15
PHILCO	11	1929	AC	CONSOLE	185
PHILCO	12-RX	1934	AC	CONSOLE REMOTE CONT. BC/SW	
PHILCO	14-B	1933	AC	CATHEDRAL BABY GRAND BC/SW	
PHILCO	14-L	1933	AC	CONSOLE LB BC/SW	
PHILCO	14-MX	1934	AC	CONSOLE	
PHILCO	14-RX	1933	AC	CONSOLE END TABLE R/C B/SW	
PHILCO	14-X	1933	AC	CONSOLE INCLINED BC/SW	
PHILCO	15-X	1932	AC	CONSOLE	95
PHILCO	16-B	1933	AC	TOMBSTONE BABY GRAND BC/SW	225
PHILCO	16-L	1933	AC	CONSOLE LB BC/SW	
PHILCO	16-PX	1934	AC	CONSOLE RADIO/PHONO	
PHILCO	16-RX	1933	AC	CONSOLE END TABLE BC/SW	
PHILCO	16-X	1933	AC	CONSOLE INCLINED BC/SW	
PHILCO	17-B	1933	AC	CATHEDRAL BABY GRAND	
PHILCO	17-D	1933	AC	CONSOLE HB BC/SW	
PHILCO	17-L	1933	AC	CONSOLE LB BC/SW	
PHILCO	17-RX	1934	AC	CONSOLE	
PHILCO	17-X	1933	AC	CONSOLE INCLINED BC/SW	
PHILCO	18-B	1933	AC	CATHEDRAL BABY GRAND BC/SW	225
PHILCO	18-H	1933	AC	CONSOLE HB BC/SW	
PHILCO	18-L	1933	AC	CONSOLE LB BC/SW	
PHILCO	18-MX	1934	AC	CONSOLE	
PHILCO	18-X	1933	AC	CONSOLE INCLINED BC/SW	
PHILCO	19-B	1933	AC	CATHEDRAL BABY GRAND	
PHILCO	19-H	1933	AC	CONSOLE HB	
PHILCO	19-LZ	1933	AC	CONSOLE LAZYBOY BC/SW	
PHILCO	19-TX	1933	AC	PORTABLE	
PHILCO	19-X	1933	AC	CONSOLE	
PHILCO	20	1930	AC	CATHEDRAL	180
PHILCO	22-L	1932	AC	CONSOLE LB RADIO/PHONO	95
PHILCO	23-L	1933	AC	CONSOLE LB BC/SW	
PHILCO	23-X	1932	AC	CONSOLE RADIO/PHONO	
PHILCO	24-L	1932	AC	CONSOLE RADIO/PHONO	
PHILCO	25-L	1933	AC	CONSOLE RADIO/PHONO BC/SW	
PHILCO	26-L	1933	AC	CONSOLE RADIO/PHONO BC/SW	
PHILCO	27-L	1933	AC	CONSOLE LB	
PHILCO	28-C	1934	AC	TABLE ORNATE	90
PHILCO	28-D	1935	AC	CONSOLE	
PHILCO	28-L	1934	AC	CONSOLE LB	
PHILCO	29	1930	AC	CONSOLE RADIO/PHONO	100
PHILCO	29-TX	1934	AC	CONSOLE RC BC/SW	

MANUFACTURER	MODEL	YEAR	POWER	STYLE	PRICE
PHILCO	29-X	1934	AC	CONSOLE BC/SW	
PHILCO	30	1930	BAT	CONSOLE HB	100
PHILCO	30	1930	BAT	CONSOLE LB	100
PHILCO	32-B	1934	32/DC	TABLE BC/SW	
PHILCO	32-F	1935	AC	CONSOLE	
PHILCO	32-L	1934	32/DC	CONSOLE HB BC/SW	
PHILCO	33-B	1936	BAT	CATHEDRAL BABY GRAND	130
PHILCO	33-F	1936	BAT	CONSOLE	
PHILCO	34-B	1934	BAT	CATHEDRAL BABY GRAND	
PHILCO	34-F	1936	2/DC	CONSOLE	
PHILCO	34-L	1934	BAT	CONSOLE BC/SW	
PHILCO	35	1931	BAT	CATHEDRAL BABY GRAND	120
PHILCO	35	1931	BAT	CONSOLE HB	90
PHILCO	35-85 B	1938	DC	TABLE AUTO. TUNING BC/SW	
PHILCO	36-B	1932	BAT	CATHEDRAL BABY GRAND	
PHILCO	36-D	1932	BAT	CONSOLE HB	
PHILCO	36-L	1932	BAT	CONSOLE LB	
PHILCO	37-10	1937	AC	CONSOLE	240
PHILCO	37-33	1937	BAT	CATHEDRAL	75
PHILCO	37-61	1937	AC	CATHEDRAL BC/SW	135
PHILCO	37-84	1936	AC	CATHEDRAL	135
PHILCO	37-116	1937	AC	CONSOLE	220
PHILCO	37-600-C	1936	AC	COMPACT	
PHILCO	37-602-C	1936	AC	COMPACT	
PHILCO	37-604-C	1937	AC	MANTLE W/DIAL ON TOP	60
PHILCO	37-610-B	1935	AC	TOMBSTONE BABY GRAND BC/SW	125
PHILCO	37-610-F	1935	AC	CONSOLE BC/SW	250
PHILCO	37-610-J	1936	AC	CONSOLE BC/SW	290
PHILCO	37-610-T	1936	AC	TABLE BC/SW	110
PHILCO	37-611-B	1935	AC	CATHEDRAL BABY GRAND BC/SW	
PHILCO	37-611-F	1936	AC	CONSOLE BC/SW	
PHILCO	37-611-J	1936	AC	CONSOLE BC/SW	
PHILCO	37-611-T	1936	AC	TABLE BC/SW	
PHILCO	37-620-B	1936	AC	CATHEDRAL BABY GRAND BC/SW	85
PHILCO	37-620-F	1935	AC	CONSOLE BC/SW	
PHILCO	37-620-J	1936	AC	CONSOLE BC/SW	
PHILCO	37-623-B	1936	2/DC	CATHEDRAL BABY GRAND BC/SW	40
PHILCO	37-623-F	1935	2/DC	CONSOLE BC/SW	
PHILCO	37-623-J	1936	BAT	CONSOLE BC/SW	
PHILCO	37-624-B	1936	6/DC	CATHEDRAL BABY GRAND BC/SW	40
PHILCO	37-624-J	1936	BAT	CONSOLE BC/SW	
PHILCO	37-630-B	1935	AC	TABLE BC/SW	
PHILCO	37-630-CSX	1935	AC	CONSOLE CHAIRSIDE BC/SW	
PHILCO	37-630-T	1936	AC	TABLE BC/SW	90
PHILCO	37-630-X	1935	AC	CONSOLE BC/SW	165

MANUFACTURER	MODEL	YEAR	POWER	STYLE	PRICE
PHILCO	37-640-B	1935	AC	CATHEDRAL BABY GRAND BC/SW	175
PHILCO	37-640-MX	1936	AC	CONSOLE BC/SW	
PHILCO	37-640-X	1935	AC	CONSOLE BC/SW	
PHILCO	37-641-B	1935	AC	CATHEDRAL BABY GRAND BC/SW	
PHILCO	37-641-X	1935	AC	CONSOLE BC/SW	
PHILCO	37-643-B	1936	2/DC	CATHEDRAL BABY GRAND BC/SW	
PHILCO	37-643-X	1935	2/DC	CONSOLE BC/SW	
PHILCO	37-650-B	1936	AC	CATHEDRAL BABY GRAND BC/SW	60
PHILCO	37-650-PX	1935	AC	CONSOLE RADIO/PHONO BC/SW	
PHILCO	37-650-X	1935	AC	CONSOLE BC/SW	
PHILCO	37-660-B	1936	AC	CATHEDRAL BABY GRAND BC/SW	
PHILCO	37-660-L	1935	AC	CONSOLE BC/SW	
PHILCO	37-660-X	1936	AC	CONSOLE BC/SW	190
PHILCO	37-665-B	1936	AC	CATHEDRAL BABY GRAND BC/SW	
PHILCO	37-665-X	1936	AC	CONSOLE BC/SW	200
PHILCO	37-670-B	1936	AC	CATHEDRAL BABY GRAND BC/SW	
PHILCO	37-670-X	1936	AC	CONSOLE BC/SW	
PHILCO	37-675-X	1936	AC	CONSOLE BC/SW	
PHILCO	37-690-X	1936	AC	CONSOLE PB BC/SW	300
PHILCO	37-2620-B	1936	AC	CATHEDRAL BABY GRAND BC/SW	
PHILCO	37-2620-J	1936	AC	CONSOLE BC/SW	
PHILCO	37-2650-B	1936	AC	CATHEDRAL BABY GRAND BC/SW	
PHILCO	37-2650-X	1936	AC	CONSOLE BC/SW	
PHILCO	37-2670-B	1936	AC	CATHEDRAL BABY GRAND BC/SW	
PHILCO	37-2670-X	1936	AC	CONSOLE BC/SW	
PHILCO	37-C	1933	AC	COMPACT	
PHILCO	38	1930	BAT	CATHEDRAL	60
PHILCO	38-1-XX	1937	AC	CONSOLE BC/SW	150
PHILCO	38-3-XX	1937	AC	CONSOLE BC/SW	
PHILCO	38-4-XX	1937	AC	CONSOLE BC/SW	
PHILCO	38-5-B	1937	AC	CATHEDRAL BC/SW	
PHILCO	38-5-X	1937	AC	CONSOLE BC/SW	
PHILCO	38-7-CS	1937	AC	CONSOLE CHAIRSIDE BC/SW	
PHILCO	38-7-T	1937	AC	TABLE BC/SW	100
PHILCO	38-7-XX	1937	AC	CONSOLE BC/SW	
PHILCO	38-8-X	1937	AC	CONSOLE BC/SW	
PHILCO	38-9-K	1937	AC	CONSOLE BC/SW	150
PHILCO	38-9-T	1937	AC	TABLE BC/SW	80
PHILCO	38-10-F	1937	AC	CONSOLE BC/SW	

MANUFACTURER	MODEL	YEAR	POWER	STYLE	PRICE
PHILCO	38-10-T	1937	AC	TABLE BC/SW	25
PHILCO	38-12-C,T	1937	AC	TABLE	45
PHILCO	38-14-CS	1937	AC	TABLE BC/SW	25
PHILCO	38-14-T	1937	AC	TABLE BC/SW	30
PHILCO	38-15-CS	1937	AC	TABLE BC/SW	20
PHILCO	38-15-T	1937	AC	TABLE BC/SW	
PHILCO	38-22-CS	1937	AC	CONSOLE CHAIRSIDE BC/SW	
PHILCO	38-22-T	1937	AC	TABLE BC/SW	
PHILCO	38-22-XX	1937	AC	CONSOLE BC/SW	
PHILCO	38-23-K	1937	AC	CONSOLE BC/SW	
PHILCO	38-23-T	1937	AC	TABLE BC/SW	40
PHILCO	38-23-X	1937	AC	CONSOLE BC/SW	
PHILCO	38-33-B	1937	2/DC	TABLE	
PHILCO	38-33-F	1937	2/DC	CONSOLE	
PHILCO	38-34-B	1937	6/DC	CATHEDRAL	
PHILCO	38-34-F	1937	6/DC	CONSOLE	
PHILCO	38-35-B	1937	6/DC	CATHEDRAL	70
PHILCO	38-35-F	1937	6/DC	CONSOLE	
PHILCO	38-38-K	1937	2/DC	CONSOLE BC/SW	
PHILCO	38-38-T	1937	2/DC	TABLE BC/SW	
PHILCO	38-38-X	1937	2/DC	CONSOLE BC/SW	
PHILCO	38-39-B	1937	AC	CATHEDRAL BC/SW	85
PHILCO	38-39-K	1937	6/DC	CONSOLE BC/SW	
PHILCO	38-39-T	1937	6/DC	TABLE BC/SW	
PHILCO	38-39-X	1937	6/DC	CONSOLE BC/SW	
PHILCO	38-40-K	1937	6/DC	CONSOLE BC/SW	
PHILCO	38-40-T	1937	6/DC	TABLE BC/SW	
PHILCO	38-40-X	1937	6/DC	TABLE BC/SW	
PHILCO	38-60-B	1937	AC	CATHEDRAL BC/SW	
PHILCO	38-60-F	1937	AC	CONSOLE BC/SW	
PHILCO	38-62-F	1937	AC	CONSOLE BC/SW	
PHILCO	38-62-T	1937	AC	TABLE BC/SW	30
PHILCO	38-89-K	1937	AC	CONSOLE BC/SW	135
PHILCO	38-93-B	1937	AC	CATHEDRAL	95
PHILCO	38-116-XX	1937	AC	CONSOLE BC/SW	150
PHILCO	38-610-B	1937	AC	CATHEDRAL BC/SW	
PHILCO	38-610-J	1937	AC	CONSOLE BC/SW	
PHILCO	38-620-T	1937	AC	TABLE BC/SW	
PHILCO	38-623-K	1937	2/DC	CONSOLE BC/SW	
PHILCO	38-623-T	1937	2/DC	TABLE BC/SW	
PHILCO	38-624-K	1937	6/DC	CONSOLE BC/SW	
PHILCO	38-624-T	1937	6/DC	TABLE BC/SW	
PHILCO	38-630-K	1937	AC	CONSOLE BC/SW	
PHILCO	38-643-B	1937	2/DC	CATHEDRAL BC/SW	
PHILCO	38-643-X	1937	2/DC	CONSOLE BC/SW	
PHILCO	38-665-B	1937	AC	CATHEDRAL BC/SW	
PHILCO	38-665-X	1937	AC	CONSOLE BC/SW	
PHILCO	38-690-XX	1937	AC	CONSOLE BC/SW	485
PHILCO	38-2620-T	1937	AC	TABLE BC/SW	
PHILCO	38-2630-K	1937	AC	CONSOLE BC/SW	

MANUFACTURER	MODEL	YEAR	POWER	STYLE	PRICE
PHILCO	38-2650-B	1937	AC	CATHEDRAL BC/SW	
PHILCO	38-2650-X	1937	AC	CONSOLE BC/SW	
PHILCO	38-2670-B	1937	AC	CATHEDRAL BC/SW	
PHILCO	38-2670-X	1937	AC	CONSOLE BC/SW	
PHILCO	38-AF	1935	2/DC	CONSOLE	
PHILCO	38-B	1933	2/DC	CATHEDRAL BABY GRAND BC/SW	95
PHILCO	38-F	1935	2/DC	CONSOLE	
PHILCO	38-H	1934	AC	CONSOLE	
PHILCO	38-J	1936	BAT	CONSOLE	
PHILCO	38-L	19 33	BAT	CONSOLE LB BC/SW	
PHILCO	39-12 CB	1938	AC	TABLE	15
PHILCO	39-12 T	1938	AC	TABLE	
PHILCO	39-14 CB	1938	AC	TABLE BC/SW	20
PHILCO	39-15 CB	1938	AC	TABLE BC/SW	15
PHILCO	39-17 F	1938	AC	CONSOLE AUTOMATIC TUNING	
PHILCO	39-17 T	1938	AC	TABLE	
PHILCO	39-18 F	1938	AC	CONSOLE AUTOMATIC TUNING	
PHILCO	39-18 T	1938	AC	TABLE AUTOMATIC TUNING	
PHILCO	39-19 F	1938	AC	CONSOLE AUTO. TUNING BC/SW	
PHILCO	39-19 T	1938	AC	TABLE AUTO. TUNING BC/SW	
PHILCO	39-25 T	1938	AC	TABLE AUTO. TUNING BC/SW	
PHILCO	39-25 XF	1938	AC	CONSOLE AUTO. TUNING BC/SW	
PHILCO	39-30 T	1938	AC	TABLE AUTO/ TUNING BC/SW	45
PHILCO	39-31-XK	1939	AC	CONSOLE PB	120
PHILCO	39-35 XX	1938	AC	CONSOLE BC/SW	
PHILCO	39-40 XX	1938	AC	CONSOLE BC/SW	150
PHILCO	39-45 XX	1938	AC	CONSOLE PB BC/SW	
PHILCO	39-70 B	1938	DC	TABLE FARM	30
PHILCO	39-70 F	1938	DC	CONSOLE FARM	
PHILCO	39-75 F	1938	DC	CONSOLE PB FARM	
PHILCO	39-75 T	1938	DC	TABLE FARM	
PHILCO	39-80 B	1938	DC	TABLE FARM	40
PHILCO	39-80 XF	1938	DC	CONSOLE FARM	
PHILCO	39-85 XF	1938	DC	CONSOLE PB BC/SW	
PHILCO	39-85-T	1939	AC	PORTABLE OSTRICH BC/SW	15
PHILCO	39-89-C	1939	BAT	PORTABLE CANDID CAMERALIKE	40
PHILCO	39-116-RX	1939	AC	CONSOLE PB RC BC/SW	115
PHILCO	39-380-X	1941	AC	CONSOLE PB BC/SW	140
PHILCO	39-B	1935	AC	TABLE	
PHILCO	39-F	1935	AC	CONSOLE	
PHILCO	40-74-T	1939	BAT	PORTABLE CLOTH	
PHILCO	40-81-T	1939	BAT	PORTABLE CLOTH	
PHILCO	40-88-T	1939	BAT	PORTABLE CLOTH BC/SW	
PHILCO	40-90-CB	1939	BAT	TABLE PLASTIC	20

MANUFACTURER	MODEL	YEAR	POWER	STYLE	PRICE
PHILCO	40-95-F	1939	BAT	CONSOLE	
PHILCO	40-95-T	1939	BAT	TABLE	25
PHILCO	40-100-F	1939	BAT	CONSOLE PB	
PHILCO	40-100-T	1939	BAT	TABLE PB	
PHILCO	40-105-B	1939	BAT	TABLE WALNUT	
PHILCO	40-105-K	1939	BAT	CONSOLE	
PHILCO	40-110-B	1939	BAT	TABLE WALNUT PB	
PHILCO	40-110-K	1939	BAT	CONSOLE PB	
PHILCO	40-115-C	1939	AC	TABLE WALNUT	
PHILCO	40-120-C	1939	AC	TABLE	
PHILCO	40-120-GI	1939	AC	TABLE	
PHILCO	40-124-C	1939	AC	TABLE PB	
PHILCO	40-125-C	1939	AC	TABLE PB	30
PHILCO	40-130-T	1939	AC	TABLE	
PHILCO	40-135-T	1939	AC	TABLE PB	
PHILCO	40-140-T	1939	AC	TABLE BC/SW	
PHILCO	40-145-T	1939	AC	TABLE PB BC/SW	
PHILCO	40-150-T	1939	AC	TABLE PB BC/SW	85
PHILCO	40-155-T	1939	AC	TABLE PB BC/SW	130
PHILCO	40-160-F	1939	AC	CONSOLE PB	
PHILCO	40-165-F	1939	AC	CONSOLE PB BC/SW	
PHILCO	40-180-XF	1939	AC	CONSOLE PB BC/SW	150
PHILCO	40-185-XX	1939	AC	CONSOLE PB BC/SW	
PHILCO	40-190-XFJ	1939	AC	CONSOLE PB BC/SW	
PHILCO	40-195-XX	1939	AC	CONSOLE PB BC/SW	
PHILCO	40-200	1939	AC	CONSOLE PB BC/SW	215
PHILCO	40-205-RX	1939	AC	CONSOLE PB	
PHILCO	40-216-RX	1939	AC	CONSOLE PB RC BC/SW	
PHILCO	40-501	1939	AC	TABLE RADIO/PHONO	
PHILCO	40-502	1939	AC	TABLE RADIO/PHONO	
PHILCO	40-503	1939	AC	TABLE RADIO/PHONE PB	
PHILCO	40-504	1939	BAT	PORTABLE RADIO/PHONO	
PHILCO	40-506	1939	AC	CONSOLE RADIO/PHONO PB	
PHILCO	40-507	1939	AC	CONSOLE R/P PB BC/SW	
PHILCO	40-508	1939	AC	CONSOLE R/P PB BC/SW	50
PHILCO	40-509	1939	AC	CONSOLE R/P PB BC/SW	
PHILCO	40-510	1939	AC	CONSOLE RADIO/PHONO PB	
PHILCO	40-516	1939	AC	CONSOLE R/P BC/SW	
PHILCO	40-527	1940	AC	CONSOLE R/P PB BC/SW	50
PHILCO	40-603-P	1941	AC	TABLE RADIO/PHONO	25
PHILCO	41-220	1941	AC	CONSOLE	150
PHILCO	41-221-C	1940	AC	TABLE W/HANDLE BC/SW	30
PHILCO	41-226-C	1940	AC	TABLE PB	30
PHILCO	41-230	1941	AC	TABLE PLASTIC	30
PHILCO	41-235	1941	AC	CONSOLE	140
PHILCO	41-258-F	1940	AC	CONSOLE BC/SW	95
PHILCO	41-260-F	1940	AC	CONSOLE PB BC/SW	95
PHILCO	41-287-X	1941	AC	CONSOLE PB BC/SW	120
PHILCO	41-300-X	1940	AC	CONSOLE PB BC/SW	185
PHILCO	41-601-P	1941	AC	TABLE RADIO/PHONO	15

MANUFACTURER	MODEL	YEAR	POWER	STYLE	PRICE
PHILCO	41-605	1940	AC	CONSOLE RADIO/PHONO	
PHILCO	41-607	1940	AC	CONSOLE RADIO/PHONO	55
PHILCO	41-608	1940	AC	CONSOLE R/P PB BC/SW	40
PHILCO	41-609	1940	AC	CONSOLE RADIO/PHONO	
PHILCO	41-610	1940	AC	CONSOLE RADIO/PHONO	
PHILCO	41-611	1940	AC	CONSOLE RADIO/PHONO	
PHILCO	41-616	1940	AC	CONSOLE RADIO/PHONO	
PHILCO	41-625-P	1941	AC	CONSOLE R/P PB BC/SW	45
PHILCO	41-695-P	1940	2/DC	CONSOLE RADIO/PHONO	30
PHILCO	41-788-T	1940	AC	TABLE HIGH PROFILE	70
PHILCO	41-844-T	1941	AC	PORTABLE WALNUT	30
PHILCO	42-345-T	1941	AC	TABLE PB BC/SW	35
PHILCO	42-350-T	1941	AC	TABLE PB BC/SW/FM	55
PHILCO	42-355-T	1941	AC	TABLE PB BC/SW/FM	55
PHILCO	42-390-X	1941	AC	CONSOLE PB BC/SW	140
PHILCO	42-1013	1941	AC	CONSOLE R/P PB BC/SW/FM	45
PHILCO	43-B	1932	AC	CATHEDRAL BABY GRAND BC/SW	
PHILCO	43-H	1932	AC	CONSOLE HB BC/SW	
PHILCO	43-X	1932	AC	CONSOLE INCLINED BC/SW	
PHILCO	44-B	1934	AC	CATHEDRAL BABY GRAND	200
PHILCO	44-H	1934	AC	CONSOLE	
PHILCO	45-C	1934	AC	TABLE ORNATE BC/SW	90
PHILCO	45-F	1935	AC	CONSOLE	
PHILCO	45-L	1934	AC	CONSOLE LB BC/SW	
PHILCO	46	1931	DC	CATHEDRAL BABY GRAND	175
PHILCO	46	1931	DC	CONSOLE HB	
PHILCO	47-B	1932	DC	CATHEDRAL BABY GRAND	
PHILCO	47-D	1932	DC	CONSOLE HB	
PHILCO	47-H	1933	AC	CONSOLE HB	
PHILCO	47-X	1932	DC	CONSOLE INCLINED	
PHILCO	48-C	1933	AC	COMPACT	
PHILCO	48-L	1933	AC	COMPACT	
PHILCO	49-B	1934	DC	TABLE BC/SW	
PHILCO	49-D	1934	DC	CONSOLE HB BC/SW	
PHILCO	49-H	1934	DC	CONSOLE HB BC/SW	
PHILCO	49-X	1934	DC	CONSOLE BC/SW	
PHILCO	50	1931	BAT	CATHEDRAL BABY GRAND	120
PHILCO	50	1931	AC	CONSOLE LB	80
PHILCO	52-B	1932	AC	CATHEDRAL BABY GRAND	220
PHILCO	52-C	1932	AC	COMPACT	50
PHILCO	52-L	1932	AC	CONSOLE	155
PHILCO	54-C	1933	AC	COMPACT BC/SW	55
PHILCO	54-S	1934	AC	COMPACT BC/SW	
PHILCO	57-C	1933	AC	TABLE BC/SW	60
PHILCO	59-C	1935	AC	COMPACT	50
PHILCO	59-S	1935	AC	COMPACT	
PHILCO	60-B	1933	AC	CATHEDRAL BABY GRAND BC/SW	150
PHILCO	60-F	1935	AC	CONSOLE	

MANUFACTURER	MODEL	YEAR	POWER	STYLE	PRICE
PHILCO	60-L	1933	AC	CONSOLE LB	155
PHILCO	60-MB	1934	AC	TOMBSTONE	80
PHILCO	61-B	1936	AC	CATHEDRAL BABY GRAND BC/SW	
PHILCO	61-F	1936	AC	CONSOLE BC/SW	
PHILCO	65	1929	AC	CONSOLE DELUXE	
PHILCO	65	1929	AC	CONSOLE HB	
PHILCO	65	1929	AC	CONSOLE LB	190
PHILCO	65	1929	AC	TABLE METAL	120
PHILCO	66-B	1934	AC	CATHEDRAL BABY GRAND BC/SW	150
PHILCO	66-L	1934	AC	CONSOLE	
PHILCO	70	1932	AC	CLOCK GRANDFATHER	350
PHILCO	70-A	1931	AC	CATHEDRAL BABY GRAND	175
PHILCO	70-B	1931	BAT	CATHEDRAL BABY GRAND	170
PHILCO	70-L	1931	AC	CONSOLE HB	140
PHILCO	71-B	1932	AC	CATHEDRAL BABY GRAND	225
PHILCO	71-D	1932	AC	CONSOLE HB	
PHILCO	71-H	1932	AC	CONSOLE HB	
PHILCO	71-L	1932	AC	CONSOLE LB	
PHILCO	71-X	1933	AC	CONSOLE	180
PHILCO	76	1930	AC	TABLE	
PHILCO	76	1930	AC	CONSOLE LB	100
PHILCO	77	1930	AC	TABLE	70
PHILCO	77	1930	AC	CONSOLE LB	100
PHILCO	77	1930	AC	CONSOLE HB	
PHILCO	80-B	1933	AC	CATHEDRAL BABY GRAND	150
PHILCO	80-P	1933	AC	PORTABLE	
PHILCO	81-B	1933	AC	TABLE	160
PHILCO	82	1929	AC	CONSOLE HB	
PHILCO	83	1929	AC	CONSOLE HB	
PHILCO	83	1929	AC	CONSOLE LB	
PHILCO	84-B	1934	AC	CATHEDRAL BABY GRAND	165
PHILCO	86	1929	AC	CONSOLE HB	125
PHILCO	87	1929	AC	CONSOLE DELUXE	
PHILCO	87	1929	AC	CONSOLE LB	
PHILCO	87	1929	AC	CONSOLE HB	
PHILCO	89-B	1933	AC	CATHEDRAL BABY GRAND BC/SW	190
PHILCO	89-F	1935	AC	CONSOLE	
PHILCO	89-L	1933	AC	CONSOLE LB BC/SW	
PHILCO	90	1931	AC	CATHEDRAL BABY GRAND	350
PHILCO	90	1931	AC	CONSOLE HB	230
PHILCO	90	1931	AC	CONSOLE LB	150
PHILCO	91-B	1932	AC	CATHEDRAL BABY GRAND	
PHILCO	91-D	1932	AC	CONSOLE HB	
PHILCO	91-H	1933	AC	CONSOLE HB	
PHILCO	91-L	1932	AC	CONSOLE LB	180
PHILCO	91-X	1932	AC	CONSOLE W/RECESSED SPEAKER	250

MANUFACTURER	MODEL	YEAR	POWER	STYLE	PRICE
PHILCO	92	1930	AC	CONSOLE LB	
PHILCO	92	1930	AC	TABLE	
PHILCO	95	1930	AC	CONSOLE LB	
PHILCO	95	1930	AC	TABLE	
PHILCO	95	1930	AC	CONSOLE HB	
PHILCO	96	1930	AC	TABLE	70
PHILCO	96	1930	AC	CONSOLE HB	175
PHILCO	96	1930	AC	CONSOLE HB	
PHILCO	112	1934	AC	CONSOLE	200
PHILCO	112	1931	AC	CONSOLE HB	
PHILCO	112	1931	AC	CONSOLE LB	200
PHILCO	116-B	1935	AC	TOMBSTONE BC/SW	85
PHILCO	116-X	1935	AC	CONSOLE BC/SW	150
PHILCO	118-B	1934	AC	CATHEDRAL BABY GRAND BC/SW	
PHILCO	118-D	1934	AC	CONSOLE HB BC/SW	
PHILCO	118-RX	1934	AC	CONSOLE RC BC/SW	
PHILCO	118-X	1934	AC	CONSOLE BC/SW	
PHILCO	130	1934	AC	CONSOLE	170
PHILCO	144-B	1934	AC	CATHEDRAL BABY GRAND BC/SW	
PHILCO	144-H	1934	AC	CONSOLE HB BC/SW	
PHILCO	144-X	1934	AC	CONSOLE BC/SW	
PHILCO	200-X	1934	AC	CONSOLE	195
PHILCO	201-X	1934	AC	CONSOLE	
PHILCO	212	1931	AC	CONSOLE RADIO/PHONO	
PHILCO	270	1931	AC	CONSOLE RADIO/PHONO	
PHILCO	296	1930	AC	CONSOLE RADIO/PHONO	100
PHILCO	370	1931	AC	CONSOLE END TABLE	
PHILCO	500	1934	AC	CONSOLE RADIO/PHONO BC/SW	
PHILCO	501	1934	AC	CONSOLE RADIO/PHONO BC/SW	
PHILCO	503-L	1934	AC	CONSOLE LOW PROFILE	
PHILCO	505	1934	AC	CONSOLE HB BC/SW	
PHILCO	506-L	1934	AC	CONSOLE LOW PROFILE BC/SW	
PHILCO	507-L	1934	AC	CONSOLE RADIO/PHONO	85
PHILCO	509-X	1934	AC	CONSOLE RADIO/PHONO	
PHILCO	511	1928	AC	TABLE 7 TUBE 1 DIAL	80
PHILCO	512	1928	AC	TABLE 7 TUBE 1 DIAL	
PHILCO	513	1928	AC	TABLE 7 TUBE 1 DIAL	170
PHILCO	514	1928	AC	TABLE 7 TUBE 1 DIAL	150
PHILCO	515	1928	AC	TABLE 7 TUBE 1 DIAL	
PHILCO	521	1928	AC	TABLE 7 TUBE 1 DIAL	
PHILCO	522	1928	AC	TABLE 7 TUBE 1 DIAL	
PHILCO	523	1928	AC	TABLE 7 TUBE 1 DIAL	
PHILCO	524	1928	AC	TABLE 7 TUBE 1 DIAL	
PHILCO	525	1928	AC	TABLE 7 TUBE 1 DIAL	
PHILCO	531	1928	AC	CONSOLE LB 7 TUBE 1 DIAL	

MANUFACTURER	MODEL	YEAR	POWER	STYLE	PRICE
PHILCO	541	1928	AC	CONSOLE LB 7 TUBE 1 DIAL	
PHILCO	551	1928	AC	CONSOLE HB 7 TUBE 1 DIAL	
PHILCO	561	1928	AC	CONSOLE HB 7 TUBE 1 DIAL	
PHILCO	642-B	1935	32/DC	TABLE BC/SW	
PHILCO	642-F	1935	32/DC	CONSOLE BC/SW	
PHILCO	650	1933	AC	CONSOLE	140
PHILCO	680-X	1935	AC	CONSOLE BC/SW	145
PHILCO	BABY GRAND	1930	AC	CATHEDRAL	100
PHILCO	CONCERT GRAND	1930	AC	CONSOLE LB	130
PHILCO	PT-2	1940	AC	TABLE PLASTIC	20
PHILCO	PT-4	1941	AC/DC	TABLE PLASTIC	20
PHILCO	PT-6	1940	AC	TABLE	30
PHILCO	PT-10	1941	AC/DC	TABLE PLASTIC	17
PHILCO	PT-25	1939	AC	MANTLE PLASTIC	20
PHILCO	PT-26	1940	AC	MANTLE PLASTIC	20
PHILCO	PT-28	1939	AC	MANTLE PLASTIC	20
PHILCO	PT-43	1939	AC	MANTLE PLASTIC	20
PHILCO	PT-91	1941	AC	MANTLE PLASTIC	15
PHILCO	TH-4	1939	AC	MANTLE	15
PHILCO	TH-5	1939	AC	MANTLE PLASTIC PB	20
PHILCO	TH-14	1939	AC	MANTLE	
PHILCO	TH-15	1939	AC	TABLE PB	
PHILCO	TH-16	1939	AC	MANTLE PLASTIC	15
PHILCO	TH-17	1939	AC	MANTLE PLASTIC PB	20
PHILCO	TH-18	1939	AC	MANTLE PLASTIC BC/SW	
PHILCO	TP-4-I,-W	1939	AC	MANTLE PLASTIC	20
PHILCO	TP-5-I,-W	1939	AC	MANTLE PLASTIC PB	25
PHILCO	TP-10	1939	AC	MANTLE PLASTIC	20
PHILCO	TP-11	1939	AC	MANTLE PLASTIC PB	25
PHILCO	TP-12	1939	AC	MANTLE	
PHILCO	TP-20	1939	AC	MANTLE PLASTIC	20
PHILCO	TP-21	1939	AC	TABLE PLASTIC PB	25
PHILMORE	P-4	1932	AC	TABLE BC/SW	
PHILMORE	P-6	1932	DC	TABLE BC/SW	
PHILMORE	P-7	1932	AC	TABLE BC/SW/LW	
PHILMORE	P-8	1932	AC	TABLE BC/SW	
PHILMORE	PD-4	1932	DC	TABLE BC/SW	
PICARD RADIO	AMES	1922	BAT	TABLE LP 2 BOXES 2 DIAL	200
PIERCE AIRO	524	1930	AC	CATHEDRAL MODERN	
PIERSON	740	1929	AC	CONSOLE	
PIERSON	750	1929	AC	CONSOLE RADIO/PHONO	
PIERSON	780	1929	AC	CONSOLE	
PILOT RADIO & TUBE	41	1935	AC	TABLE	
PILOT RADIO & TUBE	43	1935	AC	TABLE BC/SW	
PILOT RADIO & TUBE	53	1934	AC	TABLE BC/SW	
PILOT RADIO & TUBE	55	1934	AC	TABLE BC/SW/LW	
PILOT RADIO & TUBE	63	1934	AC	TABLE BC/SW	
PILOT RADIO & TUBE	65	1934	AC	TABLE BC/SW/LW	
PILOT RADIO & TUBE	68	1934	DC	TABLE BC/SW	
PILOT RADIO & TUBE	69	1934	AC	TABLE BC/SW/LW	

MANUFACTURER	MODEL	YEAR	POWER	STYLE	PRICE
PILOT RADIO & TUBE	73	1934	BAT	TABLE BC/SW	
PILOT RADIO & TUBE	93	1934	AC	TABLE BC/SW	
PILOT RADIO & TUBE	103	1934	AC	TABLE BC/SW	
PILOT RADIO & TUBE	108	1935	AC	TABLE BC/SW	
PILOT RADIO & TUBE	114	1934	AC	TABLE BC/SW	
PILOT RADIO & TUBE	123	1935	AC	TABLE BC/SW	
PILOT RADIO & TUBE	183	1935	AC	TABLE BC/SW	
PILOT RADIO & TUBE	193	1936	AC	TABLE BC/SW	
PILOT RADIO & TUBE	203	1936	AC	TABLE BC/SW	
PILOT RADIO & TUBE	213	1935	AC	TABLE BC/SW	
PILOT RADIO & TUBE	253	1936	6/DC	TABLE BC/SW	
PILOT RADIO & TUBE	293	1936	AC	TABLE BC/SW	
PILOT RADIO & TUBE	303	1935	AC	TABLE BC/SW	
PILOT RADIO & TUBE	304	1936	AC	TABLE BC/SW	
PILOT RADIO & TUBE	364	1936	AC	TABLE BC/SW	
PILOT RADIO & TUBE	393	1936	AC	TABLE BC/SW	
PILOT RADIO & TUBE	423	1937	AC	TABLE BC/SW	
PILOT RADIO & TUBE	1010	1932	AC	TABLE DRAGON MONARCH BC/SW	
PILOT RADIO & TUBE	1011-R	1932	DC	TABLE DRAGON MONARCH BC/SW	
PILOT RADIO & TUBE	7639	1932	AC	TABLE TWENTIETH CENTURY	
PILOT RADIO & TUBE	7641-R	1932	DC	TABLE TWENTIETH CENTURY	
PILOT RADIO & TUBE	8239	1932	AC	CONSOLE GOLDEN ARROW	
PILOT RADIO & TUBE	8241-R	1932	DC	CONSOLE GOLDEN ARROW	
PILOT RADIO & TUBE	8643	1932	AC	CONSOLE ARMADA	
PILOT RADIO & TUBE	8645-R	1932	DC	CONSOLE ARMADA	
PILOT RADIO & TUBE	8810	1932	AC	CONSOLE DRAGON EMPEROR B/S	
PILOT RADIO & TUBE	8811-R	1932	DC	CONSOLE DRAGON EMPEROR B/S	
PILOT RADIO & TUBE	9255	1932	AC	TABLE CHEST CORSAIR	
PILOT RADIO & TUBE	9257-R	1932	DC	TABLE CHEST CORSAIR	
PILOT RADIO & TUBE	ARH-224	1939	AC	CONSOLE PERIOD BC/SW	
PILOT RADIO & TUBE	ASH-224	1939	AC	CONSOLE RADIO/PHONO BC/SW	
PILOT RADIO & TUBE	B-2	1933	AC	MANTLE	
PILOT RADIO & TUBE	B-1	1940	AC	MANTLE PLASTIC	25
PILOT RADIO & TUBE	BX-203	1937	AC	TABLE BAKELITE BC/SW	
PILOT RADIO & TUBE	C-63	1934	AC	CONSOLE BC/SW	
PILOT RADIO & TUBE	C-114	1934	AC	CONSOLE BC/SW	
PILOT RADIO & TUBE	C-151	1931	AC	CONSOLE	
PILOT RADIO & TUBE	C-152	1931	DC	CONSOLE	
PILOT RADIO & TUBE	C-153	1931	AC	CONSOLE	
PILOT RADIO & TUBE	C-154	1931	AC	CONSOLE	
PILOT RADIO & TUBE	C-162	1931	AC	CONSOLE	
PILOT RADIO & TUBE	C-163	1931	DC	CONSOLE	
PILOT RADIO & TUBE	C-183	1936	AC	CONSOLE BC/SW	
PILOT RADIO & TUBE	C-193	1937	AC	CONSOLE BC/SW	
PILOT RADIO & TUBE	C-293	1936	AC	CONSOLE BC/SW	

MANUFACTURER	MODEL	YEAR	POWER	STYLE	PRICE
PILOT RADIO & TUBE	C-293	1937	AC	CONSOLE BC/SW	
PILOT RADIO & TUBE	C-300	1937	AC	CONSOLE BC/SW	
PILOT RADIO & TUBE	CG-56	1937	6/DC	CONSOLE BC/SW	
PILOT RADIO & TUBE	CG-184	1938	AC	CONSOLE BC/SW	
PILOT RADIO & TUBE	CG-508	1938	AC	CONSOLE BC/SW	
PILOT RADIO & TUBE	CG-528	1938	AC	CONSOLE BC/SW	
PILOT RADIO & TUBE	CG-674	1938	AC	CONSOLE BC/SW	
PILOT RADIO & TUBE	CG-5206	1937	32/DC	CONSOLE BC/SW	
PILOT RADIO & TUBE	CH-134	1939	AC	CONSOLE BC/SW	
PILOT RADIO & TUBE	CH-594	1938	AC	CONSOLE AUTO. TUNING BC/SW	
PILOT RADIO & TUBE	CH-714	1939	AC	CONSOLE PB BC/SW	
PILOT RADIO & TUBE	CH-864	1939	AC	CONSOLE BC/SW	
PILOT RADIO & TUBE	CX-23	1937	AC	CONSOLE BC/SW	
PILOT RADIO & TUBE	CX-63	1935	AC	CONSOLE BC/SW	
PILOT RADIO & TUBE	CX-68	1935	DC	CONSOLE BC/SW	
PILOT RADIO & TUBE	CX-114	1935	AC	CONSOLE BC/SW	
PILOT RADIO & TUBE	CX-304	1937	AC	CONSOLE BC/SW	
PILOT RADIO & TUBE	CX-474	1939	AC	CONSOLE PB BC/SW	
PILOT RADIO & TUBE	CX-664	1939	AC	CONSOLE PB BC/SW	
PILOT RADIO & TUBE	CY-63	1936	AC	CONSOLE BC/SW	
PILOT RADIO & TUBE	D-3	1933	AC	COMPACT	
PILOT RADIO & TUBE	E-20	1933	AC	COMPACT	
PILOT RADIO & TUBE	F-12	1933	AC	TABLE	
PILOT RADIO & TUBE	H-7	1933	AC	TABLE	
PILOT RADIO & TUBE	H-34	1933	AC	TABLE BC/SW	
PILOT RADIO & TUBE	H-36	1933	DC	TABLE BC/SW	
PILOT RADIO & TUBE	H-38	1933	DC	TABLE	
PILOT RADIO & TUBE	H-393	1937	AC	TABLE BC/SW	
PILOT RADIO & TUBE	HA-191	1940	AC	CONSOLE RADIO/PHONO	
PILOT RADIO & TUBE	HA-192	1940	AC	CONSOLE RADIO/PHONO	
PILOT RADIO & TUBE	HA-1424	1940	AC	CONSOLE RADIO/PHONO	
PILOT RADIO & TUBE	HA-1524	1940	AC	CONSOLE RADIO/PHONO	
PILOT RADIO & TUBE	HD-191	1940	AC	CONSOLE	45
PILOT RADIO & TUBE	HD-192	1940	AC	CONSOLE R/P W/RECORDER	
PILOT RADIO & TUBE	HFM-12	1940	AC	CONSOLE RADIO/PHONO	
PILOT RADIO & TUBE	HR-192	1940	AC	CONSOLE RADIO/PHONO	
PILOT RADIO & TUBE	HX-304	1938	AC	TABLE BC/SW	
PILOT RADIO & TUBE	L-8	1933	AC	TABLE	
PILOT RADIO & TUBE	L-28	1933	DC	TABLE	
PILOT RADIO & TUBE	L-30	1933	AC	TABLE BC/SW	
PILOT RADIO & TUBE	L-32	1933	DC	TABLE BC/SW	
PILOT RADIO & TUBE	LA-1424	1940	AC	CONSOLE RADIO/PHONO	
PILOT RADIO & TUBE	LA-1524	1940	AC	CONSOLE RADIO/PHONO	
PILOT RADIO & TUBE	LD-192	1940	AC	CONSOLE R/P W/RECORDER	
PILOT RADIO & TUBE	LFM-12	1940	AC	CONSOLE RADIO/PHONO	
PILOT RADIO & TUBE	M-7	1933	AC	CONSOLE RADIO/PHONO	
PILOT RADIO & TUBE	M-8	1933	AC	CONSOLE	
PILOT RADIO & TUBE	M-30	1933	AC	CONSOLE BC/SW/LW	
PILOT RADIO & TUBE	M-34	1933	AC	CONSOLE R/P BC/SW/LW	

MANUFACTURER	MODEL	YEAR	POWER	STYLE	PRICE
PILOT RADIO & TUBE	M-206	1941	AC	CONSOLE R/P BC/SW/FM	45
PILOT RADIO & TUBE	P-63	1934	AC	CONSOLE BC/SW	
PILOT RADIO & TUBE	P-393	1936	AC	CONSOLE RADIO/PHONO BC/SW	
PILOT RADIO & TUBE	PA-1424	1940	AC	CONSOLE RADIO/PHONO	
PILOT RADIO & TUBE	PA-1524	1940	AC	CONSOLE RADIO/PHONO	
PILOT RADIO & TUBE	PG-508	1939	AC	CONSOLE RADIO/PHONO BC/SW	
PILOT RADIO & TUBE	PG-528	1938	AC	CONSOLE RADIO/PHONO BC/SW	
PILOT RADIO & TUBE	PH-393	1937	AC	CONSOLE RADIO/PHONO BC/SW	
PILOT RADIO & TUBE	PH-474	1938	AC	CONSOLE RP AU. TUN. BC/SW	
PILOT RADIO & TUBE	PH-484	1938	AC	CONSOLE RP AU. TUN. BC/SW	
PILOT RADIO & TUBE	PH-664	1938	AC	CONSOLE RP AUTO TUN. BC/SW	
PILOT RADIO & TUBE	PH-874	1938	AC	CONSOLE RP AU. TUN. BC/SW	
PILOT RADIO & TUBE	PX-23	1937	AC	CONSOLE RADIO/PHONO BC/SW	
PILOT RADIO & TUBE	PX-63	1935	AC	CONSOLE RADIO/PHONO BC/SW	
PILOT RADIO & TUBE	PX-68	1935	DC	CONSOLE RADIO/PHONO BC/SW	
PILOT RADIO & TUBE	PX-114	1935	AC	CONSOLE RADIO/PHONO BC/SW	
PILOT RADIO & TUBE	PX-304	1937	AC	CONSOLE RADIO/PHONO BC/SW	
PILOT RADIO & TUBE	QG-184	1938	AC	CONSOLE RADIO/PHONO BC/SW	
PILOT RADIO & TUBE	QG-584	1938	AC	CONSOLE RADIO/PHONO BC/SW	
PILOT RADIO & TUBE	QG-674	1938	AC	CONSOLE RADIO/PHONO BC/SW	
PILOT RADIO & TUBE	QX-304	1938	AC	CONSOLE RADIO/PHONO BC/SW	
PILOT RADIO & TUBE	RG-184	1937	AC	CONSOLE RADIO/PHONO BC/SW	
PILOT RADIO & TUBE	RG-674	1938	AC	CONSOLE LB BC/SW	
PILOT RADIO & TUBE	RH-224	1939	AC	CONSOLE PERIOD BC/SW	
PILOT RADIO & TUBE	RX-304	1938	AC	CONSOLE RADIO/PHONO BC/SW	
PILOT RADIO & TUBE	S-148	1931	AC	TABLE	
PILOT RADIO & TUBE	S-149	1931	DC	TABLE	
PILOT RADIO & TUBE	S-150	1931	BAT	TABLE	
PILOT RADIO & TUBE	S-164	1931	AC	TABLE	
PILOT RADIO & TUBE	S-165	1931	DC	TABLE	
PILOT RADIO & TUBE	S-166	1931	BAT	TABLE	
PILOT RADIO & TUBE	S-167	1931	AC	TABLE	
PILOT RADIO & TUBE	S-168	1931	DC	TABLE	
PILOT RADIO & TUBE	S-169	1931	BAT	TABLE	

MANUFACTURER	MODEL	YEAR	POWER	STYLE	PRICE
PILOT RADIO & TUBE	SG-184	1938	AC	CONSOLE RADIO/PHONO BC/SW	
PILOT RADIO & TUBE	SG-674	1938	AC	CONSOLE RADIO/PHONO BC/SW	
PILOT RADIO & TUBE	SH-224	1939	AC	CONSOLE RADIO/PHONO BC/SW	
PILOT RADIO & TUBE	T-71	1940	AC	PORTABLE FEATHERWEIGHT	15
PILOT RADIO & TUBE	T-186	1940	AC	PORTABLE BC/SW	15
PILOT RADIO & TUBE	T-301	1941	AC	TABLE BC/FM	30
PILOT RADIO & TUBE	T-341	1941	AC	TABLE BC/SW	35
PILOT RADIO & TUBE	T-1021	1939	BAT	PORTABLE	
PILOT RADIO & TUBE	T-1252	1939	AC	PORTABLE WALNUT BC/SW	
PILOT RADIO & TUBE	T-1451	1939	AC	PORTABLE CLOTH	
PILOT RADIO & TUBE	T-1452	1939	AC	PORTABLE CLOTH BC/SW	
PILOT RADIO & TUBE	TG-56	1937	6/DC	TABLE BC/SW	
PILOT RADIO & TUBE	TG-162	1937	AC	TABLE BC/SW	
PILOT RADIO & TUBE	TG-184	1937	AC	TABLE BC/SW	
PILOT RADIO & TUBE	TG-284	1939	AC	TABLE WALNUT BC/SW	
PILOT RADIO & TUBE	TG-462	1938	AC	TABLE BC/SW	
PILOT RADIO & TUBE	TG-508	1938	AC	TABLE BC/SW	
PILOT RADIO & TUBE	TG-528	1937	AC	TABLE BC/SW	
PILOT RADIO & TUBE	TG-584	1937	AC	TABLE BC/SW	
PILOT RADIO & TUBE	TG-674	1938	AC	TABLE BC/SW	
PILOT RADIO & TUBE	TG-752	1937	AC	TABLE BC/SW	
PILOT RADIO & TUBE	TG-5206	1937	32/DC	TABLE BC/SW	
PILOT RADIO & TUBE	TH-11	1939	BAT	PORTABLE CLOTH	
PILOT RADIO & TUBE	TH-12	1939	BAT	PORTABLE CLOTH BC/SW	
PILOT RADIO & TUBE	TH-64	1939	AC	TABLE WALNUT BC/SW	
PILOT RADIO & TUBE	TH-134	1939	AC	TABLE BC/SW	
PILOT RADIO & TUBE	TH-151	1938	AC	TABLE	
PILOT RADIO & TUBE	TH-224	1939	AC	TABLE WALNUT BC/SW	
PILOT RADIO & TUBE	TH-372	1938	AC	TABLE BC/SW	
PILOT RADIO & TUBE	TH-384	1939	AC	TABLE BC/SW	
PILOT RADIO & TUBE	TH-454	1938	AC	TABLE BC/SW	
PILOT RADIO & TUBE	TH-474	1938	AC	TABLE AUTO. TUNING BC/SW	
PILOT RADIO & TUBE	TH-484	1938	AC	TABLE AUTO. TUNING BC/SW	
PILOT RADIO & TUBE	TH-554	1938	AC	TABLE BC/SW	
PILOT RADIO & TUBE	TH-594	1938	AC	TABLE AUTO. TUNING BC/SW	
PILOT RADIO & TUBE	TH-651	1938	AC	TABLE	
PILOT RADIO & TUBE	TH-664	1938	AC	TABLE AUTO. TUNING BC/SW	
PILOT RADIO & TUBE	TH-714	1939	AC	TABLE PB BC/SW	
PILOT RADIO & TUBE	TH-762	1938	AC	TABLE BC/SW	
PILOT RADIO & TUBE	TH-864	1939	AC	TABLE WALNUT BC/SW	
PILOT RADIO & TUBE	TH-874	1938	AC	TABLE AUTO. TUNING BC/SW	
PILOT RADIO & TUBE	TH-974	1939	AC	TABLE BC/SW	
PILOT RADIO & TUBE	TH-6204	1939	AC	TABLE BC/SW	
PILOT RADIO & TUBE	TP-32	1941	AC	TABLE RADIO/PHONO BC/SW	15
PILOT RADIO & TUBE	TP-423	1937	AC	TABLE RADIO/PHONO BC/SW	
PILOT RADIO & TUBE	TX-42	1941	AC	TABLE RADIO/PHONO	15
PILOT RADIO & TUBE	TX-474	1939	AC	TABLE PB BC/SW	

MANUFACTURER	MODEL	YEAR	POWER	STYLE	PRICE
PILOT RADIO & TUBE	TX-484	1939	AC	TABLE PB BC/SW	
PILOT RADIO & TUBE	TX-508	1939	AC	TABLE BC/SW	
PILOT RADIO & TUBE	TX-664	1939	AC	TABLE PB BC/SW	
PILOT RADIO & TUBE	TX-874	1939	AC	TABLE PB BC/SW	
PILOT RADIO & TUBE	V-191	1931	AC	TABLE	
PILOT RADIO & TUBE	V-192	1931	DC	TABLE	
PILOT RADIO & TUBE	V-193	1931	BAT	TABLE	
PILOT RADIO & TUBE	VG-352	1937	AC	TABLE BAKELITE BC/SW	
PILOT RADIO & TUBE	VX-201	1937	AC	TABLE BAKELITE	
PILOT RADIO & TUBE	VX-203	1937	AC	TABLE BAKELITE BC/SW	
PILOT RADIO & TUBE	WG-352	1937	AC	TABLE BAKELITE BC/SW	
PILOT RADIO & TUBE	WH-141	1938	AC	TABLE PLASTIC	
PILOT RADIO & TUBE	WX-201	1937	AC	TABLE BAKELITE	
PILOT RADIO & TUBE	WX-203	1937	AC	TABLE BAKELITE BC/SW	
PILOT RADIO & TUBE	X-23	1937	AC	TABLE BC/SW	
PILOT RADIO & TUBE	X-41	1936	AC	TABLE	
PILOT RADIO & TUBE	X-43	1936	AC	TABLE BC/SW	
PILOT RADIO & TUBE	X-63	1935	AC	TABLE BC/SW	
PILOT RADIO & TUBE	X-68	1935	DC	TABLE BC/SW	
PILOT RADIO & TUBE	X-73	1935	2/DC	TABLE BC/SW	
PILOT RADIO & TUBE	X-114	1935	AC	TABLE BC/SW	
PILOT RADIO & TUBE	X-1451	1940	AC	PORTABLE AIRPLANE LUGGAGE	20
PILOT RADIO & TUBE	XG-56	1938	6/DC	TABLE BC/SW	
PILOT RADIO & TUBE	XG-184	1938	AC	TABLE UPRIGHT BC/SW	
PILOT RADIO & TUBE	XG-674	1938	AC	TABLE UPRIGHT BC/SW	
PILOT RADIO & TUBE	XG-5206	1938	32/DC	TABLE BC/SW	
PINTO SPEAKING CLOCK	D-X-12A	1924	BAT	CONSOLE/CLOCK/ PHONOGRAPH	
PINTO SPEAKING CLOCK	D-X-12B	1924	BAT	CONSOLE/CLOCK/ PHONOGRAPH	
PIONEER	62	1929	AC	TABLE	
PIONEER	66	1929	AC	CONSOLE	
PIONEER	68	1929	AC	CONSOLE	
PIONEER	70	1929	AC	CONSOLE	
PIONEER	72	1929	AC	CONSOLE	
PIONEER	74	1929	AC	CONSOLE	
PIONEER	76	1929	AC	CONSOLE	
PIONEER	101	1930	AC	CONSOLE LB	
PIONEER	102	1930	AC	CONSOLE HB	
PLYMOUTH RADIO & PHONOGRAPH	3	1930	AC	CATHEDRAL MODERN	90
PORTOLA RADIO	101	1923	BAT	PORTABLE 6 TUBE 2 DIAL	175
PORTS MFG.	105	1925	BAT	TABLE LP 6 TUBE 1 DIAL	
PORTS MFG.	110	1925	BAT	CONSOLE 6 TUBE 1 DIAL	
POWELL RADIO	CATHEDRAL	1930	AC	CATHEDRAL MODERN	95
POWEROLA RADIO	POWEROLA	1925	AC	TABLE LP SLOPE PAN. 5T 3DL	
PRECEL RADIO	SUPERFIVE	1923	BAT	TABLE HP SLOPE PAN. 5T 3DL	
PRECISION PRODUCTS	ARBORPHONE	1926	BAT	TABLE LP 2 DIAL	125
PREMIER ELECTRIC	HOME-PAL	1930	AC	CATHEDRAL	

MANUFACTURER	MODEL	YEAR	POWER	STYLE	PRICE
PREMIER RADIO	7-A	1924	BAT	CONSOLE 5 TUBE 2 DIAL	
PREMIER RADIO	7-B	1924	BAT	TABLE LP 5 TUBE 2 DIAL	
PRIESS RADIO	9 STRAIGHT	1926	BAT	TABLE LP LOOP ANT 9 T 1 DL	140
PRIESS RADIO	9 STRAIGHT	1926	BAT	CONSOLE LB 9 TUBE 1 DIAL	150
PRIESS RADIO	PR-3 TELOFLEX	1925	BAT	TABLE LP 5 TUBE 2 DIAL	100
PRIESS RADIO	PR-4	1925	BAT	TABLE LP 8 TUBE 2 DIAL	
PRIESS RADIO	PR-5	1925	BAT	CONSOLE 5 TUBE 2 DIAL	
PRIESS RADIO	PR-6	1925	BAT	CONSOLE 8 TUBE 2 DIAL	
R					
R-B RADIO	BELLE CLAIRE	1925	BAT	TABLE LP 3 DIAL	140
R-B RADIO	BELLE CLAIRE	1925	BAT	CONSOLE 3 DIAL	
R-B RADIO	PAGE-FIVE	1924	BAT	TABLE LP 5 TUBE 3 DIAL	
R-B RADIO	PAGE-SIX	1925	BAT	TABLE LP 6 TUBE 3 DIAL	80
R. K. RADIO LABS	421	1934	AC	TABLE	45
R. K. RADIO LABS	633	1934	AC	CONSOLE	90
RADIO GRAPH	11	1924	BAT	TABLE LP 5 TUBE 3 DIAL	
RADIO GRAPH	11	1924	BAT	CONSOLE 5 TUBE 3 DIAL	
RADIO INDUSTRIES	RICO-DYNE	1924	BAT	TABLE LP 5 TUBE 3 DIAL	65
RADIO INDUSTRIES	RIVOLI CONSOLE	1925	BAT	CONSOLE 5 TUBE 3 DIAL	
RADIO INDUSTRIES	RIVOLI DELUXE	1925	BAT	TABLE LP SPEAKER 5 T 3 DL	
RADIO INDUSTRIES	RIVOLI JUNIOR	1925	BAT	TABLE LP 5 TUBE 3 DIAL	
RADIO MASTER	5-T-1	1924	BAT	TABLE LP 5 TUBE 3 DIAL	
RADIO MASTER	5-T-2	1924	BAT	CONSOLE HB 5 TUBE 3 DIAL	
RADIO MASTER	5-T-3	1924	BAT	CONSOLE HB 5 TUBE 3 DIAL	
RADIO MASTER	100	1923	BAT	TABLE HP 4 TUBE 2 DIAL	125
RADIO MASTER	218	1929	AC	CONSOLE LB RADIO/PHONO	150
RADIO MASTER	275	1923	BAT	CONSOLE 4 TUBE 1 DIAL	
RADIO MASTER	375	1926	BAT	CONSOLE HB 5 TUBE 2 DIAL	160
RADIO PANEL	FLIVER	1922	BAT	MANTLE 1 TUBE 1 DIAL	175
RADIO PRODUCTS	52	1923	BAT	TABLE LP 6 TUBE 3 DIAL	
RADIO PRODUCTS	RPM ODYNE	1923	BAT	BREADBOARD 6 TUBE 3 DIAL	
RADIO RECEPTOR	M-3	1925	AC	TABLE LP 4 TUBE 2 DIAL	
RADIO RECEPTOR	R-M 1	1924	BAT	TABLE LP 4 TUBE 2 DIAL	
RADIO RECEPTOR	R-M 2	1924	DC	TABLE LP 4 TUBE 2 DIAL	
RADIO RECEPTOR	R-M 3	1925	BAT	CONSOLE 4 TUBE 2 DIAL	
RADIO SHOP				See also ECHOPHONE RADIO	
RADIO SHOP	3 ECHOPHONE	1923	BAT	TABLE HP SLOPE PAN. 3T 2DL	400
RADIO SHOP	4 ECHOPHONE	1924	BAT	TABLE HP SLOPE PAN. 4T 2DL	400
RADIO SHOP	5 ECHOPHONE	1924	BAT	TABLE HP SLOPE PAN. 6T 3DL	
RADIO SHOP	A ECHOPHONE	1923	BAT	TABLE HP 3 TUBE 2 DIAL	300
RADIO SHOP	B ECHOPHONE	1923	BAT	TABLE HP 4 TUBE 2 DIAL	325
RADIO SHOP	F ECHOPHONE	1925	BAT	CONSOLE GRAND 2 DIAL	300
RADIO SHOP	F ECHOPHONE	1924	BAT	TABLE HP 2 DIAL	300
RADIO SHOP	F ECHOPHONE	1925	BAT	TABLE LP STANDARD 2 DIAL	250
RADIO SHOP	J ECHOPHONE	1923	BAT	TABLE HP 1 TUBE 2 DIAL	300
RADIO SHOP	METEOR	1924	BAT	TABLE SLOPE PANEL 3 T 2 DL	275
RADIO SHOP	R-3	1925	BAT	TABLE LP SLOPE PAN. 3T 1DL	
RADIO SHOP	R-5	1925	BAT	TABLE LP SLOPE PAN. 5T 2DL	
RADIO SHOP	R-5	1925	BAT	CONSOLE 5 TUBE 2 DIAL	

MANUFACTURER	MODEL	YEAR	POWER	STYLE	PRICE
RADIO SHOP	S SPECIAL	1922	BAT	TABLE LP 3 TUBE 2 DIAL	350
RADIO SHOP	S/2-STEP AMP.	1922	BAT	TABLE LP 2 BOXES 3 T 2 DL	350
RADIO SHOP	STANDARD	1922	BAT	TABLE LP 3 TUBE 2 DIAL	275
RADIO SHOP	V-3	1924	BAT	PORTABLE 3 TUBE 2 DIAL	
RADIO SHOP	V-3 EARLY	1924	BAT	TABLE LP 3 TUBE 2 DIAL	250
RADIO SHOP	V-3 ECHOPHONE	1925	BAT	CONSOLE 3 TUBE 2 DIAL	250
RADIO SUPPLY	V PARAMOUNT	1925	BAT	TABLE LP 5 TUBE 3 DIL	
RADIO SUPPLY	VI PARAMOUNT	1926	BAT	TABLE LP 6 TUBE 3 DIAL	
RADIOBAR	RP-200	1939	AC	CONSOLE R/P PB W/BAR	250
RADISTO	RADISTO	1924	BAT	TABLE HP 4 TUBE 1 DIAL	
RADOLEK	MARVELO	1925	BAT	TABLE LP 3 DIAL	95
RANDOLPH RADIO	5 SYMPHONIC	1926	BAT	TABLE LP SLOPE PAN. 5T 3DL	85
RANDOLPH RADIO	6 AMPLIPHONIC	1926	BAT	CONSOLE W/DOORS 6 T 2 DL	140
RANDOLPH RADIO	6 COLUMBIA GRND	1926	BA	CONSOLE LB 6 TUBE 3 DIAL	125
RANDOLPH RADIO	6 COLUMBIA SR.	1926	BA	TABLE LP SLOPE PAN. 6T 3DL	120
RANDOLPH RADIO	6 SENIOR	1927	AC	TABLE	75
RANDOLPH RADIO	7 SENIOR	1927	AC	TABLE LP	85
RANDOLPH RADIO	AMPLIPHONIC	1927	AC	CONSOLE	95
RCA	4-T	1936	AC	CATHEDRAL COMPACT	135
RCA	4-X	1936	AC	TOMBSTONE COMPACT	50
RCA	4-X-3	1936	AC	TOMBSTONE COMPACT	45
RCA	4-X-4	1936	AC	TOMBSTONE COMPACT	45
RCA	5-Q-55	1939	AC	MANTLE PLASTIC BC/SW	20
RCA	5-T	1936	AC	TOMBSTONE BC/SW	40
RCA	5-X	1936	AC	TABLE BC/SW	40
RCA	5-X-3	1936	AC	TABLE BC/SW	40
RCA	5-X-4	1936	AC	TABLE BC/SW	40
RCA	5-X-5	1939	AC	MANTLE PLASTIC	
RCA	6-BK	1936	2/DC	CONSOLE BC/SW	35
RCA	6-BT	1936	2/DC	TOMBSTONE BC/SW	40
RCA	6-K	1936	AC	CONSOLE BC/SW	65
RCA	6-K-1	1937	AC	CONSOLE BC/SW	65
RCA	6-T	1936	AC	TOMBSTONE BC/SW	40
RCA	7-K	1936	AC	CONSOLE BC/SW	70
RCA	7-T	1936	AC	TOMBSTONE BC/SW	65
RCA	7-U	1936	AC	CONSOLE R/P TUN. EYE BC/SW	60
RCA	7-X	1936	AC	TOMBSTONE BC/SW	40
RCA	8-BK	1936	2/DC	CONSOLE BC/SW	45
RCA	8-BK-6	1936	6/DC	CONSOLE BC/SW	45
RCA	8-BT	1936	2/DC	TOMBSTONE BC/SW	30
RCA	8-K	1936	AC	CONSOLE TUNING EYE BC/SW	75
RCA	8-QB	1938	6/DC	TABLE BC/SW	65
RCA	8-QBK	1938	6/DC	CONSOLE GRAND	90
RCA	8-T	1936	AC	TOMBSTONE TUNING EYE BC/SW	45
RCA	8-U	1936	AC	CONSOLE R/P TE BC/SW/LW	70
RCA	9-K	1936	AC	CONSOLE TUN. EYE BC/SW/LW	75

MANUFACTURER	MODEL	YEAR	POWER	STYLE	PRICE
RCA	9-T	1936	AC	TOMBSTONE TUN EYE BC/SW/LW	45
RCA	9-TX-23	1939	AC	TABLE	50
RCA	9-TX-31	1939	AC	MANTLE PLASTIC	20
RCA	9-TX-32	1939	AC	MANTLE PLASTIC	
RCA	9-TX-50	1939	AC	TABLE	
RCA	9-U	1936	AC	CONSOLE R/P TE BC/SW/LW	50
RCA	9-U-2	1936	AC	CONSOLE R/P TE BC/SW/LW	50
RCA	9-X	1938	AC	COMPACT	40
RCA	10-K	1936	AC	CONSOLE TUN. EYE BC/SW/LW	80
RCA	10-T	1936	AC	TOMBSTONE TUN EYE BC/SW/LW	50
RCA	10-X	1940	AC	TABLE PLASTIC	20
RCA	15-BP- 6	1940	2/DC	TABLE	15
RCA	15-BT	1940	2/DC	TABLE	30
RCA	15-K	1936	AC	CONSOLE TUN. EYE BC/SW/LW	90
RCA	15-X	1940	AC	TABLE PLASTIC WALNUT	20
RCA	16-X- 2	1940	AC	TABLE PLASTIC IVORY	25
RCA	16-X- 3	1940	AC	TABLE	30
RCA	16-X-11	1940	AC	TABLE PLASTIC BC/SW	25
RCA	16-X-13	1940	AC	TABLE BC/SW	30
RCA	16-X-14	1940	AC	TABLE BC/SW	35
RCA	18-T	1940	AC	TABLE PB BC/SW	40
RCA	19-K	1940	AC	CONSOLE PB BC/SW	155
RCA	25-BP	1941	AC	PORTABLE TAN LEATHERETTE	15
RCA	26-X- 1	1941	AC	TABLE BC/SW	25
RCA	26-X-3	1941	AC	TABLE BC/SW	30
RCA	26-X-4	1941	AC	TABLE PB BC/SW	30
RCA	28-T	1941	AC	TABLE PB BC/SW	35
RCA	28-X	1941	AC	TABLE BC/SW	30
RCA	28-X-5	1941	AC	TABLE PB BC/SW	35
RCA	29-K	1941	AC	CONSOLE PB BC/SW	135
RCA	29-K-2	1941	AC	CONSOLE PB BC/SW	150
RCA	36-X	1941	AC	TABLE	35
RCA	40-X-30, 31	1940	AC	MANTLE PLASTIC	20
RCA	40-X-50,51,52	1939	AC	TABLE	20
RCA	40-X-53,54,55	1939	AC	TABLE	20
RCA	40-X-56,57	1939	AC	TABLE	20
RCA	45-X-1,11,13	1940	AC	MANTLE PLASTIC	20
RCA	46-X-3	1939	AC	TABLE	35
RCA	46-X-11,13	1940	AC	MANTLE PLASTIC	25
RCA	46-X-12	1939	AC	MANTLE PLASTIC IVORY BC/SW	30
RCA	84BT	1937	BAT	CATHEDRAL MODERN	40
RCA	85BK	1937	BAT	CONSOLE BC/SW	55
RCA	85BT	1937	BAT	TOMBSTONE BC/SW	35
RCA	85T	1937	AC	TABLE	30
RCA	85T-1	1937	AC	TABLE BC/SW	30

MANUFACTURER	MODEL	YEAR	POWER	STYLE	PRICE
RCA	86BK	1937	2/DC	CONSOLE BC/SW	55
RCA	86BT	1937	2/DC	TABLE BC/SW	40
RCA	86E	1937	AC	CONSOLE END TABLE BC/SW	60
RCA	86K	1937	AC	CONSOLE BC/SW	60
RCA	86K-7	1937	AC	CONSOLE BC/SW	60
RCA	86T	1937	AC	TABLE BC/SW	30
RCA	86T-1	1937	AC	TABLE BC/SW	50
RCA	86X	1937	AC	TABLE BC/SW	30
RCA	87EY	1937	AC	CONSOLE END TABLE BC/SW	55
RCA	87K	1937	AC	CONSOLE BC/SW	85
RCA	87T	1937	AC	TABLE BC/SW	40
RCA	87X	1937	AC	TABLE BC/SW	30
RCA	87Y	1937	AC	CONSOLE BC/SW	50
RCA	88K	1937	AC	CONSOLE TUNING EYE BC/SW	95
RCA	94-BK	1938	2/DC	CONSOLE	70
RCA	94-BP-1	1939	BAT	PORTABLE	25
RCA	94-BP-80	1939	BAT	PORTABLE	15
RCA	94-BT	1938	2/DC	TABLE	20
RCA	94-BT-1	1938	BAT	TABLE	25
RCA	94-BT-6	1938	6/DC	TABLE	20
RCA	95T-1	1938	AC	TABLE	30
RCA	95T-5	1938	AC	TABLE PB ET BC/SW	50
RCA	95X	1938	AC	TABLE	35
RCA	95X-1	1938	AC	TABLE ET	35
RCA	96-E-2	1939	AC	CONSOLE PERIOD PB BC/SW	95
RCA	96-T-4	1939	AC	TABLE PB BC/SW	45
RCA	96-XI	1939	AC	MANTLE PLASTIC BC/SW	
RCA	96-XII	1939	AC	TABLE PB BC/SW	20
RCA	96E	1938	AC	CONSOLE END TABLE ET	45
RCA	96K	1938	AC	CONSOLE ET BC/SW	60
RCA	96K-2	1938	AC	CONSOLE ET BC/SW	85
RCA	96T	1938	AC	TABLE ET	50
RCA	96T-1	1938	AC	TABLE PB ET BC/SW	55
RCA	96T-2	1938	AC	TABLE ET BC/SW	55
RCA	96T-3	1938	AC	TABLE ET BC/SW	55
RCA	97-T-2	1939	AC	TABLE PB BC/SW	60
RCA	97E	1938	AC	CONSOLE END TABLE ET BC/SW	80
RCA	97KG	1938	AC	CONSOLE GRAND ET BC/SW	95
RCA	97T	1938	AC	TABLE ET BC/SW	55
RCA	97X	1938	AC	TABLE ET	50
RCA	98K	1938	AC	CONSOLE ET BC/SW	95
RCA	98X	1938	AC	TABLE ET BC/SW	55
RCA	99K	1938	AC	CONSOLE ET BC/SW	95
RCA	99T	1938	AC	TABLE ET BC/SW	65
RCA	100	1933	AC	CATHEDRAL BC/SW	155
RCA	101	1934	AC	TOMBSTONE	70
RCA	102	1934	AC	TABLE	40
RCA	103	1935	AC	TOMBSTONE	60

MANUFACTURER	MODEL	YEAR	POWER	STYLE	PRICE
RCA	110	1933	AC	CATHEDRAL BC/SW	160
RCA	110-K	1940	AC	CONSOLE PB BC/SW	140
RCA	111	1933	AC	TABLE CHEST BC/SW	70
RCA	114	1934	AC	COMPACT	60
RCA	115	1934	AC	TABLE BC/SW	70
RCA	117	1935	AC	CATHEDRAL MODERN BC/SW	70
RCA	118	1934	AC	TOMBSTONE BC/SW	50
RCA	119	1935	AC	CATHEDRAL MODERN	45
RCA	120	1933	AC	CATHEDRAL BC/SW	195
RCA	121	1933	AC	CATHEDRAL BC/SW	190
RCA	122	1933	AC	TOMBSTONE BC/SW	
RCA	124	1934	AC	TOMBSTONE BC/SW	90
RCA	125	1935	AC	TOMBSTONE MODERN BC/SW	55
RCA	126-B	1934	BAT	CATHEDRAL	135
RCA	128	1934	AC	CATHEDRAL BC/SW	120
RCA	135-B	1934	BAT	TOMBSTONE BC/SW	55
RCA	140	1933	AC	CATHEDRAL BC/SW	150
RCA	142-B	1934	BAT	CATHEDRAL	115
RCA	143	1934	AC	CATHEDRAL BC/SW/LW	175
RCA	210	1934	AC	CONSOLE BC/SW	70
RCA	211	1934	AC	CONSOLE BC/SW	70
RCA	211-K	1941	AC	CONSOLE PB BC/SW	150
RCA	214	1935	AC	CONSOLE BC/SW	75
RCA	220	1934	AC	CONSOLE BC/SW	125
RCA	221	1934	AC	CONSOLE BC/SW	90
RCA	222	1934	AC	CONSOLE BC/SW	135
RCA	223	1934	AC	CONSOLE BC/SW	65
RCA	224	1934	AC	CONSOLE BC/SW	130
RCA	225	1935	AC	CONSOLE BC/SW	85
RCA	226	1935	AC	CONSOLE BC/SW	120
RCA	235-B	1934	BAT	CONSOLE BC/SW	70
RCA	240	1933	AC	CONSOLE BC/SW	140
RCA	241-B	1934	BAT	CONSOLE	145
RCA	242	1934	AC	CONSOLE BC/SW/LW	150
RCA	243	1935	AC	CONSOLE BC/SW/LW	130
RCA	260	1933	AC	CONSOLE BC/SW	145
RCA	261	1934	AC	CONSOLE BC/SW	145
RCA	262	1934	AC	CONSOLE BC/SW/LW	160
RCA	263	1935	AC	CONSOLE BC/SW/LW	150
RCA	280	1933	AC	CONSOLE BC/SW	160
RCA	281	1934	AC	CONSOLE BC/SW/LW	180
RCA	300	1933	AC	TABLE RADIO/PHONO	35
RCA	301	1934	AC	TABLE RADIO/PHONO BC/SW	25
RCA	310	1933	AC	CONSOLE RADIO/PHONO BC/SW	85
RCA	320	1933	AC	CONSOLE HB BC/SW	
RCA	321	1934	AC	CONSOLE RADIO/PHONO BC/SW	65
RCA	322	1934	AC	CONSOLE RADIO/PHONO BC/SW	65

MANUFACTURER	MODEL	YEAR	POWER	STYLE	PRICE
RCA	330	1933	AC	CONSOLE RADIO/PHONO BC/SW	55
RCA	331	1933	AC	CONSOLE RADIO/PHONO BC/SW	65
RCA	341	1934	AC	CONSOLE RP/ BC/SW/LW	80
RCA	380	1934	AC	CONSOLE RADIO/PHONO	95
RCA	381	1934	AC	CONSOLE R/P RC BC/SW/LW	110
RCA	510	1940	AC	TABLE PLASTIC	20
RCA	515	1941	AC	TABLE BC/SW	35
RCA	675	1935	AC	TOMBSTONE BC/SW	
RCA	810K	1937	AC	CONSOLE TUNING EYE BC/SW	120
RCA	810K- 1	1937	AC	CONSOLE TUNING EYE BC/SW	150
RCA	810T	1937	AC	TOMBSTONE TUN. EYE BC/SW	75
RCA	811K	1937	AC	CONSOLE PB ET TE BC/SW	110
RCA	812K	1937	AC	CONSOLE PB ET TE BC/SW	115
RCA	813K	1937	AC	CONSOLE PB ET TE BC/SW	150
RCA	816K	1937	AC	CONSOLE PB ET TE BC/SW	160
RCA	910KG	1938	AC	CONSOLE GRAND ET BC/SW	125
RCA	911K	1938	AC	CONSOLE ET BC/SW	90
RCA	AR-1300/AA-1400	1922	BAT	MANTLE (2 BOXES) 3 T 2 DL	
RCA	BC6-4	1935	2/DC	CONSOLE BC/SW	40
RCA	BC6-6	1935	2/DC	CONSOLE BC/SW	45
RCA	BC7-9	1935	2/DC	CONSOLE BC/SW	40
RCA	BP-10	1941	BAT	PORTABLE MINIATURE LEATHER	40
RCA	BP-56	1940	AC	PORTABLE LUGGAGE CLOTH	15
RCA	BT-40	1939	BAT	TABLE	20
RCA	BT-42	1940	AC	TABLE	35
RCA	BT6-3	1935	2/DC	CATHEDRAL MODERN BC/SW	50
RCA	BT6-5	1935	2/DC	TOMBSTONE BC/SW	30
RCA	BT7-8	1935	2/DC	TOMBSTONE BC/SW	30
RCA	C 7-6	1935	AC	CONSOLE BC/SW	85
RCA	C 8-15	1935	AC	CONSOLE BC/SW	90
RCA	C 9-4	1935	AC	CONSOLE TUNING EYE BC/SW	90
RCA	C11-1	1935	AC	CONSOLE TUNING EYE BC/SW	115
RCA	C13-2	1935	AC	CONSOLE TUN. EYE BC/SW/LW	145
RCA	C15-3	1935	AC	CONSOLE TUN. EYE BC/SW/LW	160
RCA	D 9-19	1935	AC	CONSOLE R/P TUN. EYE BC/SW	65
RCA	D11-2	1935	AC	CONSOLE R/P TUN. EYE BC/SW	65
RCA	D22-1	1935	AC	CONSOLE R/P TUN. EYE BC/SW	220
RCA	HF-1	1938	AC	CONSOLE	100
RCA	HF-2	1938	AC	CONSOLE GRAND ET BC/SW	95

MANUFACTURER	MODEL	YEAR	POWER	STYLE	PRICE
RCA	HF-4	1938	AC	CONSOLE 18TH CENT ET BC/SW	110
RCA	HF-6	1938	AC	CONSOLE GRAND ET BC/SW	155
RCA	HF-8	1938	AC	CONSOLE 18TH CENT ET BC/SW	165
RCA	K-50	1939	AC	CONSOLE	50
RCA	K-60	1939	AC	CONSOLE PB BC/SW	110
RCA	K-61	1940	AC	CONSOLE PB BC/SW	120
RCA	K-80	1939	AC	CONSOLE PB BC/SW	130
RCA	K-130	1939	AC	CONSOLE PB BC/SW	150
RCA	P-5	1940	AC	PORTABLE	15
RCA	R-4	1932	AC	TABLE	
RCA	R-5	1931	AC	CATHEDRAL	95
RCA	R-6	1932	AC	CONSOLE	
RCA	R-7	1931	AC	TABLE	
RCA	R-9	1931	AC	CONSOLE	
RCA	R-11	1931	AC	CONSOLE	
RCA	R-17	1933	AC	TABLE	
RCA	R-18-W	1933	AC	COMPACT	
RCA	R-20 VICTOR	1927	BAT	CONSOLE 5 TUBE 3 DIAL	
RCA	R-22	1933	AC	MANTLE	60
RCA	R-27	1933	AC	COMPACT	
RCA	R-28	1933	AC	CATHEDRAL	225
RCA	R-28-B	1933	AC	TABLE	40
RCA	R-32 VICTOR	1929	AC	CONSOLE 10 TUBE 1 DIAL	
RCA	R-35 VICTOR	1930	AC	CONSOLE LB 8 TUBE 1 DIAL	
RCA	R-37	1933	AC	CATHEDRAL	270
RCA	R-38	1933	AC	CONSOLE	160
RCA	R-39 VICTOR	1930	AC	CONSOLE 8 TUBE 1 DIAL	
RCA	R-43	1931	BAT	CONSOLE	
RCA	R-50	1931	AC	CONSOLE	
RCA	R-51-B	1933	BAT	TABLE	
RCA	R-52 VICTOR	1930	AC	CONSOLE 10 TUBE 1 DIAL	
RCA	R-53-B	1933	BAT	CONSOLE	
RCA	R-55	1931	AC	CONSOLE	
RCA	R-70	1933	AC	TABLE	
RCA	R-71	1932	AC	TABLE	
RCA	R-71-B	1932	BAT	TABLE	
RCA	R-72	1932	AC	CONSOLE	
RCA	R-74	1932	AC	CATHEDRAL	120
RCA	R-76	1932	AC	CONSOLE	170
RCA	R-77	1932	AC	CONSOLE	
RCA	R-78	1932	AC	CONSOLE	150
RCA	R-90	1933	AC	CONSOLE	
RCA	RADIOLA 16	1927	BAT	TABLE LP 5 TUBE 1 DIAL	95
RCA	RADIOLA 17	1927	AC	TABLE LP 7 TUBE 1 DIAL	80
RCA	RADIOLA 18	1927	AC	TABLE LP 7 TUBE 1 DIAL	75
RCA	RADIOLA 20	1925	BAT	TABLE LP SLOPE PAN 5T 2 DL	150
RCA	RADIOLA 21	1929	BAT	TABLE LP 5 TUBE 1 DIAL	95
RCA	RADIOLA 22	1929	BAT	CONSOLE LB 5 TUBE 1 DIAL	110

MANUFACTURER	MODEL	YEAR	POWER	STYLE	PRICE
RCA	RADIOLA 24	1925	BAT	PORTABLE 6 TUBE 2 DIAL	300
RCA	RADIOLA 25	1925	BAT	TABLE LP LOOP ANT 6 T 2 DL	160
RCA	RADIOLA 26	1925	BAT	PORTABLE 6 TUBE 2 DIAL	300
RCA	RADIOLA 28	1925	BAT	CONSOLE LB LOOP ANT 8T 2 DIAL	225
RCA	RADIOLA 30	1926	AC	CONSOLE HB(RAD. 28)10 T 1D	200
RCA	RADIOLA 32	1927	AC	CONSOLE (RADIOLA 28)	200
RCA	RADIOLA 33	1929	AC	TABLE METAL 7 TUBE 1 DIAL	65
RCA	RADIOLA 33	1929	AC	CONSOLE-TABLE MODEL W/LEGS	90
RCA	RADIOLA 41	1928	AC	CONSOLE HP 6 TUBE 1 DIAL	200
RCA	RADIOLA 44	1929	AC	TABLE 5 TUBE 1 DIAL	75
RCA	RADIOLA 46	1929	AC	CONSOLE LB 5 TUBE 1 DIAL	85
RCA	RADIOLA 47	1929	AC	CONSOLE RADIO/PHONO 5T 1 DIAL	500
RCA	RADIOLA 48	1930	AC	CONSOLE 7 TUBE 1 DIAL	300
RCA	RADIOLA 51	1927	AC	CONSOLE 7 TUBE 1 DIAL	
RCA	RADIOLA 60	1928	AC	TABLE LP 9 TUBE 1 DIAL	100
RCA	RADIOLA 62	1928	AC	CONSOLE HB 9 TUBE 1 DIAL	200
RCA	RADIOLA 64	1928	AC	CONSOLE HB 11 TUBE 1 DIAL	490
RCA	RADIOLA 66	1929	AC	CONSOLE HB 8 TUBE 1 DIAL	125
RCA	RADIOLA 67	1929	AC	CONSOLE RADIO/PHONO 10T 1 DIAL	
RCA	RADIOLA 80	1930	AC	CONSOLE 9 TUBE 1 DIAL	120
RCA	RADIOLA 82	1930	AC	CONSOLE 9 TUBE 1 DIAL	
RCA	RADIOLA 86	1930	AC	CONSOLE RADIO/PHONO 9T 1 DIAL	
RCA	RADIOLA GRAND	1922	BAT	TABLE HP 4 TUBE 1 DIAL	140
RCA	RADIOLA II	1923	BAT	PORTABLE 2 TUBE 2 DIAL	250
RCA	RADIOLA III	1924	BAT	TABLE 2 TUBE 1 DIAL	80
RCA	RADIOLA III-A	1924	BAT	TABLE 4 TUBE 1 DIAL	110
RCA	RADIOLA IV	1923	BAT	TABLE HP SPEAKER 3 T 3 DL	300
RCA	RADIOLA IX	1924	BAT	CONSOLE IX IN VICTROLA LID	
RCA	RADIOLA V	1923	BAT	TABLE LP 3 TUBE 2 DIAL	400
RCA	RADIOLA VI	1923	BAT	TABLE HP 6 TUBE 2 DIAL	250
RCA	RADIOLA VII	1923	BAT	TABLE HP 5 TUBE 2 DIAL	
RCA	RADIOLA VII-A	1923	BAT	CONSOLE LB 5 TUBE 2 DIAL	
RCA	RADIOLA VII-B	1924	BAT	TABLE HP SPEAKER 5 T 2 DL	
RCA	RADIOLA VIII	1924	BAT	CONSOLE HB SUPER 6 T 2 DL	285
RCA	RADIOLA X	1924	BAT	TABLE HP SPEAKER 4 T 2 DL	340
RCA	RAE-26	1931	AC	CONSOLE RADIO/PHONO	150
RCA	RAE-59	1931	AC	CONSOLE RADIO/PHONO	
RCA	RAE-79	1931	AC	CONSOLE R/P W/RECORDER	180
RCA	RAE-84	1932	AC	CONSOLE RADIO/PHONO	80
RCA	RC (RA/DA)	1922	BAT	TABLE 3 TUBE 1 DIAL	175
RCA	RE-40	1933	AC	TABLE CHEST RADIO/PHONO	
RCA	RE-45 VICTOR	1929	AC	CONSOLE RADIO/PHONO 10T 1 DIAL	195
RCA	RE-57 VICTOR	1930	AC	CONSOLE RADIO/PHONO 8T 1 DIAL	

MANUFACTURER	MODEL	YEAR	POWER	STYLE	PRICE
RCA	RE-73	1931	AC	CONSOLE RADIO/PHONO	
RCA	RE-75 VICTOR	1930	AC	CONSOLE HB R/P 10 T 1 DIAL	225
RCA	RE-81	1932	AC	CONSOLE	190
RCA	RE-154 VICTOR	1930	AC	CONSOLE RADIO/PHONO 10T 1 DIAL	
RCA	RE-156 VICTOR	1930	AC	CONSOLE RADIO/PHONO 10T 1 DIAL	
RCA	REGENOFLEX	1924	BAT	TABLE MODEL X NO SPEAKER	175
RCA	RF AERIOLA SR.	1922	BAT	MANTLE 1 TUBE 1 DIAL	140
RCA	RG AERIOLA GRND	1922	BAT	CONSOLE 4 TUBE 1 DIAL	
RCA	RS RADIOLA	1923	BAT	MANTLE 2 TUBE 1 DIAL	290
RCA	SUPERHETERODYNE	1924	BAT	TABLE/PORTABLE SUPER VIII	
RCA	T 4-8	1935	AC	CATHEDRAL	85
RCA	T 4-9	1935	AC	TABLE	35
RCA	T 4-10	1935	AC	TABLE	45
RCA	T 5-2	1935	AC	TABLE BC/SW	50
RCA	T 8-14	1935	AC	TOMBSTONE BC/SW	150
RCA	T-9-10	1935	AC	TOMBSTONE TUNING EYE BC/SW	
RCA	T-10-1	1935	AC	TABLE BC/SW	85
RCA	T-63	1940	AC	TABLE PB BC/SW	40
RCA	T-64	1939	AC	TABLE PB BC/SW	50
RCA	T-80	1939	AC	TABLE PB BC/SW	50
RCA	U-8	1939	AC	TABLE	15
RCA	U-10	1939	AC	TABLE RADIO/PHONO PB	20
RCA	U-20	1939	AC	CONSOLE RADIO/PHONO BC/SW	115
RCA	U-25	1939	AC	CONSOLE PB BC/SW	50
RCA	U-26	1939	AC	CONSOLE R/P PB BC/SW	45
RCA	U-30	1939	AC	CONSOLE PB BC/SW	50
RCA	U-42	1939	AC	CONSOLE RADIO/PHONO BC/SW	45
RCA	U-43	1940	AC	CONSOLE RADIO/PHONO BC/SW	45
RCA	U-45	1940	AC	CONSOLE R/P PB TE BC/SW	55
RCA	U-46	1940	AC	CONSOLE R/P PB BC/SW	50
RCA	U-50	1939	AC	TABLE BC/SW	15
RCA	U-111	1938	AC	TABLE RADIO/PHONO	15
RCA	U-112	1939	AC	TABLE	15
RCA	U-119	1938	AC	CONSOLE RADIO/PHONO ET B/S	20
RCA	U-121	1939	AC	CONSOLE PB	45
RCA	U-122E	1938	AC	CONSOLE RADIO/PHONO ET B/S	45
RCA	U-123	1939	AC	CONSOLE PB BC/SW	45
RCA	U-124	1938	AC	CONSOLE RADIO/PHONO ET B/S	45
RCA	U-125	1939	AC	CONSOLE PB BC/SW	55

MANUFACTURER	MODEL	YEAR	POWER	STYLE	PRICE
RCA	U-126	1938	AC	CONSOLE RADIO/PHONO ET B/S	45
RCA	U-127E	1939	AC	CONSOLE END TABLE R/P	60
RCA	U-128	1938	AC	CONSOLE RADIO/PHONO ET B/S	50
RCA	U-129	1939	AC	CONSOLE R/P PB BC/SW	50
RCA	U-130	1938	AC	CONSOLE RADIO/PHONO ET B/S	55
RCA	U-132	1938	AC	CONSOLE RADIO/PHONO ET B/S	60
RCA	U-134	1938	AC	CONSOLE RADIO/PHONO ET B/S	60
RCA	V-101	1941	AC	TABLE RADIO/PHONO	15
RCA	V-170	1940	AC	CONSOLE RADIO/PHONO BC/SW	45
RCA	V-200	1940	AC	CONSOLE RADIO/PHONO	50
RCA	V-201	1940	AC	CONSOLE RADIO/PHONO	40
RCA	V-205	1940	AC	CONSOLE R/P PB BC/SW	50
RCA	V-210	1941	AC	CONSOLE RADIO/PHONO BC/SW	50
RCA	V-225	1941	AC	CONSOLE R/P PB BC/SW	50
RCA	V-300	1940	AC	CONSOLE RADIO/PHONO	50
RCA	V-301	1940	AC	CONSOLE RADIO/PHONO	50
RCA	V-302	1940	AC	CONSOLE PB TUN EYE BC/SW	95
RCA	V-405	1940	AC	CONSOLE RADIO/PHONO	
RCA	VHR-202	1940	AC	CONSOLE R/P W/RECORDER	55
RCA	VHR-207	1940	AC	CONSOLE R/P W/RECORDER	55
RCA	VHR-407	1940	AC	CONSOLE R/P W/RECORDER	55
RCA	VICTOR 7-1	1926	BAT	CONSOLE ALHAMBRA I 5T 3 DIAL	
RCA	VICTOR 7-2	1926	BAT	CONSOLE ALHAMBRA II 6T 2 DIAL	
RCA	VICTOR 7-10	1928	BAT	CONSOLE 6 TUBE 1 DIAL	
RCA	VICTOR 7-11	1928	AC	CONSOLE VICTROLA 7 T 1 DL	
RCA	VICTOR 7-25	1928	AC	CONSOLE 7 TUBE 1 DIAL	
RCA	VICTOR 7-26	1928	AC	CONSOLE 7 TUBE 1 DIAL	
RCA	VICTOR 7-30	1927	BAT	CONSOLE 5 TUBE 3 DIAL	
RCA	VICTOR 9-1	1926	BAT	CONSOLE FLORENZA 8 T 2 DL	
RCA	VICTOR 9-2	1926	AC	CONSOLE BORGIA II 10T 2 DL	
RCA	VICTOR 9-3	1926	BAT	CONSOLE BORGIA I 8 T 2 DL	
RCA	VICTOR 9-15	1927	BAT	CONSOLE 8 TUBE 2 DIAL	
RCA	VICTOR 9-16	1928	AC	CONSOLE 8 TUBE 2 DIAL	
RCA	VICTOR 9-18	1929	AC	CONSOLE 11 TUBE 2 DIAL	
RCA	VICTOR 9-25	1927	AC	CONSOLE 10 TUBE 2 DIAL	
RCA	VICTOR 9-40	1927	AC	CONSOLE BORGIA 10 T 2 DIAL	
RCA	VICTOR 9-54	1928	AC	CONSOLE RADIO/PHONO 11T 2 DIAL	
RCA	VICTOR 9-55	1927	AC	CONSOLE RADIO/PHONO 10T 2 DIAL	500
RCA	VICTOR 9-56	1929	AC	CONSOLE RADIO/PHONO 11T 2 DIAL	

MANUFACTURER	MODEL	YEAR	POWER	STYLE	PRICE
RCA	VICTOR 15-1	1926	AC	CONSOLE HYPERION 10 T 2 DIAL	
RECEPTON RADIO	RECEPTON	1924	BAT	TABLE LP 5 TUBE 3 DIAL	
REGENT RADIO	R-200	1925	BAT	TABLE LP 5 TUBE 3 DIAL	
REICHMANN	50 THOROLA	1925	BAT	TABLE 5 TUBE 3 DIAL	140
REICHMANN	52 THOROLA	1925	BAT	CONSOLE HB 5 TUBE 3 DIAL	150
REICHMANN	52 THOROLA	1925	BAT	TABLE 5 TUBE 3 DIAL	
REICHMANN	55 THOROLA	1925	BAT	TABLE 5 TUBE 3 DIAL	115
REICHMANN	57 THOROLA	1926	BAT	TABLE LP 5 TUBE 2 DIAL	95
REICHMANN	58 THOROLA	1926	BAT	CONSOLE LB 5 TUBE 2 DIAL	125
REICHMANN	59 THOROLA	1926	BAT	CONSOLE LB 5 TUBE 2 DIAL	140
REICHMANN	60 THOROLA	1927	AC	CONSOLE LB ENCLOSED	200
REICHMANN	61 THOROLA	1927	AC	CONSOLE HB	165
REICHMANN	65 THOROLA	1927	BAT	TABLE 2 DIAL	
REICHMANN	66 THOROLA	1927	BAT	TABLE 2 DIAL	
REMLER	10-3	1932	AC	TABLE BC/SW	
REMLER	11	1931	AC	TABLE	
REMLER	12-3	1932	AC	TABLE BC/SW	
REMLER	14	1938	AC	MANTLE PLASTIC	
REMLER	15	1931	AC	TABLE	
REMLER	15-3	1932	AC	TABLE BC/SW	
REMLER	15-3-C	1932	AC	CONSOLE BC/SW	
REMLER	15-C	1931	AC	CONSOLE LB	
REMLER	19-C	1931	AC	CONSOLE LB	
REMLER	21	1931	AC	TABLE	
REMLER	21-3	1932	AC	TABLE BC/SW	
REMLER	26	1933	AC	MANTLE BC/SW	
REMLER	27	1933	AC	MANTLE BC/SW	
REMLER	28	1936	AC	MANTLE PLASTIC BC/SW	
REMLER	40	1935	AC	MANTLE BC/SW	
REMLER	41	1935	AC	MANTLE BC/SW	
REMLER	43	1935	AC	TABLE BC/SW	
REMLER	45	1936	AC	TABLE BC/SW	
REMLER	46	1936	AC	MANTLE PLASTIC BC/SW	25
REMLER	47	1937	AC	TABLE PLASTIC BC/SW	
REMLER	51	1936	AC	TABLE	
REMLER	52	1937	AC	TABLE PLASTIC BC/SW	
REMLER	54	1938	AC	TABLE PLASTIC	20
REMLER	55	1938	AC	TABLE	
REMLER	56	1938	AC	MANTLE PLASTIC BC/SW	
REMLER	60	1936	AC	TABLE BC/SW	
REMLER	61	1938	AC	TABLE PB	30
REMLER	62	1935	AC	TABLE BC/SW	
REMLER	64	1936	AC	TABLE BC/SW	
REMLER	65	1937	AC	TABLE FLAT BC/SW	
REMLER	71	1936	AC	TABLE BC/SW	
REMLER	72	1937	AC	TABLE FLAT BC/SW	
REMLER	88	1935	AC	TABLE BC/SW	
REMLER	89	1936	AC	TABLE BC/SW	
REMLER	91	1936	2/DC	TABLE BC/SW	

MANUFACTURER	MODEL	YEAR	POWER	STYLE	PRICE
REMLER	471	1940	AC	TABLE RADIO/PHONO	15
REMLER	479	1938	AC	TABLE	
REPUBLIC RADIO	1 PC	1930	AC	CONSOLE RADIO/PHONO	
REPUBLIC RADIO	31-H	1930	AC	CONSOLE HB	
REPUBLIC RADIO	31-J	1930	AC	CATHEDRAL	
REPUBLIC RADIO	31-L	1930	AC	CONSOLE LB	
REPUBLIC RADIO	31-M	1930	AC	CATHEDRAL	
RESAS	5-T	1924	BAT	TABLE LP 5 TUBE 3 DIAL	105
REX RADIO	MASTERDYNE	1924	BAT	TABLE LP 5 TUBE 3 DIAL	70
RICHARDS	HEMCO	1923	BAT	PORTABLE 6 TUBE 2 DIAL	
RICHARDSON RADIO	2-A	1924	BAT	TABLE LP 5 TUBE 3 DIAL	65
ROBERTS RADIO	ORIOLE FIVE	1924	BAT	TABLE LP 5 TUBE 3 DIAL	65
RUCKELSHAUS	RUTIC SUPER 5	1925	BAT	TABLE LP SLANT PANEL 3 DL	95
S					
SCOTT TRANSFORMER	ABBEY	1930	AC	CONSOLE W/REMOTE CONTROL	200
SCOTT TRANSFORMER	CAVALIER	1930	AC	CONSOLE	160
SCOTT TRANSFORMER	CORTEZ	1930	AC	CONSOLE HB DELUXE	300
SCOTT TRANSFORMER	LANCASTER	1930	AC	CONSOLE	160
SCOTT TRANSFORMER	MODERNE	1930	AC	CONSOLE	150
SCOTT TRANSFORMER	ORLEANS	1930	AC	CONSOLE	175
SCOTT, E. H., RADIO LABS	ALLWAVE 2 VOLT	1934	BAT	CHASSIS AND SPEAKER	150
SCOTT, E. H., RADIO LABS	ALLWAVE 12	1931	AC	CHASSIS AND SPEAKER	200
SCOTT, E. H., RADIO LABS	ALLWAVE 12 DLUX	1932	AC	CHASSIS AND SPEAKER	250
SCOTT, E. H., RADIO LABS	ALLWAVE 15	1934	AC	CHASSIS AND SPEAKER	300
SCOTT, E. H., RADIO LABS	ALLWAVE IMPER.	1935	AC	CHASSIS AND SPEAKER	300
SCOTT, E. H., RADIO LABS	AUTOTROPE	1937	AC	CONSOLE RADIO/PHONO	450
SCOTT, E. H., RADIO LABS	LAUREATE	1941	AC	CHASSIS AND SPEAKER	200
SCOTT, E. H., RADIO LABS	MASTERPIECE	1938	AC	CHASSIS AND SPEAKER	275
SCOTT, E. H., RADIO LABS	PHANTOM	1938	AC	CHASSIS AND SPEAKER	300
SCOTT, E. H., RADIO LABS	PHANTOM DELUXE	1939	AC	CHASSIS AND SPEAKER	325
SCOTT, E. H., RADIO LABS	PHANTOM DELUXE	1941	AC	CHASSIS AND SPEAKER BC/FM	350
SCOTT, E. H., RADIO LABS	PHILHARMONIC	1938	AC	CHASSIS AND SPEAKER	450
SCOTT, E. H., RADIO LABS	PHILHARMONIC	1941	AC	CHASSIS AND SPEAKER	450
SCOTT, E. H., RADIO LABS	QUARANTA	1935	AC	CONSOLE RADIO/PHONO	3500
SCOTT, E. H., RADIO LABS	SHIELD GR. SYMP	1929	AC	CHASSIS	400
SCOTT, E. H., RADIO LABS	SHIELD GRID 9	1928	AC	TABLE	350
SCOTT, E. H., RADIO LABS	SHIELD GRID 10	1929	AC	TABLE	300
SCOTT, E. H., RADIO LABS	SIXTEEN-18	1937	AC	CHASSIS AND SPEAKER	275
SCOTT, E. H., RADIO LABS	SUPER 8	1925	BAT	TABLE	400
SCOTT, E. H., RADIO LABS	SUPER 9	1925	BAT	TABLE	400
SCOTT, E. H., RADIO LABS	SUPER 10	1927	BAT	TABLE	450
SCOTT, E. H., RADIO LABS	SUPER XII	1938	AC	CHASSIS AND SPEAKER	250
SCOTT, E. H., RADIO LABS	TELEMATIC	1938	AC	CHASSIS AND SPEAKER	250
SEARS, MANUFACTURING	A	1924	BAT	TABLE LP METER 4 TUBE 1 DL	110
SEARS, MANUFACTURING	B	1924	BAT	TABLE LP METER 5 TUBE 2 DL	125
SEARS, MANUFACTURING	T TORODYNE	1924	BAT	TABLE LP 5 TUBE 3 DIAL	85
SEARS, ROEBUCK	24	1929	BAT	TABLE METAL	40
SEARS, ROEBUCK	25	1929	BAT	CONSOLE LB	55

MANUFACTURER	MODEL	YEAR	POWER	STYLE	PRICE
SEARS, ROEBUCK	26	1929	BAT	CONSOLE LB	60
SEARS, ROEBUCK	27	1929	BAT	TABLE METAL	40
SEARS, ROEBUCK	28	1929	BAT	CONSOLE LB	50
SEARS, ROEBUCK	29	1929	BAT	CONSOLE HB	65
SEARS, ROEBUCK	30	1929	AC	TABLE METAL	40
SEARS, ROEBUCK	31	1929	AC	CONSOLE HB	65
SEARS, ROEBUCK	32	1929	AC	CONSOLE HB	70
SEARS, ROEBUCK	33	1929	AC	TABLE METAL	40
SEARS, ROEBUCK	34	1929	AC	CONSOLE LB	75
SEARS, ROEBUCK	35	1929	AC	CONSOLE HB	80
SEARS, ROEBUCK	40	1929	AC	CONSOLE HB	
SEARS, ROEBUCK	41	1929	AC	CONSOLE HB	
SEARS, ROEBUCK	42	1929	AC	CONSOLE HB	
SEARS, ROEBUCK	44	1929	BAT	TABLE METAL	35
SEARS, ROEBUCK	45	1929	BAT	CONSOLE LB	45
SEARS, ROEBUCK	46	1929	BAT	TABLE METAL	40
SEARS, ROEBUCK	47	1929	BAT	CONSOLE LB	60
SEARS, ROEBUCK	49	1929	AC	TABLE METAL	40
SEARS, ROEBUCK	50	1929	AC	CONSOLE HB	75
SEARS, ROEBUCK	51	1929	AC	CONSOLE HB	
SEARS, ROEBUCK	52	1929	AC	TABLE LOW PROFILE	50
SEARS, ROEBUCK	53	1929	AC	CONSOLE LB	145
SEARS, ROEBUCK	54	1929	AC	CONSOLE HB	160
SEARS, ROEBUCK	56	1929	AC	CONSOLE HB	
SEARS, ROEBUCK	57	1929	AC	CONSOLE HB	95
SEARS, ROEBUCK	58	1929	AC	CONSOLE HB	110
SEARS, ROEBUCK	59	1929	AC	TABLE METAL	35
SEARS, ROEBUCK	60	1929	AC	CONSOLE LB RADIO/PHONO	60
SEARS, ROEBUCK	80	1929	AC	CONSOLE RADIO/PHONO	140
SEARS, ROEBUCK	106A	1931	BAT	CONSOLE	50
SEARS, ROEBUCK	107A	1931	BAT	CONSOLE	60
SEARS, ROEBUCK	952	1930	AC	TABLE LOW PROFILE	
SEARS, ROEBUCK	1050	1930	AC	TABLE METAL	
SEARS, ROEBUCK	1060	1930	AC	TABLE METAL	35
SEARS, ROEBUCK	1064	1930	AC	CONSOLE LB	
SEARS, ROEBUCK	1069	1931	AC	CONSOLE LB	
SEARS, ROEBUCK	1070	1930	AC	TABLE METAL	35
SEARS, ROEBUCK	1074	1930	AC	CONSOLE LB	
SEARS, ROEBUCK	1076	1931	AC	CONSOLE LB	
SEARS, ROEBUCK	1080	1930	AC	TABLE METAL	
SEARS, ROEBUCK	1090	1930	AC	CONSOLE LB	
SEARS, ROEBUCK	1096	1931	AC	CONSOLE	65
SEARS, ROEBUCK	1100	1930	AC	CONSOLE LB	
SEARS, ROEBUCK	1110	1930	AC	CONSOLE HB	95
SEARS, ROEBUCK	1114	1930	AC	CONSOLE TABLE	
SEARS, ROEBUCK	1118	1931	AC	CONSOLE	65
SEARS, ROEBUCK	1120	1930	AC	CONSOLE LB W/CLOCK	
SEARS, ROEBUCK	1140	1931	AC	CATHEDRAL	95
SEARS, ROEBUCK	1150	1931	AC	CATHEDRAL	145
SEARS, ROEBUCK	1160	1931	AC	CONSOLE	50

MANUFACTURER	MODEL	YEAR	POWER	STYLE	PRICE
SEARS, ROEBUCK	1174	1931	AC	CONSOLE	
SEARS, ROEBUCK	1250	1931	AC	CATHEDRAL	
SEARS, ROEBUCK	1260	1931	AC	CONSOLE RADIO/PHONO	
SEARS, ROEBUCK	1291	1931	BAT	CATHEDRAL	110
SEARS, ROEBUCK	1310	1931	AC	CONSOLE	
SEARS, ROEBUCK	1320	1931	AC	CONSOLE	
SEARS, ROEBUCK	1330	1931	AC	CONSOLE RADIO/PHONO	
SEARS, ROEBUCK	1371	1931	AC	CATHEDRAL	
SEARS, ROEBUCK	1463	1932	AC	CONSOLE	
SEARS, ROEBUCK	1471	1932	AC	CATHEDRAL	110
SEARS, ROEBUCK	1481	1932	AC	CATHEDRAL	140
SEARS, ROEBUCK	1483	1932	AC	CONSOLE	
SEARS, ROEBUCK	1485	1932	AC	CLOCK GRANDFATHER	130
SEARS, ROEBUCK	1511	1932	AC	CONSOLE	70
SEARS, ROEBUCK	1531	1932	AC	CONSOLE RADIO/PHONO	80
SEARS, ROEBUCK	1571	1931	BAT	CATHEDRAL	65
SEARS, ROEBUCK	1575	1933	BAT	CONSOLE	80
SEARS, ROEBUCK	1576	1933	BAT	CATHEDRAL	
SEARS, ROEBUCK	1585	1932	AC	CATHEDRAL BC/SW	130
SEARS, ROEBUCK	1587	1932	AC	CONSOLE BC/SW	80
SEARS, ROEBUCK	1588	1933	AC	CONSOLE BC/SW	
SEARS, ROEBUCK	1591	1932	AC	CATHEDRAL	90
SEARS, ROEBUCK	1593	1932	AC	CONSOLE	65
SEARS, ROEBUCK	1597	1933	AC	CONSOLE	
SEARS, ROEBUCK	1641	1933	AC	CONSOLE	
SEARS, ROEBUCK	1660	1933	AC	CATHEDRAL	75
SEARS, ROEBUCK	1670	1933	AC	CONSOLE	
SEARS, ROEBUCK	1703	1933	AC	MANTLE	35
SEARS, ROEBUCK	1704	1933	AC	MANTLE	40
SEARS, ROEBUCK	1705	1933	AC	MANTLE	40
SEARS, ROEBUCK	1706	1933	AC	CATHEDRAL	90
SEARS, ROEBUCK	1707	1933	AC	CONSOLE	45
SEARS, ROEBUCK	1708	1933	AC	TABLE BC/SW	60
SEARS, ROEBUCK	1710	1933	BAT	PORTABLE	15
SEARS, ROEBUCK	1711	1933	BAT	TABLE	80
SEARS, ROEBUCK	1712	1933	BAT	CONSOLE	40
SEARS, ROEBUCK	1713	1933	BAT	CATHEDRAL	65
SEARS, ROEBUCK	1714	1933	BAT	CONSOLE	70
SEARS, ROEBUCK	1715	1933	32/DC	CONSOLE	
SEARS, ROEBUCK	1716	1933	32/DC	TABLE	
SEARS, ROEBUCK	1720	1933	AC	CONSOLE	70
SEARS, ROEBUCK	1721	1933	AC	CONSOLE BC/SW	65
SEARS, ROEBUCK	1722	1933	AC	CONSOLE BC/SW	95
SEARS, ROEBUCK	1724	1933	AC	TABLE	
SEARS, ROEBUCK	1725	1933	AC	CONSOLE	
SEARS, ROEBUCK	1726	1933	AC	CONSOLE BC/SW	
SEARS, ROEBUCK	1728	1933	AC	MANTLE	
SEARS, ROEBUCK	1729	1933	AC	CONSOLE	
SEARS, ROEBUCK	1732	1933	AC	CONSOLE BC/SW	
SEARS, ROEBUCK	1800	1934	AC	CATHEDRAL BC/SW	120

MANUFACTURER	MODEL	YEAR	POWER	STYLE	PRICE
SEARS, ROEBUCK	1801	1934	AC	MANTLE BC/SW	40
SEARS, ROEBUCK	1802	1934	AC	TABLE BC/SW	40
SEARS, ROEBUCK	1804	1934	AC	TABLE BC/SW	45
SEARS, ROEBUCK	1809	1935	AC	TABLE BC/SW	45
SEARS, ROEBUCK	1810	1935	AC	TABLE	45
SEARS, ROEBUCK	1820	1934	AC	CONSOLE BC/SW	60
SEARS, ROEBUCK	1821	1934	AC	CONSOLE BC/SW	65
SEARS, ROEBUCK	1822	1935	AC	CONSOLE BC/SW	65
SEARS, ROEBUCK	1823	1934	AC	CONSOLE BC/SW	60
SEARS, ROEBUCK	1824	1934	AC	CONSOLE BC/SW	70
SEARS, ROEBUCK	1833	1935	AC	CONSOLE BC/SW	45
SEARS, ROEBUCK	1835	1935	AC	CONSOLE BC/SW	60
SEARS, ROEBUCK	1850	1935	BAT	TABLE	45
SEARS, ROEBUCK	1852	1935	BAT	TABLE	35
SEARS, ROEBUCK	1853	1935	BAT	CONSOLE	50
SEARS, ROEBUCK	1854	1934	AC	CONSOLE BC/SW	65
SEARS, ROEBUCK	1855	1934	32/DC	TABLE	40
SEARS, ROEBUCK	1856	1934	AC	TABLE BC/SW	40
SEARS, ROEBUCK	1857	1934	AC	CONSOLE BC/SW	65
SEARS, ROEBUCK	1863	1935	32/DC	TABLE	30
SEARS, ROEBUCK	1865	1935	32/DC	TABLE	30
SEARS, ROEBUCK	1866	1935	BAT	CONSOLE BC/SW	55
SEARS, ROEBUCK	1867	1935	BAT	TABLE BC/SW	30
SEARS, ROEBUCK	1903	1936	AC	TABLE	40
SEARS, ROEBUCK	1904	1936	AC	TABLE	40
SEARS, ROEBUCK	1905	1936	AC	TABLE	50
SEARS, ROEBUCK	1911	1936	AC	CONSOLE BC/SW	
SEARS, ROEBUCK	1914	1936	AC	CONSOLE	45
SEARS, ROEBUCK	1915	1936	AC	CONSOLE BC/SW	50
SEARS, ROEBUCK	1917	1936	AC	CONSOLE	50
SEARS, ROEBUCK	1918	1936	AC	CONSOLE	60
SEARS, ROEBUCK	1923	1936	BAT	TABLE BC/SW	60
SEARS, ROEBUCK	1924	1936	6/DC	TABLE BC/SW	
SEARS, ROEBUCK	1925	1936	32/DC	TABLE BC/SW	30
SEARS, ROEBUCK	1928	1936	BAT	TABLE	
SEARS, ROEBUCK	1932	1936	BAT	CONSOLE	45
SEARS, ROEBUCK	1933	1936	BAT	CONSOLE BC/SW	50
SEARS, ROEBUCK	1934	1936	6/DC	CONSOLE BC/SW	
SEARS, ROEBUCK	1935	1936	32/DC	CONSOLE BC/SW	45
SEARS, ROEBUCK	1936	1936	BAT	CONSOLE	45
SEARS, ROEBUCK	1938	1936	AC	TABLE BC/SW	
SEARS, ROEBUCK	1939	1936	AC	TOMBSTONE BC/SW	
SEARS, ROEBUCK	1942	1936	AC	TOMBSTONE TUNING EYE BC/SW	
SEARS, ROEBUCK	1944	1936	AC	CONSOLE TUNING EYE BC/SW	
SEARS, ROEBUCK	1945	1936	AC	CONSOLE TUNING EYE BC/SW	
SEARS, ROEBUCK	1946	1936	AC	CONSOLE TUNING EYE BC/SW	
SEARS, ROEBUCK	1947	1936	BAT	TABLE BC/SW	
SEARS, ROEBUCK	1948	1936	BAT	CONSOLE BC/SW	
SEARS, ROEBUCK	3061-A, B, C	1941	AC	TABLE PLASTIC	20

MANUFACTURER	MODEL	YEAR	POWER	STYLE	PRICE
SEARS, ROEBUCK	3161-A, B, C	1941	AC	TABLE PLASTIC	25
SEARS, ROEBUCK	3261-A, B, C	1941	AC	TABLE PLASTIC	45
SEARS, ROEBUCK	3361	1941	AC	TABLE PLASTIC	20
SEARS, ROEBUCK	3561	1941	AC	TABLE PLASTIC	45
SEARS, ROEBUCK	4414	1936	BAT	TABLE	
SEARS, ROEBUCK	4415	1936	BAT	TABLE	
SEARS, ROEBUCK	4418	1937	BAT	MANTLE BC/SW	25
SEARS, ROEBUCK	4419	1937	BAT	TABLE BC/SW	30
SEARS, ROEBUCK	4420	1936	BAT	TABLE	
SEARS, ROEBUCK	4421	1936	BAT	TABLE BC/SW	20
SEARS, ROEBUCK	4422	1936	2/DC	TABLE BC/SW	
SEARS, ROEBUCK	4423	1936	2/DC	TABLE BC/SW	
SEARS, ROEBUCK	4424	1936	2/DC	TABLE BC/SW	
SEARS, ROEBUCK	4426	1936	BAT	TABLE BC/SW	
SEARS, ROEBUCK	4427	1936	2/DC	TABLE BC/SW	
SEARS, ROEBUCK	4429	1937	32/DC	TABLE BC/SW	30
SEARS, ROEBUCK	4437	1937	BAT	TABLE	30
SEARS, ROEBUCK	4438	1937	2/DC	TABLE	30
SEARS, ROEBUCK	4439	1937	BAT	TABLE BC/SW	30
SEARS, ROEBUCK	4440	1937	2/DC	TABLE BC/SW	30
SEARS, ROEBUCK	4441	1937	2/DC	TABLE TUNING EYE BC/SW	40
SEARS, ROEBUCK	4442	1936	BAT	CONSOLE	
SEARS, ROEBUCK	4443	1936	2/DC	CONSOLE	
SEARS, ROEBUCK	4444	1936	BAT	CONSOLE BC/SW	
SEARS, ROEBUCK	4445	1936	2/DC	CONSOLE BC/SW	
SEARS, ROEBUCK	4446	1936	BAT	CONSOLE BC/SW	
SEARS, ROEBUCK	4447	1936	2/DC	CONSOLE BC/SW	
SEARS, ROEBUCK	4448	1936	BAT	CONSOLE BC/SW	
SEARS, ROEBUCK	4449	1937	32/DC	CONSOLE BC/SW	45
SEARS, ROEBUCK	4450	1937	BAT	CONSOLE TUNING EYE BC/SW	
SEARS, ROEBUCK	4451	1937	BAT	CONSOLE BC/SW	60
SEARS, ROEBUCK	4455	1937	BAT	CONSOLE BC/SW	55
SEARS, ROEBUCK	4456	1937	4/DC	CONSOLE BC/SW	55
SEARS, ROEBUCK	4459	1937	BAT	CONSOLE BC/SW	60
SEARS, ROEBUCK	4461	1936	AC	TABLE	
SEARS, ROEBUCK	4462	1936	AC	TABLE BC/SW	
SEARS, ROEBUCK	4463	1937	AC	TABLE	35
SEARS, ROEBUCK	4464	1937	AC	TABLE BC/SW	
SEARS, ROEBUCK	4465	1937	AC	CONSOLE TUNING EYE BC/SW	45
SEARS, ROEBUCK	4466	1937	AC	TABLE TUNING EYE BC/SW	35
SEARS, ROEBUCK	4467	1937	AC	TABLE BC/SW	
SEARS, ROEBUCK	4469	1937	AC	TABLE	30
SEARS, ROEBUCK	4470	1937	AC	TABLE TUNING EYE BC/SW	35
SEARS, ROEBUCK	4472	1937	BAT	TABLE	25
SEARS, ROEBUCK	4473	1937	4/DC	TABLE	25
SEARS, ROEBUCK	4477	1937	BAT	CONSOLE	45
SEARS, ROEBUCK	4478	1937	4/DC	CONSOLE	45
SEARS, ROEBUCK	4484	1936	AC	CONSOLE BC/SW	
SEARS, ROEBUCK	4485	1936	AC	CONSOLE TUNING EYE BC/SW	55
SEARS, ROEBUCK	4486	1937	AC	CONSOLE TUNING EYE BC/SW	55

MANUFACTURER	MODEL	YEAR	POWER	STYLE	PRICE
SEARS, ROEBUCK	4487	1937	AC	CONSOLE SELF-TUNING BC/SW	70
SEARS, ROEBUCK	4488	1936	AC	CONSOLE BC/SW	
SEARS, ROEBUCK	4490	1937	AC	CONSOLE TUNING EYE BC/SW	50
SEARS, ROEBUCK	4498	1937	BAT	TABLE BC/SW	30
SEARS, ROEBUCK	4499	1937	2/DC	TABLE BC/SW	
SEARS, ROEBUCK	4500	1938	AC	MANTLE	15
SEARS, ROEBUCK	4505	1938	AC	MANTLE	20
SEARS, ROEBUCK	4532	1938	BAT	TABLE	
SEARS, ROEBUCK	4612	1938	32/DC	TABLE	35
SEARS, ROEBUCK	4620	1938	BAT	TABLE	20
SEARS, ROEBUCK	4638	1938	BAT	TABLE BC/SW	30
SEARS, ROEBUCK	4641	1938	BAT	TABLE	20
SEARS, ROEBUCK	4644	1938	BAT	TABLE	25
SEARS, ROEBUCK	4660	1937	AC	TABLE TUNING EYE	30
SEARS, ROEBUCK	4662	1938	AC	TABLE	30
SEARS, ROEBUCK	4663	1938	AC	TABLE PB BC/SW	45
SEARS, ROEBUCK	4664	1937	AC	TABLE TUNING EYE	35
SEARS, ROEBUCK	4665	1937	AC	TABLE TUNING EYE	50
SEARS, ROEBUCK	4666	1937	AC	TABLE TUNING EYE	40
SEARS, ROEBUCK	4667	1938	AC	TABLE	35
SEARS, ROEBUCK	4668	1938	AC	TABLE BC/SW	15
SEARS, ROEBUCK	4675	1938	AC	TOMBSTONE W/TILTBACK DIAL	35
SEARS, ROEBUCK	4677	1938	AC	CONSOLE BC/SW	50
SEARS, ROEBUCK	4680	1938	AC	CONSOLE	50
SEARS, ROEBUCK	4681	1938	AC	CONSOLE BC/SW	
SEARS, ROEBUCK	4684	1938	AC	CONSOLE ARMCHAIR	45
SEARS, ROEBUCK	4685	1937	AC	CONSOLE	50
SEARS, ROEBUCK	4686	1937	AC	CONSOLE TUNING EYE	65
SEARS, ROEBUCK	4687	1937	AC	CONSOLE TUNING EYE	65
SEARS, ROEBUCK	4688	1937	AC	CONSOLE TUN. EYE ET BC/SW	80
SEARS, ROEBUCK	4980	1931	AC	CONSOLE RADIO/PHONO	90
SEARS, ROEBUCK	4982	1931	BAT	CONSOLE RADIO/PHONO	75
SEARS, ROEBUCK	4983	1930	AC	CONSOLE LB RADIO/PHONO	
SEARS, ROEBUCK	4984	1931	BAT	CONSOLE RADIO/PHONO	85
SEARS, ROEBUCK	4985	1930	AC	CONSOLE LB RADIO/PHONO	
SEARS, ROEBUCK	4988	1931	AC	CONSOLE RADIO/PHONO	90
SEARS, ROEBUCK	5661	1941	AC	CONSOLE RADIO/PHONO	
SEARS, ROEBUCK	6038	1939	AC	CONSOLE PB BC/SW	85
SEARS, ROEBUCK	6102	1939	AC	MANTLE BLACK PLASTIC PB	20
SEARS, ROEBUCK	6103	1939	AC	MANTLE IVORY PLASTIC PB	20
SEARS, ROEBUCK	6109	1939	AC	MANTLE WALNUT PLASTIC	
SEARS, ROEBUCK	6110	1939	AC	MANTLE BLACK PLASTIC	25
SEARS, ROEBUCK	6111	1939	AC	MANTLE IVORY PLASTIC	
SEARS, ROEBUCK	6250	1939	AC	TABLE PB BC/SW	30
SEARS, ROEBUCK	6251	1939	AC	TABLE PB BC/SW	30
SEARS, ROEBUCK	6252	1939	AC	TABLE PB BC/SW	30
SEARS, ROEBUCK	6253	1939	AC	CONSOLE PB BC/SW	50
SEARS, ROEBUCK	6254	1939	AC	CONSOLE PB BC/SW	55
SEARS, ROEBUCK	6256	1940	BAT	PORTABLE	15

MANUFACTURER	MODEL	YEAR	POWER	STYLE	PRICE
SEARS, ROEBUCK	6261	1939	BAT	TABLE PB	25
SEARS, ROEBUCK	6262	1939	BAT	TABLE PB BC/SW	30
SEARS, ROEBUCK	6264	1939	BAT	TABLE PB BC/SW	30
SEARS, ROEBUCK	6265	1939	BAT	CONSOLE PB BC/SW	50
SEARS, ROEBUCK	6270	1939	6/DC	TABLE	30
SEARS, ROEBUCK	6271	1939	6/DC	TABLE	40
SEARS, ROEBUCK	6272	1939	6/DC	CONSOLE	70
SEARS, ROEBUCK	6274	1940	BAT	PORTABLE	15
SEARS, ROEBUCK	6320	1940	AC	CONSOLE PB BC/SW	40
SEARS, ROEBUCK	6321	1940	AC	TABLE PB	35
SEARS, ROEBUCK	6324	1940	AC	TABLE PB	35
SEARS, ROEBUCK	6325	1940	AC	TABLE PB TUN EYE BC/SW	
SEARS, ROEBUCK	6326	1940	AC	TABLE RADIO/PHONO	15
SEARS, ROEBUCK	6335	1939	AC	CONSOLE	110
SEARS, ROEBUCK	6336	1940	AC	CONSOLE	120
SEARS, ROEBUCK	6337	1940	AC	CONSOLE TUNING EYE BC/SW	130
SEARS, ROEBUCK	6346	1940	AC	CONSOLE RADIO/PHONO BC/SW	45
SEARS, ROEBUCK	6354	1940	BAT	TABLE	
SEARS, ROEBUCK	6356	1940	BAT	TABLE	30
SEARS, ROEBUCK	6359	1940	BAT	TABLE PB BC/SW	
SEARS, ROEBUCK	6363	1940	BAT	TABLE PB	
SEARS, ROEBUCK	6368	1940	BAT	TABLE PB BC/SW	40
SEARS, ROEBUCK	6380	1940	BAT	CONSOLE PB BC/SW	
SEARS, ROEBUCK	6382	1940	BAT	CONSOLE PB BC/SW	65
SEARS, ROEBUCK	6400	1940	AC	MANTLE BLACK PLASTIC	25
SEARS, ROEBUCK	6401	1940	AC	MANTLE IVORY PLASTIC	25
SEARS, ROEBUCK	6402	1940	AC	TABLE WALNUT PLASTIC	25
SEARS, ROEBUCK	6403	1940	AC	TABLE BLACK PLASTIC	15
SEARS, ROEBUCK	6404	1940	AC	MANTLE IVORY PLASTIC	20
SEARS, ROEBUCK	6405	1940	AC	MANTLE WALNUT PLASTIC	15
SEARS, ROEBUCK	6407	1940	AC	MANTLE BLACK PLASTIC	15
SEARS, ROEBUCK	6408	1940	AC	MANTLE IVORY PLASTIC	20
SEARS, ROEBUCK	6409	1940	AC	MANTLE WALNUT PLASTIC	15
SEARS, ROEBUCK	6435	1940	AC	CONSOLE PB BC/SW	
SEARS, ROEBUCK	6436	1940	AC	CONSOLE PB BC/SW	
SEARS, ROEBUCK	6437	1940	AC	CONSOLE PB BC/SW	175
SEARS, ROEBUCK	6438	1940	AC	CONSOLE PB TUN EYE BC/SW	120
SEARS, ROEBUCK	6490	1940	AC	CONSOLE PB BC/SW	60
SEARS, ROEBUCK	6491	1940	AC	CONSOLE DUNCAN PHYFE TABLE	45
SEARS, ROEBUCK	6493	1940	AC	CONSOLE BOOKCASE PB BC/SW	
SEARS, ROEBUCK	7000	1941	AC	MANTLE WALNUT PLASTIC	
SEARS, ROEBUCK	7002	1941	AC	MANTLE ONYX PLASTIC	
SEARS, ROEBUCK	7003	1941	AC	MANTLE WALNUT PLASTIC PB	
SEARS, ROEBUCK	7006	1941	AC	MANTLE ONYX PLASTIC PB	
SEARS, ROEBUCK	7010	1941	AC	TABLE WALNUT PLASTIC	
SEARS, ROEBUCK	7014	1941	AC	TABLE BC/SW	
SEARS, ROEBUCK	7036	1941	AC	TABLE PB BC/SW	
SEARS, ROEBUCK	7037	1941	AC	TABLE PB BC/SW	20

MANUFACTURER	MODEL	YEAR	POWER	STYLE	PRICE
SEARS, ROEBUCK	7038	1941	AC	TABLE PB BC/SW	
SEARS, ROEBUCK	7039	1941	AC	TABLE PB BC/SW	40
SEARS, ROEBUCK	7046	1941	AC	CONSOLE PB BC/SW	
SEARS, ROEBUCK	7048	1941	AC	CONSOLE PB BC/SW	50
SEARS, ROEBUCK	7050	1941	AC	CONSOLE PB BC/SW	
SEARS, ROEBUCK	7054	1941	AC	TABLE RADIO/PHONO	15
SEARS, ROEBUCK	7056	1941	AC	TABLE RADIO/PHONO	15
SEARS, ROEBUCK	7057	1941	AC	TABLE RADIO/PHONO	15
SEARS, ROEBUCK	7066	1941	AC	CONSOLE RADIO/PHONO	
SEARS, ROEBUCK	7068	1941	AC	CONSOLE R/P PB BC/SW	
SEARS, ROEBUCK	7069	1941	AC	CONSOLE RADIO/PHONO PB FM	45
SEARS, ROEBUCK	7084	1941	BAT	PORTABLE	15
SEARS, ROEBUCK	7101	1941	BAT	TABLE	
SEARS, ROEBUCK	7102	1941	BAT	TABLE	
SEARS, ROEBUCK	7105	1941	BAT	TABLE	30
SEARS, ROEBUCK	7106	1941	AC	TABLE	
SEARS, ROEBUCK	7109	1941	AC	TABLE BC/SW	35
SEARS, ROEBUCK	7111	1941	AC	TABLE TUNING EYE BC/SW	
SEARS, ROEBUCK	7120	1941	32/DC	TABLE PB BC/SW	
SEARS, ROEBUCK	7127	1941	AC	TABLE PB BC/SW	
SEARS, ROEBUCK	7159	1941	BAT	CONSOLE BC/SW	
SEARS, ROEBUCK	7177	1941	AC	CONSOLE PB BC/SW	
SEARS, ROEBUCK	XIV	1928	BAT	TABLE LOW PROFILE	55
SEARS, ROEBUCK	XIX	1928	BAT	TABLE HIGH PROFILE	90
SEARS, ROEBUCK	XV	1928	BAT	TABLE LOW PROFILE	85
SEARS, ROEBUCK	XVI	1928	BAT	CONSOLE HB	120
SEARS, ROEBUCK	XVII	1928	BAT	CONSOLE	110
SEARS, ROEBUCK	XVIII	1928	AC	TABLE LOW PROFILE	55
SEARS, ROEBUCK	XX	1928	AC	CONSOLE LB	115
SEARS, ROEBUCK	XXIII	1928	BAT	CONSOLE LB	110
SECTRON RADIO	A DELUXE	1923	BAT	TABLE HP 6 TUBE 2 DIAL	
SECTRON RADIO	B	1923	BAT	PORTABLE 6 TUBE 2 DIAL	
SECTRON RADIO	C BUS. MAN'S	1923	BAT	CONSOLE DESK 6 TUBE 2 DIAL	
SEELEY ELECTRIC	5	1930	AC	CATHEDRAL	
SEELEY ELECTRIC	6	1930	AC	CATHEDRAL	
SENTINEL RADIO	14A72	1935	AC	CONSOLE BC/SW	
SENTINEL RADIO	14A86	1935	AC	TABLE BC/SW	
SENTINEL RADIO	19A68	1935	AC	TABLE BC/SW	
SENTINEL RADIO	19A71	1935	AC	CONSOLE BC/SW	
SENTINEL RADIO	20A71	1935	AC	CONSOLE BC/SW	
SENTINEL RADIO	20A73	1935	AC	TABLE BC/SW	
SENTINEL RADIO	25A72	1935	AC	CONSOLE BC/SW	
SENTINEL RADIO	25A74	1935	AC	TABLE BC/SW	
SENTINEL RADIO	30A83	1935	AC	TABLE BC/SW	
SENTINEL RADIO	31B72	1935	6/DC	CONSOLE BC/SW	
SENTINEL RADIO	31B74	1935	6/DC	TABLE BC/SW	
SENTINEL RADIO	32B69	1935	2/DC	TABLE	30
SENTINEL RADIO	33B70	1935	2/DC	TABLE BC/SW	
SENTINEL RADIO	33B71	1935	2/DC	CONSOLE BC/SW	

MANUFACTURER	MODEL	YEAR	POWER	STYLE	PRICE
SENTINEL RADIO	34B71	1935	6/DC	CONSOLE BC/SW	
SENTINEL RADIO	34B73	1935	6/DC	TABLE BC/SW	
SENTINEL RADIO	35B70	1935	2/DC	TABLE BC/SW	
SENTINEL RADIO	36L71	1935	32/DC	CONSOLE BC/SW	
SENTINEL RADIO	36L73	1935	32/DC	TABLE BC/SW	
SENTINEL RADIO	40A	1935	AC	CONSOLE BC/SW	
SENTINEL RADIO	40A	1935	AC	TABLE BC/SW	
SENTINEL RADIO	118BCT	1938	1.5DC	CONSOLE	
SENTINEL RADIO	118BT	1938	1.5DC	TABLE	
SENTINEL RADIO	119BCT	1938	6/DC	CONSOLE	
SENTINEL RADIO	119BT	1938	6/DC	TABLE	
SENTINEL RADIO	124AA	1938	AC	CONSOLE CHAIRSIDE	
SENTINEL RADIO	124ACT	1938	AC	CONSOLE	
SENTINEL RADIO	124AT	1938	AC	TABLE	
SENTINEL RADIO	125ACE	1938	AC	CONSOLE BC/SW	
SENTINEL RADIO	125ATE	1938	AC	TABLE BC/SW	
SENTINEL RADIO	127BC	1938	1.5DC	CONSOLE	
SENTINEL RADIO	127BT	1938	1.5DC	TABLE	
SENTINEL RADIO	130BC	1938	6/DC	CONSOLE BC/SW	
SENTINEL RADIO	130BT	1938	6/DC	TABLE BC/SW	
SENTINEL RADIO	137UT	1938	AC	TABLE	
SENTINEL RADIO	140BC	1938	6/DC	CONSOLE BC/SW	
SENTINEL RADIO	140BT	1938	6/DC	TABLE BC/SW	
SENTINEL RADIO	142AT	1938	AC	TABLE BC/SW	
SENTINEL RADIO	143LC	1938	32/DC	CONSOLE	
SENTINEL RADIO	143LT	1938	32/DC	TABLE BC/SW	
SENTINEL RADIO	144XC	1938	AC	CONSOLE BC/SW	
SENTINEL RADIO	144XT	1938	AC	TABLE BC/SW	
SENTINEL RADIO	168BC	1939	AC	CONSOLE BC/SW	40
SENTINEL RADIO	168BT	1939	AC	TABLE BC/SW	
SENTINEL RADIO	175BC-CB	1939	BAT	CONSOLE RADIO/PHONO	
SENTINEL RADIO	175BCT	1939	AC	CONSOLE	40
SENTINEL RADIO	175BT	1939	AC	TABLE PLASTIC	20
SENTINEL RADIO	175BT-CB	1939	BAT	TABLE RADIO/PHONO	
SENTINEL RADIO	175BTW	1940	2/DC	TABLE WALNUT	35
SENTINEL RADIO	176BC	1939	AC	CONSOLE PB BC/SW	
SENTINEL RADIO	176BT	1939	AC	TABLE PB BC/SW	30
SENTINEL RADIO	178BL	1939	BAT	PORTABLE	
SENTINEL RADIO	180XL	1939	AC	PORTABLE	
SENTINEL RADIO	186BC	1939	BAT	CONSOLE BC/SW	
SENTINEL RADIO	186BT	1939	BAT	TABLE BC/SW	
SENTINEL RADIO	189LC	1939	32/DC	CONSOLE BC/SW	
SENTINEL RADIO	189LT	1939	32/DC	TABLE BC/SW	
SENTINEL RADIO	193UC	1939	AC	CONSOLE BC/SW	
SENTINEL RADIO	193UT	1939	AC	TABLE PLASTIC BC/SW	
SENTINEL RADIO	194UT	1939	AC	MANTLE PLASTIC	25
SENTINEL RADIO	195UT	1939	AC	MANTLE PB	30
SENTINEL RADIO	196AC-CB	1939	AC	CONSOLE RADIO/PHONO	
SENTINEL RADIO	196AT	1939	AC	TABLE	
SENTINEL RADIO	196AT-CB	1939	AC	TABLE RADIO/PHONO	

MANUFACTURER	MODEL	YEAR	POWER	STYLE	PRICE
SENTINEL RADIO	198AC	1939	AC	CONSOLE PB BC/SW	50
SENTINEL RADIO	198AT	1939	AC	TABLE PB BC/SW	35
SENTINEL RADIO	199AC	1939	AC	CONSOLE PB BC/SW	100
SENTINEL RADIO	199AC-CB	1939	AC	CONSOLE R/P PB BC/SW	
SENTINEL RADIO	200XT	1939	AC	TABLE	
SENTINEL RADIO	204ACE-CB	1940	AC	CONSOLE TUNING EYE	45
SENTINEL RADIO	212W	1940	AC	MANTLE PLASTIC	20
SENTINEL RADIO	213P	1940	AC	PORTABLE STAG LEATHERETTE	15
SENTINEL RADIO	216FY	1940	AC	CONSOLE R/P W/RECORDER	
SENTINEL RADIO	216JY	1940	AC	CONSOLE R/P W/REC. BC/SW	50
SENTINEL RADIO	216MY	1940	AC	CONSOLE R/P W/RECORDER	
SENTINEL RADIO	217P	1940	AC	PORTABLE LUGGAGE STYLE CAB	15
SENTINEL RADIO	218O	1940	AC	MANTLE PLASTIDC	25
SENTINEL RADIO	220F	1940	AC	CONSOLE RADIO/PHONO	
SENTINEL RADIO	220J	1940	AC	CONSOLE RADIO/PHONO	
SENTINEL RADIO	220M	1940	AC	CONSOLE RADIO/PHONO	
SENTINEL RADIO	221F	1940	AC	CONSOLE RADIO/PHONO	
SENTINEL RADIO	221J	1940	AC	CONSOLE RADIO/PHONO	
SENTINEL RADIO	221M	1940	AC	CONSOLE RADIO/PHONO	
SENTINEL RADIO	223D & RCD	1940	AC	CONSOLE TWIN COMMODES R/P	100
SENTINEL RADIO	227P	1940	AC	PORTABLE LIGHTWEIGHT	25
SENTINEL RADIO	238V	1941	AC	TABLE BOOK RADIO/PHONO	45
SENTINEL RADIO	249I	1941	AC	TABLE PLASTIC IVORY	25
SENTINEL RADIO	412	1932	AC	TABLE	
SENTINEL RADIO	501	1933	BAT	TABLE	
SENTINEL RADIO	502	1933	BAT	CONSOLE	
SENTINEL RADIO	513	1932	AC	TABLE	
SENTINEL RADIO	599	1933	AC	TABLE BC/SW	
SENTINEL RADIO	614	1932	AC	TABLE BC/SW	
SENTINEL RADIO	623	1933	AC	TABLE BC/SW	
SENTINEL RADIO	625	1933	AC	CONSOLE BC/SW	
SENTINEL RADIO	634	1933	AC	TABLE BC/SW	
SENTINEL RADIO	635	1933	AC	CONSOLE BC/SW	
SENTINEL RADIO	815	1932	AC	TABLE	
SENTINEL RADIO	816	1932	AC	CONSOLE	
SENTINEL RADIO	817	1932	AC	CONSOLE HB	
SENTINEL RADIO	826	1932	AC	CONSOLE BC/SW/LW	
SENTINEL RADIO	827	1932	AC	CONSOLE HB BC/SW/LW	
SENTINEL RADIO	1016	1932	AC	CONSOLE	
SENTINEL RADIO	1017	1932	AC	CONSOLE HB	
SENTINEL RADIO	1046	1933	AC	CONSOLE BC/SW	
SENTINEL RADIO	4143	1934	AC	TABLE	
SENTINEL RADIO	4536	1934	AC	TABLE BC/SW	
SENTINEL RADIO	5211	1934	AC	TABLE BC/SW	
SENTINEL RADIO	5628	1934	32/DC	TABLE	
SENTINEL RADIO	5721	1934	AC	TABLE BC/SW	
SENTINEL RADIO	6232	1934	BAT	CONSOLE	
SENTINEL RADIO	6241	1934	BAT	TABLE	

MANUFACTURER	MODEL	YEAR	POWER	STYLE	PRICE
SENTINEL RADIO	6315	1934	AC	TABLE BC/SW	
SENTINEL RADIO	6317	1934	AC	CONSOLE BC/SW	
SENTINEL RADIO	6323	1934	AC	TABLE BC/SW	
SENTINEL RADIO	7732	1934	BAT	CONSOLE BC/SW	
SENTINEL RADIO	7741	1934	BAT	TABLE BC/SW	
SENTINEL RADIO	DUOLA-A	1933	AC	CONSOLE W/EXTERNAL SPEAKER	
SHAMROCK	CONSOLE	1929	AC	CONSOLE 7 TUBE 1 DIAL	
SHEPARD-POTTER	T-12	1923	BAT	TABLE LP 1 TUBE 2 DIAL	95
SHEPARD-POTTER	TF-6	1923	BAT	TABLE LP 6 TUBE 3 DIAL	160
SHERMAN	CLEARFIELD	1924	BAT	TABLE LP CLEAR CAB. 6T 3DL	300
SILVA	R-5	1923	BAT	TABLE LP 5 TUBE 3 DIAL	
SILVER-MARSHALL	34-A	1930	AC	CONSOLE LB	
SILVER-MARSHALL	35-A	1930	AC	CONSOLE LB	
SILVER-MARSHALL	37 BEARCAT	1930	AC	TABLE METAL	
SILVER-MARSHALL	38	1930	AC	TABLE METAL	
SILVER-MARSHALL	39	1930	AC	TABLE METAL	
SILVER-MARSHALL	60	1929	AC	CONSOLE LB SHERATON	175
SILVER-MARSHALL	75	1929	AC	CONSOLE LB CONCERT GRAND	195
SILVER-MARSHALL	95	1929	AC	CONSOLE HB SHERATON	175
SILVER-MARSHALL	722	1930	AC	TABLE METAL	75
SILVER-MARSHALL	4802	1933	AC	CONSOLE	
SILVER-MARSHALL	4810	1933	AC	CONSOLE	
SILVER-MARSHALL	ALL WORLD 5	1934	AC	TABLE BC/SW	
SILVER-MARSHALL	ALL WORLD DLX	1934	AC	CONSOLE BC/SW	
SILVER-MARSHALL	ALL WORLD DLX	1934	AC	TABLE BC/SW	
SILVER-MARSHALL	D-1	1931	AC	CONSOLE LB	
SILVER-MARSHALL	F-4	1931	AC	TABLE	
SILVER-MARSHALL	F-5	1931	AC	CONSOLE LB	
SILVER-MARSHALL	G-6	1931	AC	CONSOLE HB	
SILVER-MARSHALL	HA	1933	AC	CONSOLE	
SILVER-MARSHALL	HM	1933	AC	CONSOLE	
SILVER-MARSHALL	HO	1933	AC	CONSOLE	
SILVER -MARSHALL	HR	1933	AC	TABLE	
SILVER-MARSHALL	HT	1933	AC	TABLE	
SILVER-MARSHALL	HU	1933	AC	COMPACT	
SILVER-MARSHALL	HV	1933	AC	TABLE	
SILVER-MARSHALL	HW	1933	AC	TABLE	
SILVER-MARSHALL	HX	1933	AC	TABLE	
SILVER-MARSHALL	KB-21	1932	AC	CONSOLE	
SILVER-MARSHALL	Q-25-A	1932	AC	CONSOLE BC/SW	
SILVER-MARSHALL	Q-32	1932	AC	CONSOLE BC/SW	
SILVER-MARSHALL	Q-34	1932	AC	CONSOLE DELUXE BC/SW	
SILVER-MARSHALL	V-30	1932	AC	CONSOLE BC/SW	
SILVER-MARSHALL	Y-39	1932	AC	CONSOLE	
SILVER-MARSHALL	Z-10	1933	AC	CONSOLE	
SILVER-MARSHALL	Z-13	1933	AC	CONSOLE HB	
SIMON RADIO	6-A	1925	BAT	TABLE LP 6 TUBE 3 DIAL	
SIMON RADIO	6-B	1925	BAT	TABLE LP SPEAKER 6 T 3 DL	

MANUFACTURER	MODEL	YEAR	POWER	STYLE	PRICE
SIMPLEX RADIO	4Z	1935	AC	MANTLE	
SIMPLEX RADIO	5C	1935	AC	MANTLE BC/SW	
SIMPLEX RADIO	5D	1935	AC	CONSOLE BC/SW	
SIMPLEX RADIO	5D	1935	AC	TABLE BC/SW	
SIMPLEX RADIO	5DA	1935	2/DC	CONSOLE BC/SW	
SIMPLEX RADIO	5DA	1935	2/DC	TABLE BC/SW	
SIMPLEX RADIO	5DB	1935	6/DC	CONSOLE BC/SW	
SIMPLEX RADIO	5DB	1935	6/DC	TABLE BC/SW	
SIMPLEX RADIO	5DF	1935	32/DC	CONSOLE BC/SW	
SIMPLEX RADIO	5DF	1935	32/DC	TABLE BC/SW	
SIMPLEX RADIO	6-A	1927	AC	TABLE 7 TUBE 1 DIAL	
SIMPLEX RADIO	6-A	1927	AC	CONSOLE LB 7 TUBE 1 DIAL	260
SIMPLEX RADIO	6G	1935	AC	CONSOLE BC/SW	
SIMPLEX RADIO	6G	1935	AC	TABLE BC/SW	
SIMPLEX RADIO	6GH	1935	AC	CONSOLE BC/SW	
SIMPLEX RADIO	6GH	1935	AC	TABLE BC/SW	
SIMPLEX RADIO	6GM	1935	AC	CONSOLE BC/SW	
SIMPLEX RADIO	6GM	1935	AC	TABLE BC/SW	
SIMPLEX RADIO	8J	1935	AC	CONSOLE BC/SW	
SIMPLEX RADIO	8J	1935	AC	TABLE BC/SW	
SIMPLEX RADIO	8JM	1935	AC	CONSOLE BC/SW	
SIMPLEX RADIO	8JM	1935	AC	TABLE BC/SW	
SIMPLEX RADIO	AA	1936	BAT	PORTABLE	
SIMPLEX RADIO	D	1936	AC	CONSOLE BC/SW	
SIMPLEX RADIO	D	1936	AC	TABLE BC/SW	
SIMPLEX RADIO	D	1928	AC	TABLE METAL 8 TUBE 1 DIAL	
SIMPLEX RADIO	D DELUXE	1936	AC	TABLE BC/SW	
SIMPLEX RADIO	DA	1936	2/DC	CONSOLE BC/SW	
SIMPLEX RADIO	DA	1936	2/DC	TABLE BC/SW	
SIMPLEX RADIO	F	1930	AC	TABLE METAL	60
SIMPLEX RADIO	G	1936	AC	CONSOLE BC/SW	
SIMPLEX RADIO	G	1936	AC	TABLE BC/SW	
SIMPLEX RADIO	G	1930	AC	CONSOLE LB	
SIMPLEX RADIO	G	1930	AC	CONSOLE HB	
SIMPLEX RADIO	GB	1936	6/DC	CONSOLE BC/SW	
SIMPLEX RADIO	GB	1936	6/DC	TABLE BC/SW	
SIMPLEX RADIO	GH	1936	AC	CONSOLE BC/SW	
SIMPLEX RADIO	GH	1936	AC	TABLE BC/SW	
SIMPLEX RADIO	H	1930	AC	CATHEDRAL	115
SIMPLEX RADIO	I	1930	DC	CATHEDRAL	115
SIMPLEX RADIO	J	1930	BAT	CATHEDRAL	115
SIMPLEX RADIO	J	1931	AC	CONSOLE	
SIMPLEX RADIO	J	1931	AC	TABLE	
SIMPLEX RADIO	K	1931	AC	TABLE	
SIMPLEX RADIO	NT	1936	AC	CONSOLE BC/SW	
SIMPLEX RADIO	NT	1936	AC	TABLE BC/SW	
SIMPLEX RADIO	P	1932	AC	TABLE BC/SW	
SIMPLEX RADIO	R	1932	AC	TABLE BC/SW	
SIMPLEX RADIO	RF	1923	BAT	TABLE LP 4 TUBE 2 DIAL	125
SIMPLEX RADIO	RFB	1923	BAT	TABLE HP 4 TUBE 2 DIAL	100

MANUFACTURER	MODEL	YEAR	POWER	STYLE	PRICE
SIMPLEX RADIO	RJ	1923	BAT	PORTABLE 4 TUBE 2 DIAL	
SIMPLEX RADIO	RK	1936	AC	MANTLE	
SIMPLEX RADIO	RK DELUXE	1936	AC	TABLE	
SIMPLEX RADIO	RX	1924	BAT	TABLE HP 4 TUBE 2 DIAL	75
SIMPLEX RADIO	S	1932	AC	TABLE BC/SW/LW	
SIMPLEX RADIO	SR-5	1924	BAT	TABLE LP 5 TUBE 3 DIAL	65
SIMPLEX RADIO	SR-8	1925	BAT	TABLE LP SLOPE PAN. 5T 3DL	
SIMPLEX RADIO	SR-9	1926	BAT	CONSOLE HB 5TUBE 3 DIAL	
SIMPLEX RADIO	SR-10	1927	BAT	CONSOLE HB 5 TUBE 2 DIAL	
SIMPLEX RADIO	SR-10	1927	BAT	TABLE 5 TUBE 2 DIAL	
SIMPLEX RADIO	SR-11	1927	BAT	TABLE 6 TUBE 1 DIAL	
SIMPLEX RADIO	SR-11	1927	BAT	CONSOLE LB 6 TUBE 1 DIAL	
SIMPLEX RADIO	SR-11	1927	BAT	CONSOLE HB 6 TUBE 1 DIAL	
SIMPLEX RADIO	U	1933	AC	MANTLE BC/SW	
SIMPLEX RADIO	V	1933	AC	MANTLE BC/SW	
SIMPLEX RADIO	V ALL WAVE	1934	AC	MANTLE BC/SW/LW	
SIMPLEX RADIO	W	1933	AC	CONSOLE BC/SW/LW	
SIMPLEX RADIO	W	1933	AC	TABLE BC/SW/LW	
SIMPLEX RADIO	X	1934	AC	TABLE	
SIMPLEX RADIO	Y	1935	AC	CONSOLE BC/SW	
SIMPLEX RADIO	Y	1935	AC	TABLE BC/SW	
SIMPLEX RADIO	Z	1936	AC	MANTLE	
SIMPLEX RADIO	Z DELUXE	1936	AC	MANTLE	
SINGER	SINGER FIVE	1925	BAT	TABLE LP 5 TUBE 3 DIAL	
SLEEPER RADIO	52 MONOTROL	1923	BAT	TABLE LP 6 TUBE 1 DIAL	175
SLEEPER RADIO	54 MONOTROL	1925	BAT	TABLE LP SLOPE PAN 6 T 1DL	325
SLEEPER RADIO	56	1926	BAT	TABLE 5 TUBE 2 DIAL	
SLEEPER RADIO	57 SCOUT	1925	BAT	TABLE 5 TUBE 2 DIAL	
SLEEPER RADIO	58 SERENADER	1925	BAT	TABLE 5 TUBE 2 DIAL	150
SLEEPER RADIO	59	1925	BAT	TABLE 5 TUBE 2 DIAL	
SLEEPER RADIO	64 SCOUT	1927	AC	TABLE LP 2 DIAL	
SLEEPER RADIO	65 SERENADER	1927	AC	TABLE W/SPEAKER 2 DIAL	
SLEEPER RADIO	66	1927	AC	TABLE LP	110
SLEEPER RADIO	67	1927	AC	CONSOLE LB 2 DIAL	165
SLEEPER RADIO	69 MONOTROL	1927	AC	CONSOLE	
SLEEPER RADIO	78	1927	AC	CONSOLE 2 DIAL	
SLEEPER RADIO	MONOTROL	1923	BAT	TABLE LP LOOP ANT 1 DIAL	175
SONORA PHONOGRAPH	241	1924	BAT	CONSOLE RADIO/PHONO 5T 3 DIAL	250
SONORA PHONOGRAPH	277	1927	BAT	CONSOLE	
SONORA PHONOGRAPH	A-31	1928	AC	CONSOLE LB	140
SONORA PHONOGRAPH	A-32	1928	AC	CONSOLE RADIO/PHONO	
SONORA PHONOGRAPH	A-33	1928	AC	CONSOLE HB	
SONORA PHONOGRAPH	A-35	1928	AC	CONSOLE	
SONORA PHONOGRAPH	A-36	1928	AC	CONSOLE	
SONORA PHONOGRAPH	A-40	1928	AC	CONSOLE RADIO/PHONO	
SONORA PHONOGRAPH	A-44	1928	AC	CONSOLE RADIO/PHONO	
SONORA PHONOGRAPH	C	1924	BAT	TABLE LP 5 TUBE 3 DIAL	75
SONORA PHONOGRAPH	C-22	1938	AC	TABLE PLASTIC	
SONORA PHONOGRAPH	D-12	1938	AC	TABLE BC/SW	35

MANUFACTURER	MODEL	YEAR	POWER	STYLE	PRICE
SONORA PHONOGRAPH	D-800	1926	BAT	TABLE 6 TUBE 2 DIAL	170
SONORA PHONOGRAPH	D-820	1926	BAT	CONSOLE 6 TUBE 2 DIAL	
SONORA PHONOGRAPH	D-830	1926	BAT	CONSOLE 6 TUBE 2 DIAL	
SONORA PHONOGRAPH	E-300	1927	BAT	CONSOLE 6 TUBE 2 DIAL	
SONORA PHONOGRAPH	E-850	1927	BAT	TABLE 6 TUBE 2 DIAL	
SONORA PHONOGRAPH	E-860	1927	BAT	CONSOLE HB 6 TUBE 2 DIAL	
SONORA PHONOGRAPH	E-865	1927	BAT	CONSOLE HB 6 TUBE 2 DIAL	
SONORA PHONOGRAPH	E-870	1927	BAT	CONSOLE HB 6 TUBE 2 DIAL	
SONORA PHONOGRAPH	F-875	1927	BAT	CONSOLE LB 7 TUBE 2 DIAL	
SONORA PHONOGRAPH	FSA-96	1940	AC	CONSOLE R/P W/RECORDER	
SONORA PHONOGRAPH	KD-75	1939	AC	PORTABLE LUG. STYLE CASE	15
SONORA PHONOGRAPH	KM-138	1940	AC	TABLE PLASTIC	40
SONORA PHONOGRAPH	KNF-99	1940	AC	TABLE RADIO/PHONO	15
SONORA PHONOGRAPH	KNF-148	1941	AC	TABLE RADIO/PHONO	15
SONORA PHONOGRAPH	KT-87	1940	AC	MANTLE PLASTIC	25
SONORA PHONOGRAPH	KW-78	1940	AC	TABLE PLASTIC RADIO/PHONO	15
SONORA PHONOGRAPH	KXF-95	1940	AC	CONSOLE RADIO/PHONO BC/SW	45
SONORA PHONOGRAPH	KZ-111	1940	2/DC	TABLE	20
SONORA PHONOGRAPH	PL-37	1939	BAT	PORTABLE	
SONORA PHONOGRAPH	TH-46	1940	2/DC	TABLE	30
SONORA PHONOGRAPH	TJ-62	1939	AC	MANTLE PLASTIC WALNUT	25
SONORA PHONOGRAPH	TJ-63	1939	AC	TABLE	40
SONORA PHONOGRAPH	TK-44	1939	AC	MANTLE	
SONORA PHONOGRAPH	TN-45	1939	AC	MANTLE	
SONORA PHONOGRAPH	TP-108	1939	AC	MANTLE PLASTIC	25
SONORA PHONOGRAPH	TR-51	1939	AC	TABLE PB BC/SW	
SONORA PHONOGRAPH	TSA-105	1939	AC	MANTLE PLASTIC WALNUT	25
SONORA PHONOGRAPH	TT-52	1939	AC	TABLE PB	
SONORA PHONOGRAPH	TV-48	1939	AC	MANTLE PLASTIC	40
SONORA PHONOGRAPH	TW-49	1939	AC	MANTLE PLASTIC PB	35
SONORA PHONOGRAPH	TX-51	1939	AC	TABLE BC/SW	
SONORA PHONOGRAPH	TX-58	1939	AC	CONSOLE PB BC/SW	
SONORA PHONOGRAPH	TY-54	1939	AC	TABLE PB BC/SW	
SONORA PHONOGRAPH	TZ-56	1939	AC	CONSOLE PB BC/SW	120
SONORA PHONOGRAPH	XL-28	1939	BAT	PORTABLE	
SOUTHEASTERN RADIO	5	1924	BAT	TABLE LP 5 TUBE 2 DIAL	
SOUTHEASTERN RADIO	5-C	1924	BAT	CONSOLE 5 TUBE 3 DIAL	
SOVEREIGN	238	1927	AC	CONSOLE 2 DIAL	
SPARKS-WITHINGTON	5	1931	AC	TABLE	
SPARKS-WITHINGTON	5-15	1925	BAT	TABLE 5 TUBE 3 DIAL	100
SPARKS-WITHINGTON	5-26	1925	BAT	TABLE 5 TUBE 3 DIAL	125
SPARKS-WITHINGTON	9	1931	AC	CONSOLE	
SPARKS-WITHINGTON	9A	1931	BAT	CONSOLE	
SPARKS-WITHINGTON	10	1931	AC	TABLE	
SPARKS-WITHINGTON	14	1932	AC	CONSOLE LB	
SPARKS-WITHINGTON	15	1931	AC	CONSOLE LB	
SPARKS-WITHINGTON	18	1932	AC	CONSOLE HB	
SPARKS-WITHINGTON	25	1931	AC	CONSOLE HB	

MANUFACTURER	MODEL	YEAR	POWER	STYLE	PRICE
SPARKS-WITHINGTON	26	1931	AC	CONSOLE HB	
SPARKS-WITHINGTON	27-A	1932	AC	CONSOLE HB	
SPARKS-WITHINGTON	28	1932	AC	CONSOLE HB	
SPARKS-WITHINGTON	30	1931	AC	CONSOLE RADIO/PHONO	
SPARKS-WITHINGTON	31	1930	BAT	CONSOLE 6 TUBE 1 DIAL	
SPARKS-WITHINGTON	32	1930	BAT	CONSOLE 6 TUBE 1 DIAL	
SPARKS-WITHINGTON	35	1931	AC	CONSOLE RADIO/PHONO AUTO	
SPARKS-WITHINGTON	39	1928	BAT	TABLE 9 TUBE 1 DIAL	
SPARKS-WITHINGTON	49	1929	BAT	TABLE 8 TUBE 1 DIAL	
SPARKS-WITHINGTON	51	1931	BAT	TABLE	
SPARKS-WITHINGTON	52	1931	BAT	CONSOLE	
SPARKS-WITHINGTON	53	1934	AC	TABLE BC/SW	
SPARKS-WITHINGTON	57	1934	AC	TABLE BC/SW	
SPARKS-WITHINGTON	58	1933	BAT	CONSOLE	
SPARKS-WITHINGTON	61	1933	AC	TABLE BC/SW	
SPARKS-WITHINGTON	61-A	1933	AC	CONSOLE BOOKSTAND	
SPARKS-WITHINGTON	61-B	1933	AC	CONSOLE END TABLE	
SPARKS-WITHINGTON	62	1933	AC	MANTLE BC/SW	
SPARKS-WITHINGTON	65	1934	AC	TABLE BC/SW	50
SPARKS-WITHINGTON	66	1934	AC	CONSOLE BC/SW	
SPARKS-WITHINGTON	67	1934	AC	TABLE BC/SW	45
SPARKS-WITHINGTON	68	1934	AC	CONSOLE BC/SW	
SPARKS-WITHINGTON	69	1928	AC	TABLE 8 TUBE 1 DIAL	
SPARKS-WITHINGTON	71	1933	AC	MANTLE	
SPARKS-WITHINGTON	71-B	1933	AC	CONSOLE	
SPARKS-WITHINGTON	72	1933	AC	CONSOLE	
SPARKS-WITHINGTON	74	1933	AC	CONSOLE BC/SW	
SPARKS-WITHINGTON	75	1934	AC	TABLE	
SPARKS-WITHINGTON	75A	1934	AC	CONSOLE BC/SW	85
SPARKS-WITHINGTON	76	1934	AC	CONSOLE	
SPARKS-WITHINGTON	78	1934	AC	CONSOLE BC/SW	
SPARKS-WITHINGTON	79	1928	AC	CONSOLE 8 TUBE 1 DIAL	95
SPARKS-WITHINGTON	80	1934	AC	CONSOLE BC/SW	
SPARKS-WITHINGTON	83	1934	AC	CONSOLE BC/SW	
SPARKS-WITHINGTON	84	1934	AC	CONSOLE	
SPARKS-WITHINGTON	86-X	1934	AC	TABLE	
SPARKS-WITHINGTON	89-A	1928	AC	CONSOLE 9 TUBE 1 DIAL	
SPARKS-WITHINGTON	99	1929	AC	CONSOLE RADIO/PHONO	
SPARKS-WITHINGTON	101	1929	AC	CONSOLE	
SPARKS-WITHINGTON	103	1930	BAT	CONSOLE 13 TUBE 1 DIAL	
SPARKS-WITHINGTON	104	1934	AC	CONSOLE	
SPARKS-WITHINGTON	109	1928	AC	CONSOLE 10 TUBE 1 DIAL	135
SPARKS-WITHINGTON	110	1929	AC	CONSOLE 12 TUBE 1 DIAL	
SPARKS-WITHINGTON	111	1929	AC	CONSOLE 12 TUBE 1 DIAL	
SPARKS-WITHINGTON	134	1934	AC	CONSOLE BC/SW	
SPARKS-WITHINGTON	135	1935	AC	CONSOLE	
SPARKS-WITHINGTON	136	1935	AC	CONSOLE	
SPARKS-WITHINGTON	235	1930	BAT	CONSOLE RADIO/PHONO 10T 1 DIAL	
SPARKS-WITHINGTON	301	1929	AC	CONSOLE 10 TUBE 1 DIAL	

MANUFACTURER	MODEL	YEAR	POWER	STYLE	PRICE
SPARKS-WITHINGTON	410	1930	BAT	TABLE 6 TUBE 1 DIAL	
SPARKS-WITHINGTON	410	1931	AC	TABLE	170
SPARKS-WITHINGTON	420	1930	BAT	CONSOLE 6 TUBE 1 DIAL	
SPARKS-WITHINGTON	475	1934	AC	CONSOLE	
SPARKS-WITHINGTON	478A	1934	AC	CONSOLE	
SPARKS-WITHINGTON	506	1935	AC	MANTLE BC/SW	
SPARKS-WITHINGTON	510-BV,-BD,-DG	1939	AC	MANTLE PLASTIC	35
SPARKS-WITHINGTON	510-DR,-W	1939	AC	MANTLE PLASTIC	
SPARKS-WITHINGTON	516	1935	AC	TABLE BC/SW	
SPARKS-WITHINGTON	517	1936	AC	TABLE BC/SW	95
SPARKS-WITHINGTON	517-B	1936	AC	TABLE BC/SW	
SPARKS-WITHINGTON	517-W	1936	AC	TABLE BC/SW	
SPARKS-WITHINGTON	518	1937	AC	TABLE BC/SW	
SPARKS-WITHINGTON	520-M	1939	AC	TABLE PB	
SPARKS-WITHINGTON	527-2	1936	2/DC	TABLE BC/SW	
SPARKS-WITHINGTON	537	1936	AC	TABLE BC/SW	
SPARKS-WITHINGTON	538	1937	AC	TABLE BC/SW	
SPARKS-WITHINGTON	538-2	1937	2/DC	TABLE BC/SW	
SPARKS-WITHINGTON	550-M	1939	AC	TABLE PB BC/SW	
SPARKS-WITHINGTON	557	1936	AC	TABLE GLASS BLUE BC/SW	800
SPARKS-WITHINGTON	558-B	1937	AC	TABLE GLASS BLUE BC/SW	950
SPARKS-WITHINGTON	558-C	1937	AC	TABLE GLASS COPPER BC/SW	700
SPARKS-WITHINGTON	567	1936	AC	CONSOLE BC/SW	
SPARKS-WITHINGTON	568	1937	AC	CONSOLE BC/SW	
SPARKS-WITHINGTON	570-M	1939	AC	TABLE PB BC/SW	
SPARKS-WITHINGTON	577	1936	AC	CONSOLE BC/SW	
SPARKS-WITHINGTON	578	1937	AC	CONSOLE CHAIRSIDE BC/SW	
SPARKS-WITHINGTON	587-2	1936	2/DC	CONSOLE	
SPARKS-WITHINGTON	588-2	1937	2/DC	CONSOLE BC/SW	
SPARKS-WITHINGTON	589	1930	BAT	CONSOLE HB 10 TUBE 1 DIAL	
SPARKS-WITHINGTON	590-I	1939	BAT	PORTABLE CLOTH	
SPARKS-WITHINGTON	591	1930	BAT	CONSOLE LB 10 TUBE 1 DIAL	
SPARKS-WITHINGTON	593	1930	BAT	CONSOLE LB 10 TUBE 1 DIAL	
SPARKS-WITHINGTON	594	1935	AC	TABLE	
SPARKS-WITHINGTON	600	1930	BAT	CONSOLE LB 10 TUBE 1 DIAL	80
SPARKS-WITHINGTON	608	1937	AC	TABLE	
SPARKS-WITHINGTON	608-B	1937	AC	TABLE PLASTIC	
SPARKS-WITHINGTON	610	1930	BAT	CONSOLE HB 10 TUBE 1 DIAL	
SPARKS-WITHINGTON	615	1927	AC	TABLE	
SPARKS-WITHINGTON	616	1935	AC	TABLE BC/SW	
SPARKS-WITHINGTON	617	1936	AC	TABLE BC/SW	
SPARKS-WITHINGTON	620	1930	BAT	CONSOLE HB 10 TUBE 1 DIAL	
SPARKS-WITHINGTON	620-M	1939	AC	TABLE PB BC/SW	
SPARKS-WITHINGTON	626	1927	BAT	TABLE 6 TUBE 1 DIAL	95
SPARKS-WITHINGTON	628	1937	AC	TABLE BC/SW	
SPARKS-WITHINGTON	638-6	1937	6/DC	TABLE BC/SW	
SPARKS-WITHINGTON	655	1935	AC	TABLE	
SPARKS-WITHINGTON	660-M	1939	AC	CONSOLE PB BC/SW	
SPARKS-WITHINGTON	666	1935	AC	CONSOLE BC/SW	
SPARKS-WITHINGTON	667	1936	AC	CONSOLE BC/SW	

MANUFACTURER	MODEL	YEAR	POWER	STYLE	PRICE
SPARKS-WITHINGTON	668	1937	AC	CONSOLE BC/SW	
SPARKS-WITHINGTON	678	1937	AC	CONSOLE BC/SW	
SPARKS-WITHINGTON	678-A	1938	AC	CONSOLE PB BC/SW	
SPARKS-WITHINGTON	685	1935	AC	CONSOLE	
SPARKS-WITHINGTON	688-6	1937	6/DC	CONSOLE	
SPARKS-WITHINGTON	691	1934	AC	CONSOLE BC/SW	
SPARKS-WITHINGTON	716X	1935	AC	TABLE BC/SW	
SPARKS-WITHINGTON	727-X	1936	AC	TABLE BC/SW/LW	
SPARKS-WITHINGTON	740	1930	BAT	CONSOLE HB 11 TUBE 1 DIAL	
SPARKS-WITHINGTON	750	1930	BAT	CONSOLE HB 11 TUBE 1 DIAL	
SPARKS-WITHINGTON	766	1935	AC	CONSOLE BC/SW	
SPARKS-WITHINGTON	768	1937	AC	CONSOLE BC/SW	
SPARKS-WITHINGTON	770-PA	1940	AC	CONSOLE RADIO/PHONO	45
SPARKS-WITHINGTON	778	1937	AC	CONSOLE BC/SW	
SPARKS-WITHINGTON	827-X	1936	AC	TABLE BC/SW/LW	
SPARKS-WITHINGTON	835	1935	AC	CONSOLE	
SPARKS-WITHINGTON	867	1936	AC	CONSOLE BC/SW	
SPARKS-WITHINGTON	870	1930	BAT	CONSOLE HB 13 TUBE 1 DIAL	
SPARKS-WITHINGTON	880	1939	AC	CONSOLE PB BC/SW	
SPARKS-WITHINGTON	930	1929	AC	CONSOLE 9 TUBE 1 DIAL	
SPARKS-WITHINGTON	931	1929	AC	CONSOLE 9 TUBE 1 DIAL	
SPARKS-WITHINGTON	966	1935	AC	CONSOLE BC/SW	
SPARKS-WITHINGTON	968	1937	AC	CONSOLE BC/SW	
SPARKS-WITHINGTON	987	1936	AC	CONSOLE BC/SW	
SPARKS-WITHINGTON	1068	1937	AC	CONSOLE AUTO TUNING BC/SW	
SPARKS-WITHINGTON	1068	1938	AC	CONSOLE AUTO TUNING BC/W	85
SPARKS-WITHINGTON	1078	1937	AC	CONSOLE AUTO TUNING BC/SW	
SPARKS-WITHINGTON	1086	1935	AC	CONSOLE BC/SW	
SPARKS-WITHINGTON	1160	1939	AC	CONSOLE PB BC/SW	145
SPARKS-WITHINGTON	1166	1935	AC	CONSOLE BC/SW/LW	
SPARKS-WITHINGTON	1167	1936	AC	CONSOLE BC/SW	
SPARKS-WITHINGTON	1176	1935	AC	CONSOLE	
SPARKS-WITHINGTON	1186	1935	AC	CONSOLE ROUND GLASS BC/SW	
SPARKS-WITHINGTON	1268	1937	AC	CONSOLE AUTO TUNING BC/SW	
SPARKS-WITHINGTON	1288-P	1937	AC	CONSOLE RADIO/PHONO	
SPARKS-WITHINGTON	1466	1935	AC	CONSOLE BC/SW/LW	
SPARKS-WITHINGTON	1476	1935	AC	CONSOLE BC/SW	
SPARKS-WITHINGTON	1567	1936	AC	CONSOLE BC/SW--TO 60 MC.	
SPARKS-WITHINGTON	1568	1937	AC	CONSOLE AUTO TUNING BC/SW	
SPARKS-WITHINGTON	1568	1938	AC	CONSOLE PB BC/SW	
SPARKS-WITHINGTON	1867	1936	AC	CONSOLE BC/SW--TO 60 MC.	
SPARKS-WITHINGTON	5008	1938	AC	TABLE	
SPARKS-WITHINGTON	5018	1938	AC	TABLE	
SPARKS-WITHINGTON	5218	1938	AC	TABLE PB BC/SW	

MANUFACTURER	MODEL	YEAR	POWER	STYLE	PRICE
SPARKS-WITHINGTON	5518A	1938	AC	TABLE PB	
SPARKS-WITHINGTON	6218	1938	AC	TABLE BC/SW	
SPARKS-WITHINGTON	7618	1938	AC	CONSOLE PB BC/SW	
SPARKS-WITHINGTON	8618	1938	AC	CONSOLE PB BC/SW	
SPARKS-WITHINGTON	AC-6	1927	AC	TABLE 7 TUBE 1 DIAL	135
SPARKS-WITHINGTON	AC-7	1927	AC	CONSOLE 8 TUBE 1 DIAL	
SPARKS-WITHINGTON	AC-63	1927	AC	TABLE 8 TUBE 1 DIAL	
SPIELMAN ELECTRIC	AIR PILOT	1924	BAT	TABLE LP SLOPE PAN. 5T 3DL	75
SPLITDORF ELECTRICAL	ABBEY	1927	BAT	TABLE 6 TUBE 1 DIAL	125
SPLITDORF ELECTRICAL	AVON	1928	AC	CONSOLE 7 TUBE 1 DIAL	
SPLITDORF ELECTRICAL	BUCKINGHAM	1927	AC	CONSOLE HB DELUXE 6 T 1 DIAL	200
SPLITDORF ELECTRICAL	C-200	1926	BAT	CONSOLE 5 TUBE 3 DIAL	95
SPLITDORF ELECTRICAL	C-215	1926	BAT	CONSOLE 6 TUBE 2 DIAL	
SPLITDORF ELECTRICAL	CONCERTO	1927	BAT	CONSOLE 5 TUBE 1 DIAL	
SPLITDORF ELECTRICAL	DEVON	1928	AC	CONSOLE HB DELUXE 8 T 1 DIAL	250
SPLITDORF ELECTRICAL	INTERMEZZO	1927	BAT	CONSOLE 6 TUBE 1 DIAL	
SPLITDORF ELECTRICAL	LORENZO	1927	BAT	CONSOLE 6 TUBE 1 DIAL	
SPLITDORF ELECTRICAL	MAESTRO	1927	BAT	CONSOLE 6 TUBE 1 DIAL	
SPLITDORF ELECTRICAL	R-100	1924	BAT	TABLE LP 5 TUBE 3 DIAL	75
SPLITDORF ELECTRICAL	R-110D	1925	BAT	TABLE LP 5 TUBE 3 DIAL	
SPLITDORF ELECTRICAL	R-150D	1925	BAT	TABLE LP 5 TUBE 3 DIAL	
SPLITDORF ELECTRICAL	R-200	1925	BAT	TABLE LP 5 TUBE 3 DIAL	150
SPLITDORF ELECTRICAL	R-400	1925	BAT	TABLE LP 4 TUBE 2 DIAL	
SPLITDORF ELECTRICAL	R-410C	1925	BAT	CONSOLE 5 TUBE 3 DIAL	
SPLITDORF ELECTRICAL	R-425C	1925	BAT	CONSOLE 5 TUBE 3 DIAL	
SPLITDORF ELECTRICAL	R-500	1925	BAT	TABLE LP 5 TUBE 3 DIAL	60
SPLITDORF ELECTRICAL	R-560	1926	BAT	TABLE LP 5 TUBE 3 DIAL	85
SPLITDORF ELECTRICAL	RV-580	1926	BAT	TABLE 5 TUBE 3 DIAL	140
SPLITDORF ELECTRICAL	RV-695	1926	BAT	TABLE LP 6 TUBE 2 DIAL	
SPLITDORF ELECTRICAL	SALEM	1928	AC	CONSOLE 8 TUBE 1 DIAL	
SPLITDORF ELECTRICAL	VIRTUOSO	1927	BAT	CONSOLE 5 TUBE 1 DIAL	
SPLITDORF ELECTRICAL	WARWICK	1927	BAT	CONSOLE 6 TUBE 1 DIAL	
SPLITDORF ELECTRICAL	WARWICK	1928	AC	CONSOLE LB 7 TUBE 1 DIAL	135
SPLITDORF ELECTRICAL	WINTHROP	1927	BAT	CONSOLE SECRETARY 6 T 1 DIAL	300
STANDARD RADIO & ELECTRIC	GIBLIN RADIOEAR	1922	BAT	TABLE LP 3 BOXES 1 DIAL	250
STANDARD RADIO & ELECTRIC	RL	1923	BAT	TABLE LP 6 TUBE 2 DIAL	
STANDARD RADIO CO.	5 MASTER	1926	BAT	TABLE LP 3 DIAL	110
STANDARD RADIO CO.	5 MASTER	1926	BAT	TABLE HP W/SPEAKER 3 DIAL	135
STANDARD RADIO CO.	5 MASTER	1926	BAT	CONSOLE HB 3 DIAL	150
STANDARD RADIO CO.	5 SUPER	1926	BAT	TABLE LP 3 DIAL	85
STANDARD RADIO CO.	8 SUPER	1926	BAT	TABLE LB 2 DIAL	145
STANDARD RADIO CORP.	B-5 STANDARDYNE	1925	BAT	TABLE LP 3 DIAL	85
STANDARD RADIO CORP.	B-7 STANDARDYNE	1925	BAT	TABLE LP 3 DIAL	120
STANDARD RADIO CORP.	BH STANDARDYNE	1925	BAT	CONSOLE HB 3 DIAL	200
STANDARD RADIO CORP.	S-27	1927	BAT	TABLE LP 6 TUBE 1 DIAL	
STANDARD RADIO CORP.	S-600	1927	BAT	CONSOLE 6 TUBE 1 DIAL	
STANDARD RADIO CORP.	S-950	1927	BAT	CONSOLE 6 TUBE 1 DIAL	

MANUFACTURER	MODEL	YEAR	POWER	STYLE	PRICE
STANDARD RADIO CORP.	STANDARDYNE	1927	BAT	TABLE 1 MULTIVALVE 2 DIAL	400
STANRAD	STANDARD FIVE	1925	BAT	TABLE LP SLOPE PAN. 5T 3DL	70
STATES RADIO	STATES RADIO	1922	BAT	TABLE LP 3 TUBE 2 DIAL	95
STEIN, FRED	130-B	1930	BAT	TABLE	
STEIN, FRED	130-C	1930	AC	CATHEDRAL	
STEINITE RADIO	5	1925	BAT	TABLE LP 5 TUBE 3 DIAL	
STEINITE RADIO	25-1	1925	BAT	TABLE 1 TUBE 2 DIAL	
STEINITE RADIO	25-2	1925	BAT	TABLE 2 TUBE 1 DIAL	
STEINITE RADIO	26-1	1925	BAT	TABLE 1 TUBE 2 DIAL	
STEINITE RADIO	26-2	1925	BAT	TABLE 2 TUBE 1 DIAL	
STEINITE RADIO	27	1926	BAT	TABLE LP 5 TUBE 3 DIAL	
STEINITE RADIO	27-C	1926	BAT	CONSOLE 5 TUBE 3 DIAL	
STEINITE RADIO	40	1929	AC	CONSOLE HB 8 TUBE 1 DIAL	
STEINITE RADIO	45	1929	AC	CONSOLE HB 8 TUBE 1 DIAL	
STEINITE RADIO	50	1929	AC	CONSOLE HB 9 TUBE 1 DIAL	
STEINITE RADIO	60	1929	AC	CONSOLE HB 8 TUBE 1 DIAL	
STEINITE RADIO	70	1929	AC	CONSOLE 7 TUBE 1 DIAL	
STEINITE RADIO	80	1929	AC	CONSOLE HB 7 TUBE 1 DIAL	
STEINITE RADIO	85	1927	AC	TABLE LP 6 TUBE 3 DIAL	
STEINITE RADIO	95	1929	AC	CONSOLE 7 TUBE 1 DIAL	
STEINITE RADIO	100	1927	AC	TABLE LP 7 TUBE 2 DIAL	
STEINITE RADIO	102	1929	AC	CONSOLE RADIO/PHONO 9T 1 DIAL	160
STEINITE RADIO	125	1927	AC	TABLE LP 7 TUBE 2 DIAL	
STEINITE RADIO	150	1927	AC	CONSOLE 7 TUBE 2 DIAL	
STEINITE RADIO	261	1928	AC	TABLE LP 7 TUBE 1 DIAL	90
STEINITE RADIO	262	1928	AC	TABLE LP 7 TUBE 1 DIAL	
STEINITE RADIO	263	1928	AC	CONSOLE 7 TUBE 1 DIAL	
STEINITE RADIO	264	1928	AC	CONSOLE 7 TUBE 1 DIAL	
STEINITE RADIO	265	1928	AC	CONSOLE 7 TUBE 1 DIAL	
STEINITE RADIO	266	1928	AC	CONSOLE 7 TUBE 1 DIAL	
STEINITE RADIO	410	1930	AC	CATHEDRAL	80
STEINITE RADIO	420	1930	AC	CONSOLE HB	
STEINITE RADIO	642B	1931	AC	CONSOLE LB	
STEINITE RADIO	700	1931	AC	TABLE	
STEINITE RADIO	705	1931	AC	TABLE	
STEINITE RADIO	706	1931	AC	TABLE	
STEINITE RADIO	725	1931	AC	CONSOLE HB	
STEINITE RADIO	GF	1931	AC	CLOCK GRANDFATHER	
STERLING RADIO	A-2-60	1929	AC	CONSOLE TROUBADOUR	125
STERLING RADIO	A-3-60	1929	AC	CONSOLE LB SERENADER	135
STERLING RADIO	B-1-60	1929	AC	CONSOLE	
STERLING RADIO	B-2-60	1929	AC	CONSOLE IMPERIAL	145
STERLING RADIO	C	1930	AC	CONSOLE HB	
STERLING RADIO	E	1930	AC	CONSOLE LB	
STERLING RADIO	F-1-60	1930	AC	CATHEDRAL	130
STEWART-WARNER	01-521	1939	AC	TABLE PB	
STEWART-WARNER	01-531	1939	AC	TABLE PB	
STEWART-WARNER	01-5D9	1939	AC	TABLE RADIO/PHONO	15
STEWART-WARNER	01-5H7	1939	AC	CONSOLE PB	115

MANUFACTURER	MODEL	YEAR	POWER	STYLE	PRICE
STEWART-WARNER	01-611	1939	AC	TABLE PB	
STEWART-WARNER	01-6A7	1939	AC	CONSOLE PB BC/SW	130
STEWART-WARNER	01-6B9	1939	AC	CONSOLE RADIO/PHONO PB	45
STEWART-WARNER	01-6C9	1939	AC	CONSOLE RADIO/PHONO PB	50
STEWART-WARNER	01-6D4	1939	AC	CONSOLE PERIOD PB	
STEWART-WARNER	01-817	1939	AC	CONSOLE PB	130
STEWART-WARNER	01-8A7	1939	AC	CONSOLE PB BC/SW	120
STEWART-WARNER	01-9A7	1940	AC	CONSOLE PB BC/SW	90
STEWART-WARNER	02-411	1939	BAT	PORTABLE	15
STEWART-WARNER	02-421	1939	BAT	TABLE	
STEWART-WARNER	02-427	1939	BAT	CO NSOLE	
STEWART-WARNER	03-5A1	1939	AC	MANTLE PLASTIC WALNUT PB	
STEWART-WARNER	03-5A2	1939	AC	MANTLE PLASTIC IVORY PB	
STEWART-WARNER	03-5A3	1939	AC	MANTLE PLASTIC IVORY PB	
STEWART-WARNER	03-5B1,2,3	1939	AC	MANTLE PLASTIC PB	
STEWART-WARNER	03-5C1	1939	AC	TABLE	30
STEWART-WARNER	03-5E1	1939	AC	TABLE PLASTIC PB	55
STEWART-WARNER	03-5J1	1939	AC	TABLE PLASTIC PB	
STEWART-WARNER	03-5K1,3	1939	AC	TABLE PLASTIC REMOTE TUNER	20
STEWART-WARNER	05-5L1	1939	AC	PORTABLE	15
STEWART-WARNER	07-321	1939	AC	MANTLE PLASTIC	
STEWART-WARNER	07-323,-I	1939	AC	MANTLE PLASTIC	
STEWART-WARNER	07-511--513	1939	AC	MANTLE DIONNE QUINTUPLETS	45
STEWART-WARNER	07-514	1939	AC	TABLE	30
STEWART-WARNER	07-516	1939	AC	CONSOLE ARMCHAIR	85
STEWART-WARNER	07-551	1939	AC	MANTLE PLASTIC WLANUT	
STEWART-WARNER	07-552	1939	AC	MANTLE PLASTIC IVORY	
STEWART-WARNER	07-553	1939	AC	MANTLE PLASTIC IVORY	
STEWART-WARNER	07-554	1939	AC	TABLE	
STEWART-WARNER	07-556	1939	AC	CONSOLE ARMCHAIR	
STEWART-WARNER	07-5B1,2,3	1939	AC	MANTLE PLASTIC PB	25
STEWART-WARNER	07-5R6	1940	AC	TABLE	30
STEWART-WARNER	4B1	1940	2/DC	MANTLE PLASTIC STREAMLINED	15
STEWART-WARNER	4B4	1940	2/DC	TABLE	30
STEWART-WARNER	5V9	1941	AC	TABLE RADIO/PHONO	15
STEWART-WARNER	5Y1	1940	AC	PORTABLE W/ALLIGATOR CLOTH	15
STEWART-WARNER	6G1	1940	AC	TABLE PB BC/SW	35
STEWART-WARNER	6T8	1940	AC	TABLE RADIO/PHONO	15
STEWART-WARNER	7A8	1940	AC	CONSOLE RADIO/PHONO	45
STEWART-WARNER	7A9	1940	AC	CONSOLE RADIO/PHONO	
STEWART-WARNER	8A8	1940	AC	CONSOLE R/P W/PA SYSTEM	45
STEWART-WARNER	8D6	1940	AC	CONSOLE RADIO/PHONO	
STEWART-WARNER	8D7	1940	AC	CONSOLE RADIO/PHONO	
STEWART-WARNER	8D8	1940	AC	CONSOLE RADIO/PHONO	50
STEWART-WARNER	8D9	1940	AC	CONSOLE RADIO/PHONO BC/SW	50

MANUFACTURER	MODEL	YEAR	POWER	STYLE	PRICE
STEWART-WARNER	8F7	1940	AC	CONSOLE PB BC/SW	85
STEWART-WARNER	8R8	1940	AC	CONSOLE R/P RECORDER	45
STEWART-WARNER	8R9	1940	AC	CONSOLE R/P W/RECORDER	
STEWART-WARNER	10A 6	1940	AC	CONSOLE RADIO/PHONO	
STEWART-WARNER	10A 8	1940	AC	CONSOLE RADIO/PHONO	55
STEWART-WARNER	10A 9	1940	AC	CONSOLE RADIO/PHONO BC/SW	50
STEWART-WARNER	10A10	1940	AC	CONSOLE RADIO/PHONO	
STEWART-WARNER	55A	1933	AC	CONSOLE	
STEWART-WARNER	56A	1933	AC	CONSOLE	
STEWART-WARNER	65	1933	AC	CONSOLE	
STEWART-WARNER	91-511	1938	AC	TABLE PB	
STEWART-WARNER	91-512	1938	AC	TABLE PB	
STEWART-WARNER	91-513	1938	AC	TABLE PB	40
STEWART-WARNER	91-514	1938	AC	TABLE PB	
STEWART-WARNER	91-531	1938	AC	TABLE PB BC/SW	
STEWART-WARNER	91-536	1938	AC	CONSOLE ARMCHAIR PB BC/SW	
STEWART-WARNER	91-537	1938	AC	CONSOLE PB BC/SW	
STEWART-WARNER	91-561	1938	AC	TABLE PLASTIC WALNUT PB	
STEWART-WARNER	91-621	1938	AC	TABLE PB BC/SW	
STEWART-WARNER	91-627	1938	AC	CONSOLE PB BC/SW	
STEWART-WARNER	91-648	1939	AC	TABLE RADIO/PHONO PB	
STEWART-WARNER	91-717	1938	AC	CONSOLE PB BC/SW	
STEWART-WARNER	91-817	1938	AC	CONSOLE BC/SW	
STEWART-WARNER	91-1117	1938	AC	CONSOLE BC/SW	
STEWART-WARNER	97-521	1938	AC	TABLE PB	
STEWART-WARNER	97-524	1938	AC	TABLE PB	
STEWART-WARNER	97-562	1938	AC	TABLE PLASTIC IVORY PB	20
STEWART-WARNER	108	1933	AC	TABLE METAL	
STEWART-WARNER	300	1925	BAT	TABLE LP 5 TUBE 3 DIAL	95
STEWART-WARNER	305 AEROMASTER	1925	BAT	TABLE LP SLOPE PAN 5T 3 DL	110
STEWART-WARNER	310	1925	BAT	CONSOLE 5 TUBE 3 DIAL	
STEWART-WARNER	315	1925	BAT	CONSOLE 5 TUBE 3 DIAL	
STEWART-WARNER	320	1925	BAT	CONSOLE 5 TUBE 3 DIAL	
STEWART-WARNER	325	1925	BAT	TABLE LP 5 TUBE 3 DIAL	50
STEWART-WARNER	330	1926	BAT	TABLE 5 TUBE 3 DIAL	
STEWART-WARNER	340	1926	BAT	TABLE 5 TUBE 3 DIAL	
STEWART-WARNER	345	1926	BAT	TABLE 6 TUBE 1 DIAL	
STEWART-WARNER	350	1926	BAT	TABLE 6 TUBE 1 DIAL	100
STEWART-WARNER	355	1926	BAT	CONSOLE 6 TUBE 1 DIAL	
STEWART-WARNER	360	1926	BAT	CONSOLE HB 6 TUBE 1 DIAL	
STEWART-WARNER	385	1927	BAT	TABLE 6 TUBE 1 DIAL	80
STEWART-WARNER	390	1927	BAT	CONSOLE 6 TUBE 1 DIAL	
STEWART-WARNER	500A	1931	AC	TABLE	
STEWART-WARNER	500C	1931	DC	TABLE	
STEWART-WARNER	501A	1931	AC	TABLE	
STEWART-WARNER	501C	1931	DC	TABLE	
STEWART-WARNER	503A	1931	AC	CONSOLE LB	
STEWART-WARNER	503C	1931	DC	CONSOLE LB	

MANUFACTURER	MODEL	YEAR	POWER	STYLE	PRICE
STEWART-WARNER	505A	1931	AC	CONSOLE LB	
STEWART-WARNER	505C	1931	DC	CONSOLE LB	
STEWART-WARNER	507A	1931	AC	CONSOLE LB	
STEWART-WARNER	507C	1931	DC	CONSOLE LB	
STEWART-WARNER	509A	1931	AC	CONSOLE LB W/TIRE CASTORS	
STEWART-WARNER	509C	1931	DC	CONSOLE LB W/TIRE CASTORS	
STEWART-WARNER	520	1927	BAT	CONSOLE LB 6 TUBE 1 DIAL	185
STEWART-WARNER	525	1927	BAT	TABLE LP 6 TUBE 1 DIAL	70
STEWART-WARNER	530	1927	AC	CONSOLE 7 TUBE 1 DIAL	
STEWART-WARNER	535	1927	AC	TABLE LP 7 TUBE 1 DIAL	70
STEWART-WARNER	705	1927	BAT	TABLE LP 6 TUBE 1 DIAL	135
STEWART-WARNER	710	1927	BAT	CONSOLE 6 TUBE 1 DIAL	
STEWART-WARNER	715	1927	AC	TABLE 7 TUBE 1 DIAL	
STEWART-WARNER	720	1927	AC	CONSOLE 7 TUBE 1 DIAL	
STEWART-WARNER	801	1928	AC	TABLE METAL 7 TUBE 1 DIAL	160
STEWART-WARNER	802	1928	AC	TABLE METAL 7 TUBE 1 DIAL	
STEWART-WARNER	806	1928	BAT	TABLE METAL 6 TUBE 1 DIAL	
STEWART-WARNER	811	1928	AC	TABLE METAL 7 TUBE 1 DIAL	
STEWART-WARNER	812	1928	AC	TABLE METAL 7 TUBE 1 DIAL	
STEWART-WARNER	900	1929	BAT	TABLE METAL 7 TUBE 1 DIAL	95
STEWART-WARNER	900 A.C.	1929	AC	TABLE METAL 8 TUBE 1 DIAL	55
STEWART-WARNER	901	1929	AC	TABLE 8 TUBE 1 DIAL	
STEWART-WARNER	911	1929	AC	TABLE 8 TUBE 1 DIAL	
STEWART-WARNER	921	1929	AC	TABLE	
STEWART-WARNER	931	1929	AC	TABLE	
STEWART-WARNER	951	1929	AC	TABLE	
STEWART-WARNER	951 AVON	1930	AC	CONSOLE LB 8 TUBE 1 DIAL	160
STEWART-WARNER	952 GRAHAM	1930	AC	CONSOLE LB 8 TUBE 1 DIAL	
STEWART-WARNER	953 RAPHAEL	1930	DC	CONSOLE LB 7 TUBE 1 DIAL	
STEWART-WARNER	954 ST. JAMES	1930	DC	CONSOLE HB 7 TUBE 1 DIAL	
STEWART-WARNER	954 ST. JAMES	1930	AC	CONSOLE HB 8 TUBE 1 DIAL	
STEWART-WARNER	955	1930	DC	CONSOLE HB R/P 7 T 1 DIAL	
STEWART-WARNER	955	1930	AC	CONSOLE HB R/P 8 T 1 DIAL	
STEWART-WARNER	1101	1933	AC	TABLE BC/SW	
STEWART-WARNER	1102	1933	AC	CONSOLE BC/SW	
STEWART-WARNER	1103	1933	AC	CONSOLE BC/SW	
STEWART-WARNER	1104	1933	AC	CONSOLE DUN. PHYFE BC/SW	
STEWART-WARNER	1105	1933	AC	CONSOLE COMMODE BC/SW	
STEWART-WARNER	1106	1933	AC	CONSOLE BOOK CASE BC/SW	
STEWART-WARNER	1107	1933	AC	TABLE	
STEWART-WARNER	1108	1933	AC	CONSOLE	
STEWART-WARNER	1111	1933	AC	MANTLE BC/SW	
STEWART-WARNER	1116	1933	AC	CONSOLE BK/CS RED LEA B/SW	
STEWART-WARNER	1117	1933	AC	CONSOLE BK/CS GRN LEA B/SW	
STEWART-WARNER	1118	1933	AC	CONSOLE BK/CS BRN LEA B/SW	
STEWART-WARNER	1119	1933	AC	MANTLE METAL JADE BC/SW	
STEWART-WARNER	1152	1933	AC	TABLE DUN. PHYFE BC/SW	

MANUFACTURER	MODEL	YEAR	POWER	STYLE	PRICE
STEWART-WARNER	1153	1933	AC	TABLE DUN. PHYFE BC/SW	
STEWART-WARNER	1154	1933	AC	CONSOLE COMMODE BC/SW	
STEWART-WARNER	1155	1933	AC	TABLE BC/SW	
STEWART-WARNER	1161	1933	AC	TABLE	
STEWART-WARNER	1163	1933	AC	COMPACT	
STEWART-WARNER	1164	1933	AC	COMPACT	
STEWART-WARNER	1191	1933	AC	TABLE	
STEWART-WARNER	1192	1933	AC	CONSOLE	
STEWART-WARNER	1201	1933	AC	CONSOLE	
STEWART-WARNER	1231	1934	AC	TABLE	
STEWART-WARNER	1234	1934	AC	TABLE	
STEWART-WARNER	1235	1934	AC	TABLE	
STEWART-WARNER	1236	1934	A C	TABLE	
STEWART-WARNER	1251	1934	AC	TABLE	
STEWART-WARNER	1252	1934	AC	TABLE	
STEWART-WARNER	1254	1934	AC	CONSOLE	
STEWART-WARNER	1261	1934	AC	TABLE BC/SW	
STEWART-WARNER	1262	1934	AC	TABLE BC/SW	
STEWART-WARNER	1264	1934	AC	CONSOLE BC/SW	
STEWART-WARNER	1265	1934	AC	CONSOLE BC/SW	
STEWART-WARNER	1266	1934	AC	CONSOLE BC/SW	
STEWART-WARNER	1271	1934	AC	TABLE BC/SW	
STEWART-WARNER	1272	1934	AC	TABLE BC/SW	
STEWART-WARNER	1274	1934	AC	CONSOLE BC/SW	
STEWART-WARNER	1301	1935	AC	TABLE BC/SW	
STEWART-WARNER	1302	1935	AC	TABLE BC/SW	
STEWART-WARNER	1306	1935	AC	CONSOLE BC/SW	
STEWART-WARNER	1345	1935	AC	CONSOLE BC/SW	
STEWART-WARNER	1361	1935	AC	TABLE BC/SW	
STEWART-WARNER	1362	1935	AC	TABLE BC/SW	50
STEWART-WARNER	1365	1935	AC	CONSOLE BC/SW	
STEWART-WARNER	1366	1935	AC	CONSOLE BC/SW	
STEWART-WARNER	1375	1935	AC	CONSOLE BC/SW/LW	
STEWART-WARNER	1376	1935	AC	CONSOLE BC/SW/LW	
STEWART-WARNER	1385	1935	AC	CONSOLE BC/SW/LW	
STEWART-WARNER	1386	1935	AC	CONSOLE BC/SW/LW	
STEWART-WARNER	1388	1935	AC	CONSOLE BC/SW/LW	
STEWART-WARNER	1391	1935	2/DC	TABLE	
STEWART-WARNER	1395	1935	2/DC	CONSOLE	
STEWART-WARNER	1401	1935	AC	TABLE BC/SW	
STEWART-WARNER	1425	1936	AC	CONSOLE BC/SW	
STEWART-WARNER	1441	1936	AC	TABLE BC/SW	
STEWART-WARNER	1451	1936	AC	TABLE BC/SW	
STEWART-WARNER	1455	1936	AC	CONSOLE BC/SW	
STEWART-WARNER	1461	1936	AC	TABLE BC/SW	
STEWART-WARNER	1465	1936	AC	CONSOLE BC/SW	
STEWART-WARNER	1475	1936	AC	CONSOLE BC/SW	
STEWART-WARNER	1485	1936	AC	CONSOLE BC/SW	
STEWART-WARNER	1495	1936	AC	CONSOLE BC/SW	
STEWART-WARNER	1611	1936	2/DC	TABLE BC/SW	

MANUFACTURER	MODEL	YEAR	POWER	STYLE	PRICE
STEWART-WARNER	1621	1936	2/DC	TABLE BC/SW/LW	
STEWART-WARNER	1625	1936	2/DC	CONSOLE BC/SW/LW	
STEWART-WARNER	1631	1936	6/DC	TABLE BC/SW	
STEWART-WARNER	1635	1936	6/DC	CONSOLE BC/SW	
STEWART-WARNER	1641	1936	6/DC	TABLE BC/SW/LW	
STEWART-WARNER	1645	1936	6/DC	CONSOLE BC/SW/LW	
STEWART-WARNER	1711	1936	AC	TABLE BC/SW	
STEWART-WARNER	A-6	1940	AC	MANTLE PLASTIC MINATURE	20
STEWART-WARNER	AIR PAL	1939	AC	MANTLE PLASTIC IV OR WAL	25
STEWART-WARNER	BKR-84	1940	AC	TABLE RADIO/PHONO PB	15
STEWART-WARNER	R-40-A	1932	AC	MANTLE	
STEWART-WARNER	R-43-A	1932	AC	CONSOLE	
STEWART-WARNER	R-44-A	1932	AC	CONSOLE BC/SW	
STEWART-WARNER	R-45-A	1932	AC	CONSOLE	
STEWART-WARNER	R-46-A	1932	AC	CONSOLE BC/SW	
STEWART-WARNER	R-47-A	1932	AC	CONSOLE W/TIRE CASTORS	
STEWART-WARNER	R-48-A	1932	AC	CONSOLE BC/SW	
STEWART-WARNER	R-50-A	1932	AC	CONSOLE BC/SW	
STEWART-WARNER	R-51-A	1932	AC	CONSOLE BC/SW	
STEWART-WARNER	R-58-A	1932	AC	CONSOLE RADIO/PHONO BC/SW	
STEWART-WARNER	R-109-A	1932	AC	CATHEDRAL	240
STORY & CLARK	1	1929	AC	CONSOLE	
STORY & CLARK	2	1929	AC	CONSOLE	
STORY & CLARK	3	1929	AC	CONSOLE	
STORY & CLARK	36	1930	AC	CONSOLE	
STORY & CLARK	43	1930	AC	CONSOLE	
STORY & CLARK	51	1930	AC	CONSOLE	
STROMBERG-CARLSON	1-A	1924	BAT	TABLE LP 5 TUBE 3 DIAL	
STROMBERG-CARLSON	1-B	1924	BAT	TABLE LP 5 TUBE 3 DIAL	95
STROMBERG-CARLSON	2	1923	BAT	CONSOLE LB 5 TUBE 3 DIAL	
STROMBERG-CARLSON	10	1930	AC	CONSOLE LB	
STROMBERG-CARLSON	11	1930	AC	CONSOLE	
STROMBERG-CARLSON	12	1930	AC	CONSOLE LB	185
STROMBERG-CARLSON	14	1930	AC	CONSOLE RADIO/PHONO	
STROMBERG-CARLSON	19	1931	AC	CONSOLE	
STROMBERG-CARLSON	20	1931	AC	CONSOLE	
STROMBERG-CARLSON	22	1931	AC	CO NSOLE RADIO/PHONO RC	
STROMBERG-CARLSON	24	1933	AC	CONSOLE RADIO/PHONO RC	
STROMBERG-CARLSON	25	1931	AC	CONSOLE	
STROMBERG-CARLSON	26	1931	AC	CONSOLE	
STROMBERG-CARLSON	27	1931	AC	CONSOLE RADIO/PHONO RC	
STROMBERG-CARLSON	37	1932	AC	CONSOLE	
STROMBERG-CARLSON	38	1932	AC	CONSOLE	
STROMBERG-CARLSON	39	1932	AC	CONSOLE	
STROMBERG-CARLSON	40	1932	AC	CONSOLE	
STROMBERG-CARLSON	41	1932	AC	CONSOLE RADIO/PHONO	
STROMBERG-CARLSON	43	1933	AC	CONSOLE	
STROMBERG-CARLSON	48	1933	AC	CONSOLE	
STROMBERG-CARLSON	49	1933	AC	CONSOLE LB	

MANUFACTURER	MODEL	YEAR	POWER	STYLE	PRICE
STROMBERG-CARLSON	50	1933	AC	CONSOLE HB	
STROMBERG-CARLSON	51	1933	AC	CONSOLE RADIO/PHONO	
STROMBERG-CARLSON	52	1933	AC	CONSOLE REMOTE CONTROL	
STROMBERG-CARLSON	53	1933	AC	CONSOLE METAL CONCEALED	
STROMBERG-CARLSON	54	1933	AC	CONSOLE R/P REMOTE CONTROL	
STROMBERG-CARLSON	55	1933	AC	CONSOLE REMOTE CONTROL	95
STROMBERG-CARLSON	56	1933	AC	CONSOLE LB	
STROMBERG-CARLSON	58-L	1935	AC	CONSOLE BC/SW	
STROMBERG-CARLSON	58-T	1935	AC	TOMBSTONE BC/SW	70
STROMBERG-CARLSON	58-W	1935	AC	CONSOLE BC/SW	170
STROMBERG-CARLSON	60-L	1934	AC	CONSOLE BC/SW	90
STROMBERG-CARLSON	60-M	1935	AC	CONSOLE	
STROMBERG-CARLSON	60-T	1934	AC	TABLE BC/SW	45
STROMBERG-CARLSON	61-L	1935	AC	CONSOLE BC/SW	90
STROMBERG-CARLSON	61-T	1935	AC	TABLE BC/SW	
STROMBERG-CARLSON	61-W	1935	AC	CONSOLE BC/SW	
STROMBERG-CARLSON	62	1935	AC	CONSOLE BC/SW	
STROMBERG-CARLSON	63	1935	AC	CONSOLE BC/SW	
STROMBERG-CARLSON	64	1934	AC	CONSOLE	
STROMBERG-CARLSON	65	1935	AC	CONSOLE Te-lek-tor-et RC	190
STROMBERG-CARLSON	68	1934	AC	CONSOLE BC/SW	125
STROMBERG-CARLSON	68R	1935	AC	CONSOLE	
STROMBERG-CARLSON	69	1934	AC	TABLE SW CONVERTER	40
STROMBERG-CARLSON	70	1935	AC	CONSOLE RC BC/SW	45
STROMBERG-CARLSON	72	1935	AC	CONSOLE R/P RC BC/SW	50
STROMBERG-CARLSON	74	1935	AC	CONSOLE R/P RC BC/SW	50
STROMBERG-CARLSON	82	1935	AC	CONSOLE RC BC/SW	
STROMBERG-CARLSON	84	1935	AC	CONSOLE BC/SW	125
STROMBERG-CARLSON	115	1936	6/DC	CONSOLE BC/SW	
STROMBERG-CARLSON	125-H	1936	AC	TABLE BC/SW	
STROMBERG-CARLSON	126-H	1937	AC	TABLE BC/SW	
STROMBERG-CARLSON	126-L	1937	AC	CONSOLE BC/SW	
STROMBERG-CARLSON	126-P	1937	AC	CONSOLE RADIO/PHONO BC/SW	
STROMBERG-CARLSON	127-H	1937	AC	TABLE BC/SW	35
STROMBERG-CARLSON	127-R	1938	AC	CONSOLE HALF-ROUND BC/SW	
STROMBERG-CARLSON	130-	1937	AC	CONSOLE	
STROMBERG-CARLSON	130-H	1936	AC	TABLE BC/SW	
STROMBERG-CARLSON	130-J	1937	AC	TABLE BC/SW	60
STROMBERG-CARLSON	130-L	1936	AC	CONSOLE BC/SW	
STROMBERG-CARLSON	130-M	1936	AC	CONSOLE BC/SW	
STROMBERG-CARLSON	130-R	1936	AC	TABLE BC/SW	
STROMBERG-CARLSON	130-U	1936	AC	TABLE BC/SW	
STROMBERG-CARLSON	140-H	1936	AC	TABLE BC/SW	
STROMBERG-CARLSON	140-K	1936	AC	CONSOLE BC/SW	
STROMBERG-CARLSON	140-L	1936	AC	CONSOLE BC/SW	
STROMBERG-CARLSON	145-L	1936	AC	CONSOLE BC/SW	
STROMBERG-CARLSON	145-P	1936	AC	CONSOLE RADIO/PHONO BC/SW	55

MANUFACTURER	MODEL	YEAR	POWER	STYLE	PRICE
STROMBERG-CARLSON	150-L	1936	AC	CONSOLE BC/SW	
STROMBERG-CARLSON	160-L	1936	AC	CONSOLE BC/SW	
STROMBERG-CARLSON	160-P	1936	AC	CONSOLE RADIO/PHONO BC/SW	
STROMBERG-CARLSON	180-L	1936	AC	CONSOLE BC/SW	
STROMBERG-CARLSON	225-H	1937	AC	TABLE BC/SW	55
STROMBERG-CARLSON	228-H	1937	AC	TABLE BC/SW	65
STROMBERG-CARLSON	228-L	1937	AC	CONSOLE BC/SW	150
STROMBERG-CARLSON	229-P	1937	AC	CONSOLE RADIO/PHONO BC/SW	170
STROMBERG-CARLSON	230-H	1937	AC	TABLE BC/SW	70
STROMBERG-CARLSON	230-L	1937	AC	CONSOLE BC/SW	150
STROMBERG-CARLSON	231-F	1937	AC	CONSOLE COFFEE TABLE BC/SW	115
STROMBERG-CARLSON	231-P	1937	AC	CONSOLE RADIO/PHONO BC/SW	190
STROMBERG-CARLSON	231-R	1937	AC	CONSOLE END TABLE BC/SW	160
STROMBERG-CARLSON	235-H	1938	AC	TABLE BC/SW	
STROMBERG-CARLSON	235-L	1938	AC	CONSOLE BC/SW	
STROMBERG-CARLSON	240-H	1937	AC	TABLE BC/SW	75
STROMBERG-CARLSON	240-L	1937	AC	CONSOLE BC/SW	175
STROMBERG-CARLSON	240-M	1937	AC	CONSOLE BC/SW	175
STROMBERG-CARLSON	240-P	1937	AC	CONSOLE RADIO/PHONO BC/SW	
STROMBERG-CARLSON	240-R	1937	AC	CONSOLE HALF-ROUND BC/SW	215
STROMBERG-CARLSON	240-S	1937	AC	CONSOLE DROP-DOOR BC/SW	200
STROMBERG-CARLSON	240-W	1937	AC	CONSOLE GOV. WINTHROP DESK	240
STROMBERG-CARLSON	245-L	1938	AC	CONSOLE BC/SW	
STROMBERG-CARLSON	245-M	1938	AC	CONSOLE BC/SW	110
STROMBERG-CARLSON	245-R	1938	AC	CONSOLE HALF-ROUND BC/SW	
STROMBERG-CARLSON	250-L	1937	AC	CONSOLE BC/SW	
STROMBERG-CARLSON	255-L	1937	AC	CONSOLE TUNING EYE BC/SW	
STROMBERG-CARLSON	260-L	1937	AC	CONSOLE BC/SW	
STROMBERG-CARLSON	260-P	1937	AC	CONSOLE RADIO/PHONO BC/SW	
STROMBERG-CARLSON	320-H	1938	AC	TABLE BC/SW	40
STROMBERG-CARLSON	320-T	1938	AC	TABLE DUNCAN PHYFE BC/SW	65
STROMBERG-CARLSON	325-J	1938	AC	TABLE BC/SW	40
STROMBERG-CARLSON	325-N	1938	AC	CONSOLE CHIPPENDALE BC/SW	60
STROMBERG-CARLSON	325-S	1938	AC	CONSOLE BUTTERFLY TABLE	60
STROMBERG-CARLSON	335-L	1938	AC	CONSOLE BC/SW	135
STROMBERG-CARLSON	336-P	1938	AC	CONSOLE CHAIRSIDE RP BC/SW	90
STROMBERG-CARLSON	337-H	1938	AC	TABLE BC/SW	40
STROMBERG-CARLSON	337-L	1938	AC	CONSOLE BC/SW	150
STROMBERG-CARLSON	340-F	1938	AC	CONSOLE BC/SW	140

MANUFACTURER	MODEL	YEAR	POWER	STYLE	PRICE
STROMBERG-CARLSON	340-H	1938	AC	TABLE BC/SW	40
STROMBERG-CARLSON	340-M	1938	AC	CONSOLE BC/SW	130
STROMBERG-CARLSON	340-P	1938	AC	CONSOLE RADIO/PHONO BC/SW	
STROMBERG-CARLSON	340-Y	1938	AC	CONSOLE CORNER CAB. BC/SW	150
STROMBERG-CARLSON	341-P	1938	AC	CONSOLE RADIO/PHONO BC/SW	45
STROMBERG-CARLSON	341-R	1938	AC	CONSOLE ARMCHAIR BC/SW	80
STROMBERG-CARLSON	345-F	1938	AC	CONSOLE CHIPPENDALE BC/SW	130
STROMBERG-CARLSON	345-M	1938	AC	CONSOLE BC/SW	150
STROMBERG-CARLSON	350-M	1938	AC	CONSOLE BC/SW	
STROMBERG-CARLSON	350-P	1938	AC	CONSOLE RADIO/PHONO BC/SW	65
STROMBERG-CARLSON	350-R	1938	AC	CONSOLE HALF ROUND BC/SW	180
STROMBERG-CARLSON	350-V	1938	AC	CONSOLE CORNER BC/SW	150
STROMBERG-CARLSON	360-M	1938	AC	CONSOLE BC/SW	
STROMBERG-CARLSON	370-M	1938	AC	CONSOLE BC/SW	190
STROMBERG-CARLSON	400-H	1939	AC	TABLE	40
STROMBERG-CARLSON	400-N	1939	AC	CONSOLE PERIOD MAHOGONY	85
STROMBERG-CARLSON	400-S	1939	AC	CONSOLE PERIOD MAPLE	85
STROMBERG-CARLSON	402-H	1939	BAT	PORTABLE	30
STROMBERG-CARLSON	405-H	1939	AC	TABLE	40
STROMBERG-CARLSON	410-H	1939	AC	TABLE WALNUT BC/SW	
STROMBERG-CARLSON	410-J	1939	AC	TABLE MAPLE BC/SW	40
STROMBERG-CARLSON	410-T	1939	AC	CONSOLE PERIOD MAHOG.BC/SW	90
STROMBERG-CARLSON	411-PF	1939	AC	CONSOLE PERIOD PB BC/SW	85
STROMBERG-CARLSON	411-PT	1939	AC	TABLE WALNUT BC/SW	15
STROMBERG-CARLSON	412-H	1939	AC	TABLE BC/SW	45
STROMBERG-CARLSON	420-F	1939	AC	CONSOLE HUTCH CAB BC/SW	90
STROMBERG-CARLSON	420-H	1939	AC	TABLE PB BC/SW	40
STROMBERG-CARLSON	420-J	1940	AC	TABLE PB BC/SW	40
STROMBERG-CARLSON	420-L	1939	AC	CONSOLE PB BC/SW	110
STROMBERG-CARLSON	420-PL	1939	AC	CONSOLE PB BC/SW	50
STROMBERG-CARLSON	420-PR	1939	AC	CONSOLE ARMCHAIR PB BC/SW	50
STROMBERG-CARLSON	420-R	1939	AC	CONSOLE PB BC/SW	140
STROMBERG-CARLSON	430-H	1939	AC	TABLE PB BC/SW	40
STROMBERG-CARLSON	430-L,-M	1939	AC	CONSOLE PB BC/SW	120
STROMBERG-CARLSON	430-MM	1940	AC	CONSOLE PB BC/SW	95
STROMBERG-CARLSON	430-PF	1939	AC	CONSOLE PB BC/SW	50
STROMBERG-CARLSON	430-PL	1939	AC	CONSOLE PB BC/SW	50
STROMBERG-CARLSON	440-M	1939	AC	CONSOLE PB BC/SW	125
STROMBERG-CARLSON	450-M	1939	AC	CONSOLE PB BC/SW	120
STROMBERG-CARLSON	455-MM	1940	AC	CONSOLE PB BC/SW/FM	95
STROMBERG-CARLSON	455-PL	1940	AC	CONSOLE PB BC/SW/FM	45

MANUFACTURER	MODEL	YEAR	POWER	STYLE	PRICE
STROMBERG-CARLSON	460-PF	1940	AC	CONSOLE R/P PB BC/SW	45
STROMBERG-CARLSON	470-PF	1939	AC	CONSOLE PB BC/SW	
STROMBERG-CARLSON	480-M	1939	AC	CONSOLE PB BC/SW	140
STROMBERG-CARLSON	480-M	1940	AC	CONSOLE PB BC/SW/FM	125
STROMBERG-CARLSON	500-J	1940	AC	TABLE	35
STROMBERG-CARLSON	500-S	1940	AC	TABLE	35
STROMBERG-CARLSON	501	1926	BAT	TABLE LP 5 TUBE 2 DIAL	
STROMBERG-CARLSON	502	1926	BAT	CONSOLE 5 TUBE 2 DIAL	
STROMBERG-CARLSON	509-PF	1940	AC	CONSOLE RADIO/PHONO	
STROMBERG-CARLSON	509-PT	1940	AC	TABLE RADIO/PHONO BC/SW	15
STROMBERG-CARLSON	520-L	1940	AC	CONSOLE PB BC/SW	85
STROMBERG-CARLSON	520-PF	1940	AC	CONSOLE RADIO/PHONO	
STROMBERG-CARLSON	520-PG	1940	AC	CONSOLE RADIO/PHONO	
STROMBERG-CARLSON	520-PL	1940	AC	CONSOLE RADIO/PHONO	
STROMBERG-CARLSON	520-PS	1940	AC	CONSOLE R/P PB BC/SW	50
STROMBERG-CARLSON	523	1927	AC	TABLE 8 TUBE 2 DIAL	
STROMBERG-CARLSON	524	1927	AC	CONSOLE 8 TUBE 2 DIAL	
STROMBERG-CARLSON	524	1927	AC	TABLE 8 TUBE 2 DIAL	
STROMBERG-CARLSON	530-PL	1940	AC	CONSOLE RADIO/PHONO	
STROMBERG-CARLSON	535-M	1940	AC	CONSOLE PB BC/SW/FM	155
STROMBERG-CARLSON	535-PG	1940	AC	CONSOLE R/P PB BC/SW/FM	65
STROMBERG-CARLSON	535-PL	1940	AC	CONSOLE RADIO/PHONO BC/FM	
STROMBERG-CARLSON	535-PS	1940	AC	CONSOLE RADIO/PHONO BC/FM	
STROMBERG-CARLSON	585-M	1941	AC	CONSOLE PB BC/SW/FM	120
STROMBERG-CARLSON	600-H	1941	AC	TABLE PLASTIC	30
STROMBERG-CARLSON	601	1924	BAT	TABLE LP 6 TUBE 2 DIAL	120
STROMBERG-CARLSON	602	1924	BAT	CONSOLE 6 TUBE 2 DIAL	
STROMBERG-CARLSON	633	1927	AC	TABLE 9 TUBE 2 DIAL	
STROMBERG-CARLSON	634	1927	AC	CONSOLE 9 TUBE 2 DIAL	
STROMBERG-CARLSON	635	1928	AC	TABLE TREASURE CHEST 7 T 1 DIAL	140
STROMBERG-CARLSON	635 D.C.	1928	DC	TABLE TREASURE CHEST 6T 1 DIAL	
STROMBERG-CARLSON	636	1928	AC	CONSOLE TREASURE CHEST	160
STROMBERG-CARLSON	638	1929	AC	CONSOLE 9 TUBE 1 DIAL	145
STROMBERG-CARLSON	638 D.C.	1929	DC	CONSOLE 8 TUBE 1 DIAL	
STROMBERG-CARLSON	641	1929	AC	TABLE 6 TUBE 1 DIAL	75
STROMBERG-CARLSON	641-A	1929	AC	TABLE TREASURE CHEST	
STROMBERG-CARLSON	642	1929	AC	CONSOLE HB 6 TUBE 1 DIAL	175
STROMBERG-CARLSON	645	1929	DC	CONSOLE	
STROMBERG-CARLSON	652	1930	AC	CONSOLE HB	190
STROMBERG-CARLSON	654	1930	AC	CONSOLE RADIO/PHONO	200
STROMBERG-CARLSON	734	1927	AC	CONSOLE 11 TUBE 1 DIAL	
STROMBERG-CARLSON	744	1927	AC	CONSOLE RADIO/PHONO 7 T 1 DIAL	300
STROMBERG-CARLSON	846	1929	AC	CONSOLE HB 10 TUBE 1 DIAL	200
STROMBERG-CARLSON	935-PL	1941	AC	CONSOLE R/P PB BC/FM	50

MANUFACTURER	MODEL	YEAR	POWER	STYLE	PRICE
STROMBERG-CARLSON	935-PR	1941	AC	CONSOLE PB TE BC/SW/FM	50
SUN MFG.	27-A	1927	BAT	TABLE LP 5 TUBE 2 DIAL	
SUNBEAM RADIO	75	1924	BAT	TABLE LP 5 TUBE 3 DIAL	75
SUNBEAM RADIO	90	1924	BAT	TABLE HP 5 TUBE 2 DIAL	
SUNBEAM RADIO	110	1924	BAT	TABLE HP SPEAKER 5 T 2 DL	
SUNBEAM RADIO	475	1924	BAT	CONSOLE LB 5 TUBE 3 DIAL	
SUNBEAM RADIO	975	1924	BAT	CONSOLE LB 5 TUBE 3 DIAL	
SUPER-ANTENNA	FRL-SUPER	1922	BAT	TABLE HP 5 TUBE 2 DIAL	250
SUPERIOR CABINET	530	1929	AC	CONSOLE HB	180
SUPERIOR CABINET	630	1929	AC	CONSOLE HB	245
SUPERIOR CABINET	730	1929	AC	CONSOLE LB	160
SYPHER	2	1924	BAT	TABLE LP 3 TUBE 2 DIAL	70
SYPHER	10	1924	BAT	TABLE LP 4 TUBE 2 DIAL	
SYPHER	SUPER-REFLEX	1923	BAT	TABLE LP 4 TUBE 2 DIAL	
TANNER	NEUTROFLEX 4	1925	BAT	TABLE LP 4 TUBE 2 DIAL	
TELEPHONE MAINTENANCE	P-1 TELMACO	1925	BAT	PORTABLE BOX 1 DIAL	160
TELETONE	R-F 5	1924	BAT	TABLE LP SLOPE PAN. 5T 3DL	80
TELETONE	T-210	1924	BAT	CONSOLE HB 5 TUBE 3 DIAL	
TEMPLE	8	1929	BAT	CONSOLE	
TEMPLE	8-60	1929	AC	CONSOLE	
TEMPLE	8-61	1929	AC	CONSOLE	
TEMPLE	8-80	1929	AC	CONSOLE	
TEMPLE	8-81	1929	AC	CONSOLE	
TEMPLE	8-90	1929	AC	CONSOLE RADIO/PHONO	
TEMPLE	8-91	1929	AC	CONSOLE RADIO/PHONO	
TERRY ELECTRICAL	A-12	1924	BAT	CONSOLE LB 5 TUBE 3 DIAL	
THERMIODYNE	CTF-6	1925	BAT	CONSOLE 6 TUBE 1 DIAL	100

T

MANUFACTURER	MODEL	YEAR	POWER	STYLE	PRICE
THERMIODYNE	TF-5	1925	BAT	TABLE LP 5 TUBE 1 DIAL	
THERMIODYNE	TF-6	1925	BAT	TABLE LP 6 TUBE 1 DIAL	90
THERMIODYNE	TF-7	1926	BAT	TABLE 7 TUBE 1 DIAL	
THOMPSON	5-A	1923	BAT	TABLE LP 5 TUBE 3 DIAL	
THOMPSON	R-81	1925	BAT	TABLE HP SPEAKER 5 T 1 DL	
THOMPSON	S-60	1925	BAT	TABLE LP SLOPE PAN 5T 3 DL	90
THOMPSON	S-70	1925	BAT	TABLE LP SLOPE PAN 5T 3 DL	90
THOMPSON	V-50	1925	BAT	TABLE LP SLOPE PAN 5T 3 DL	80
TRANSFORMER CORPORATION OF AMERICA	40	1931	AC	TABLE	
TRANSFORMER CORPORATION OF AMERICA	84	1931	AC	CONSOLE LB	
TRANSFORMER CORPORATION OF AMERICA	85	1931	AC	TABLE	
TRANSFORMER CORPORATION OF AMERICA	94	1931	AC	CONSOLE HB	
TRANSFORMER CORPORATION OF AMERICA	95	1931	AC	CONSOLE HB	
TRANSFORMER CORPORATION OF AMERICA	220	1932	AC	TABLE	

MANUFACTURER	MODEL	YEAR	POWER	STYLE	PRICE
TRANSFORMER CORPORATION OF AMERICA	240	1932	AC	TABLE BC/SW	
TRANSFORMER CORPORATION OF AMERICA	241	1932	AC	TABLE BC/SW	
TRANSFORMER CORPORATION OF AMERICA	260	1932	AC	CONSOLE LB	
TRANSFORMER CORPORATION OF AMERICA	280	1932	AC	CONSOLE	
TRANSFORMER CORPORATION OF AMERICA	300	1932	AC	CONSOLE	
TRANSFORMER CORPORATION OF AMERICA	423	1933	AC	TABLE BC/SW	
TRANSFORMER CORPORATION OF AMERICA	451	1933	AC	TABLE BC/SW	
TRANSFORMER CORPORATION OF AMERICA	470	1933	AC	TABLE BC/SW	
TRANSFORMER CORPORATION OF AMERICA	471	1933	AC	TABLE BC/SW	
TRANSFORMER CORPORATION OF AMERICA	472	1933	AC	CONSOLE BC/SW	
TRANSFORMER CORPORATION OF AMERICA	480	1933	AC	CONSOLE BC/SW	
TRANSFORMER CORPORATION OF AMERICA	490	1933	AC	CONSOLE BC/SW	
TRANSFORMER CORPORATION OF AMERICA	AC-51	1930	AC	CONSOLE LB	
TRANSFORMER CORPORATION OF AMERICA	AC-53	1930	AC	CONSOLE HB	165
TRANSFORMER CORPORATION OF AMERICA	AC-55	1930	AC	CONSOLE RADIO/PHONO	
TRANSFORMER CORPORATION OF AMERICA	CLARION JR.	1930	AC	CATHEDRAL	
TRANSFORMER CORPORATION OF AMERICA	TC-1	1934	AC	TABLE BC/SW	
TRANSFORMER CORPORATION OF AMERICA	TC-2	1934	AC	CONSOLE RADIO/PHONO BC/SW	85
TRANSFORMER CORPORATION OF AMERICA	TC-20	1934	AC	TABLE BC/SW	
TRANSFORMER CORPORATION OF AMERICA	TC-30	1934	AC	TABLE	
TRANSFORMER CORPORATION OF AMERICA	TC-50	1934	BAT	TABLE	
TRANSFORMER CORPORATION OF AMERICA	TC-52	1934	BAT	TABLE BC/LW	
TRANSFORMER CORPORATION OF AMERICA	TC-60	1934	AC	TABLE BC/SW/LW	
TRAV-LER	6 STANDARD	1930	BAT	PORTABLE	125
TRAV-LER	7 DELUXE	1930	BAT	PORTABLE	145
TRAV-LER	10	1930	BAT	PORTABLE	125

MANUFACTURER	MODEL	YEAR	POWER	STYLE	PRICE
TRAV-LER	41	1936	AC	MANTLE DELUXE	
TRAV-LER	46R	1936	AC	TABLE	
TRAV-LER	50A	1935	AC	TABLE	
TRAV-LER	51	1935	AC	TABLE	
TRAV-LER	53A	1935	AC	TABLE	
TRAV-LER	60A	1935	AC	TABLE BC/SW	
TRAV-LER	76	1935	AC	TABLE	
TRAV-LER	100	1935	AC	CONSOLE	
TRAV-LER	163	1936	AC	TABLE BC/SW	
TRAV-LER	173	1936	6/DC	CONSOLE BC/SW	
TRAV-LER	315	1939	AC/DC	MANTLE	30
TRAV-LER	325	1939	AC	MANTLE W/CLOCK	30
TRAV-LER	442	1936	AC	TABLE OVAL	
TRAV-LER	511A	1936	AC	TABLE BC/SW	
TRAV-LER	512	1936	AC	TABLE	
TRAV-LER	539-M	1939	AC	TABLE PB TUN EYE BC/SW	45
TRAV-LER	555	1939	BAT	PORTABLE BC/SW	
TRAV-LER	556	1939	BAT	PORTABLE BC/SW	
TRAV-LER	623	1936	AC	TABLE BC/SW	
TRAV-LER	633	1936	AC	TABLE BC/SW	
TRAV-LER	634	1936	AC	TABLE BC/SW	
TRAV-LER	720	1939	AC	TABLE	
TRAV-LER	733	1936	6/DC	CONSOLE BC/SW	
TRAV-LER	802	1940	AC	TABLE PB	35
TRAV-LER	1555	1939	BAT	PORTABLE	
TRAV-LER	1556	1939	BAT	PORTABLE	
TRAV-LER	A	1930	AC	CATHEDRAL	75
TRAV-LER	A-D.C.	1930	DC	CATHEDRAL	70
TRAV-LER	B	1930	AC	CATHEDRAL	70
TRAV-LER	DX	1931	DC	TABLE	
TRAV-LER	K	1931	AC	CONSOLE LB	
TRAV-LER	K	1931	AC	TABLE	
TRAV-LER	L	1931	AC	TABLE	
TRAV-LER	S 9	1931	AC	CONSOLE HB	
TRAV-LER	S 9	1931	AC	CONSOLE LB	
TRAV-LER	S 9	1931	AC	TABLE	
TRAV-LER	T	1929	BAT	PORTABLE 5 TUBE 1 DIAL	125
TRAV-LER	TRAV-LETTE	1931	AC	TABLE	
TREGO RADIO	TREGOSIX	1927	BAT	CONSOLE LB 2 DIAL	165
TROJAN RADIO	4-A TINY	1932	AC	TABLE	
TROJAN RADIO	5-C BABY	1932	AC	TABLE	
TROJAN RADIO	8-H GIANT	1932	AC	TABLE	
TROJAN RADIO	BABY	1931	AC	TABLE	
TROJAN RADIO	GIANT	1931	AC	CONSOLE	
TROJAN RADIO	GIANT	1931	AC	TABLE	
TROJAN RADIO	JUNIOR	1931	AC	CONSOLE	
TROJAN RADIO	JUNIOR	1931	AC	TABLE	
TROJAN RADIO	TINY	1931	AC	TABLE	
TROY RADIO	4	1937	AC	TABLE	
TROY RADIO	14	1934	AC	TABLE	

MANUFACTURER	MODEL	YEAR	POWER	STYLE	PRICE
TROY RADIO	40	1934	AC	TABLE	
TROY RADIO	42	1934	AC	TABLE	
TROY RADIO	44	1934	AC	TABLE	
TROY RADIO	45	1938	AC	TABLE	40
TROY RADIO	45-M (MIRROR)	1938	AC	TABLE-PEACH,SIL.,GRN,BLUE	
TROY RADIO	52	1934	AC	TABLE	
TROY RADIO	54	1934	AC	TABLE BC/SW	
TROY RADIO	55	1937	AC	TABLE	
TROY RADIO	55CC	1937	AC	PORTABLE	
TROY RADIO	57	1937	AC	TABLE	
TROY RADIO	57B	1937	2/DC	TABLE	
TROY RADIO	57PC	1937	AC	TABLE RADIO/PHONO	
TROY RADIO	57PC-CC	1937	AC	PORTABLE RADIO/PHONO	
TROY RADIO	58PC	1938	AC	TABLE RADIO/PHONO	
TROY RADIO	59PC	1938	AC	TABLE RADIO/PHONO	
TROY RADIO	67SW	1937	AC	TABLE BC/SW	
TROY RADIO	79	1937	AC	TABLE BC/SW	
TROY RADIO	79PC	1938	AC	TABLE RADIO/PHONO BC/SW	
TROY RADIO	84	1934	AC	TABLE BC/SW	
TROY RADIO	84C	1934	AC	TABLE BC/SW	
TROY RADIO	95	1937	AC	TABLE BC/SW	
TROY RADIO	95PC	1937	AC	CONSOLE RADIO/PHONO BC/SW	
TROY RADIO	100	1937	AC	TABLE	
TROY RADIO	113	1938	AC	TABLE BC/SW	
TROY RADIO	157PC	1937	AC	CONSOLE RADIO/PHONO	
TROY RADIO	170PC	1938	AC	CONSOLE RADIO/PHONO BC/SW	
TROY RADIO	179	1937	AC	TABLE BC/SW	
TROY RADIO	825	1938	AC	TABLE PB	45
TROY RADIO	TR-4	1935	AC	TABLE	
TROY RADIO	TR-5A	1935	AC	TABLE	
TROY RADIO	TR-51B	1935	2/DC	TABLE	
TROY RADIO	TR-63	1935	AC	TABLE BC/SW	
TROY RADIO	TR-63B	1935	2/DC	TABLE BC/SW	
TROY RADIO	TR-74B	1935	2/DC	TABLE BC/SW	
TROY RADIO	TR-84	1935	AC	TABLE BC/SW	
TROY RADIO	TR-151	1935	AC	TABLE	
TROY RADIO	TR-163	1935	AC	TABLE BC/SW	
TROY RADIO	TR-184	1935	AC	TABLE BC/SW	
TROY RADIO	TR-711	1935	AC	TABLE	
TURNER RADIO	A	1924	BAT	TABLE HP SLOPE PANEL 5T 3D	
TUSKA	220	1922	BAT	TABLE LP 3 DIAL	325
TUSKA	222	1922	BAT	TABLE LP 2 DIAL	275
TUSKA	224	1922	BAT	TABLE 1 TUBE 2 DIAL	195
TUSKA	225	1923	BAT	TABLE LP 3 TUBE 2 DIAL	275
TUSKA	227	1922	BAT	TABLE 3 TUBE 2 DIAL	275
TUSKA	228 SUPERDYNE	1924	BAT	TABLE LP 4 TUBE 2 DIAL	195
TUSKA	305 SUPERDYNE	1925	BAT	TABLE LP 4 TUBE 2 DIAL	160
TWITCHELL	F-403	1924	BAT	TABLE LP 3 TUBE 2 DIAL	

MANUFACTURER	MODEL	YEAR	POWER	STYLE	PRICE
U					
UNION CONSTRUCTION	MASTER	1922	BAT	TABLE HP 3 TUBE 5 DIAL	275
UNITED AIR CLEANER	8	1930	AC	CONSOLE HB	
UNITED AIR CLEANER	9	1930	AC	CONSOLE RADIO/PHONO	
UNITED AIR CLEANER	10	1930	AC	CONSOLE LB	
UNITED AIR CLEANER	11	1930	AC	CONSOLE LB	
UNITED AIR CLEANER	12	1930	AC	CONSOLE RADIO/PHONO	
UNITED AIR CLEANER	15	1930	AC	CATHEDRAL	
UNITED AIR CLEANER	16	1930	AC	CATHEDRAL	
UNITED AIR CLEANER	106	1931	AC	TABLE SEMI-PORTABLE	
UNITED AIR CLEANER	108A	1931	AC	CONSOLE LB	
UNITED AIR CLEANER	108A	1931	AC	TABLE	
UNITED AIR CLEANER	109	1931	AC	CONSOLE	
UNITED AIR CLEANER	111	1931	AC	TABLE	
UNITED AIR CLEANER	115	1931	AC	TABLE	
UNITED AIR CLEANER	116	1931	AC	TABLE	
UNITED AIR CLEANER	444	1929	AC	CONSOLE	
UNITED AIR CLEANER	445	1929	AC	TABLE	
UNITED AIR CLEANER	555	1929	AC	CONSOLE	
UNITED AIR CLEANER	666	1929	AC	CONSOLE	
UNITED AIR CLEANER	666-C	1929	AC	CONSOLE RADIO/PHONO	
UNITED ENGINE	K	1925	BAT	TABLE LP 5 TUBE 3 DIAL	75
UNITED ENGINE	M	1925	BAT	TABLE HP 5 TUBE 3 DIAL	
UNITED ENGINE	S	1925	BAT	CONSOLE HP 5 TUBE 3 DIAL	
UNITED ENGINE	T	1925	BAT	CONSOLE R/P 5 TUBE 3 DIAL	
UNITED ENGINE	W	1925	BAT	TABLE HP SPEAKER 5 T 3 DL	
UNITED MANUFACTURING	UR	1923	BAT	TABLE HP 4 TUBE 2 DIAL	
UNITED MOTOR SERVICE	1101	1935	AC	TABLE	
UNITED MOTOR SERVICE	1105	1935	AC	TABLE	
UNITED MOTOR SERVICE	1106	1935	AC	TABLE BC/SW	
UNITED MOTOR SERVICE	1107	1935	AC	TABLE BC/SW	
UNITED MOTOR SERVICE	1108	1935	AC	CONSOLE BC/SW	
UNITED MOTOR SERVICE	1109	1935	AC	CONSOLE BC/SW	
UNITED MOTOR SERVICE	1110	1935	AC	CONSOLE BC/SW	
UNITED MOTOR SERVICE	3205	1935	32/DC	TABLE BC/SW	
UNITED MOTOR SERVICE	3206	1935	32/DC	TABLE BC/SW	
UNITED MOTOR SERVICE	3207	1935	32/DC	CONSOLE BC/SW	
UNITED MOTOR SERVICE	4048	1933	AC	TABLE	
UNITED MOTOR SERVICE	4049	1933	AC	TABLE BC/SW	
UNITED MOTOR SERVICE	4050	1933	AC	CONSOLE BC/SW	
UNITED MOTOR SERVICE	6010	1935	6/DC	TABLE BC/SW	
UNITED MOTOR SERVICE	R-1125	1937	AC	TABLE	
UNITED MOTOR SERVICE	R-1126	1937	AC	TABLE BC/SW	
UNITED MOTOR SERVICE	R-1127	1937	AC	TABLE BC/SW	
UNITED MOTOR SERVICE	R-1128	1937	AC	TABLE BC/SW	
UNITED MOTOR SERVICE	R-1129	1937	AC	CONSOLE BC/SW	
UNITED MOTOR SERVICE	R-1130	1937	AC	CONSOLE BC/SW	
UNITED MOTOR SERVICE	R-1131	1937	AC	CONSOLE BC/SW	
UNITED MOTOR SERVICE	R-1132	1937	AC	CONSOLE BC/SW	
UNITED MOTOR SERVICE	R-2055	1937	2/DC	TABLE	

MANUFACTURER	MODEL	YEAR	POWER	STYLE	PRICE
UNITED MOTOR SERVICE	R-3210	1937	32/DC	TABLE BC/SW	
UNITED MOTOR SERVICE	R-3212	1937	32/DC	CONSOLE	
UNITED MOTOR SERVICE	R-6015	1937	6/DC	TABLE BC/SW	
UNITED RADIO	VICTORY	1925	BAT	TABLE LP 6 TUBE 3 DIAL	125
UNITED REPRODUCERS	21 PEERLESS	1929	AC	CONSOLE	
UNITED REPRODUCERS	22 PEERLESS	1929	AC	CONSOLE	
UNITED REPRODUCERS	23 PEERLESS	1929	AC	CONSOLE	
UNITED REPRODUCERS	24 PEERLESS	1929	AC	CONSOLE DELUXE	
UNITED REPRODUCERS	25 PEERLESS	1929	AC	CONSOLE RADIO/PHONO	
UNITED REPRODUCERS	65 COURIER	1929	AC	TABLE METAL	
UNITED REPRODUCERS	651 COURIER	1929	AC	CONSOLE	
UNITED REPRODUCERS	652 COURIER	1929	AC	CONSOLE	
UNITED REPRODUCERS	653 COURIER	1929	AC	CONSOLE	
UNITED STATES RADIO & TV	5A	1932	AC	TABLE BC/SW	
UNITED STATES RADIO & TV	7D	1932	AC	TABLE BC/SW	
UNITED STATES RADIO & TV	8A	1931	AC	TABLE	
UNITED STATES RADIO & TV	8B	1931	AC	CONSOLE	
UNITED STATES RADIO & TV	9A	1932	AC	TABLE	
UNITED STATES RADIO & TV	9B	1932	AC	CONSOLE LB	
UNITED STATES RADIO & TV	10B	1931	AC	CONSOLE	
UNITED STATES RADIO & TV	12B	1932	AC	CONSOLE HB	
UNITED STATES RADIO & TV	19B	1932	AC	CONSOLE LB	
UNITED STATES RADIO & TV	21	1929	BAT	CONSOLE	
UNITED STATES RADIO & TV	22	1929	BAT	TABLE METAL	
UNITED STATES RADIO & TV	25A	1932	AC	TABLE	
UNITED STATES RADIO & TV	26B	1931	AC	CONSOLE	
UNITED STATES RADIO & TV	26P	1931	AC	TABLE	90
UNITED STATES RADIO & TV	27	1930	AC	CATHEDRAL MODERN	160
UNITED STATES RADIO & TV	28-A	1930	AC	CONSOLE LB	
UNITED STATES RADIO & TV	31-B	1930	AC	CONSOLE LB	
UNITED STATES RADIO & TV	31-C	1930	AC	CONSOLE RADIO/PHONO	
UNITED STATES RADIO & TV	31-D	1930	AC	CONSOLE LB	
UNITED STATES RADIO & TV	36	1929	AC	TABLE METAL	
UNITED STATES RADIO & TV	37	1929	AC	CONSOLE	
UNITED STATES RADIO & TV	46	1929	AC	CONSOLE	
UNITED STATES RADIO & TV	47	1929	AC	TABLE METAL	
UNITED STATES RADIO & TV	50	1929	AC	CONSOLE	
UNITED STATES RADIO & TV	54	1929	BAT	CONSOLE LB	
UNITED STATES RADIO & TV	55	1930	BAT	TABLE METAL	
UNITED STATES RADIO & TV	60	1929	AC	CONSOLE	
UNITED STATES RADIO & TV	70	1929	AC	CONSOLE	
UNITED STATES RADIO & TV	80	1929	AC	CONSOLE	
UNITED STATES RADIO & TV	89	1929	AC	TABLE	
UNITED STATES RADIO & TV	120B	1932	AC	CONSOLE HB	
USL RADIO	AC-7	1927	AC	TABLE 1 DIAL	
USL RADIO	DC-6	1927	BAT	TABLE 1 DIAL	65
USL RADIO	DC-7	1927	BAT	TABLE 1 DIAL	65
USL RADIO	RC-5	1924	BAT	TABLE LP 5 TUBE 2 DIAL	75

MANUFACTURER	MODEL	YEAR	POWER	STYLE	PRICE
V					
VALLEY ELECTRIC	5	1924	BAT	TABLE LP 5 TUBE 3 DIAL	80
VALLEY ELECTRIC	52	1926	BAT	TABLE LP 2 DIAL	135
VALLEY ELECTRIC	71	1926	BAT	TABLE LP 7 TUBE 1 DIAL	
VALLEY ELECTRIC	VALLEYTONE	1925	BAT	TABLE LP 3 DIAL	
VERNON RADIO	VERNON	1927	AC	TABLE LP 3 TUBE 2 DIAL	175
VIBROPLEX	1	1923	BAT	MANTLE 1 TUBE 1 DIAL	
VIBROPLEX	2	1923	BAT	TABLE LP 4 TUBE 2 DIAL	75
VIBROPLEX	5	1923	BAT	TABLE LP 5 TUBE 2 DIAL	85
VICTOREEN RADIO	345	1930	AC	CONSOLE LB	
VICTOREEN RADIO	395	1930	AC	CONSOLE HB	
VICTOREEN RADIO	595	1930	AC	CONSOLE RADIO/PHONO	
W					
W-K ELECTRIC	5 ORIOLE	1925	BAT	TABLE LP 5 TUBE 2 DIAL	80
W-K ELECTRIC	6 ORIOLE	1925	BAT	TABLE LP 5 TUBE 2 DIAL	90
W-K ELECTRIC	7 ORIOLE	1924	BAT	TABLE LP 5 TUBE 2 DIAL	90
W-K ELECTRIC	60 ORIOLE	1925	BAT	TABLE LP 5 TUBE 2 DIAL	95
W-K ELECTRIC	71 WARWICK	1926	BAT	TABLE LP 6 TUBE 2 DIAL	95
W-K ELECTRIC	75 CANTERBURY	1926	BAT	CONSOLE LB 6 TUBE 2 DIAL	165
W-K ELECTRIC	78 MAYFAIR	1926	BAT	CONSOLE LB 6 TUBE 2 DIAL	160
W-K ELECTRIC	90 ORIOLE	1927	BAT	TABLE LP 5 TUBE 2 DIAL	
WALBERT	26 ISOFARAD	1927	AC	TABLE 2 DIAL	
WALTHAM RADIO	31	1930	AC	CATHEDRAL	
WARE RADIO	60	1925	BAT	TABLE LP 3 DIAL	85
WARE RADIO	175	1925	BAT	TABLE HP SLOPE PANEL 2 DL	125
WARE RADIO	AD-2	1922	BAT	TABLE 4 TUBE 1 DIAL	200
WARE RADIO	T	1924	BAT	TABLE HP 3 TUBE 3 DIAL	130
WARE RADIO	W	1924	BAT	TABLE 5 TUBE 3 DIAL	100
WELLS RADIO	5 BEARCAT	1926	BAT	TABLE HP SPEAKER 5T 3 DL	
WELLS RADIO	24	1923	BAT	TABLE LP 4 TUBE 1 DIAL	
WELLS RADIO	50 BEARCAT	1926	BAT	TABLE LP 5 TUBE 3 DIAL	
WELLS RADIO	55 BEARCAT	1926	BAT	CONSOLE LB 5 TUBE 3 DIAL	75
WELLS-GARDNER	60	1929	BAT	TABLE METAL	
WELLS-GARDNER	70	1929	BAT	TABLE METAL	
WELLS-GARDNER	72	1929	BAT	CONSOLE	
WELLS-GARDNER	92	1929	AC	CONSOLE	
WELLS-GARDNER	94	1929	AC	CONSOLE	
WESTBURR	WB- 6	1923	BAT	PORTABLE 6 TUBE 1 DIAL	
WESTERN COIL & ELECTRICAL	WC-5 RADIODYNE	1924	BAT	TABLE LP 3 DIAL	95
WESTERN COIL & ELECTRICAL	WC-5B	1924	BAT	TABLE LP 4 TUBE 2 DIAL	115
WESTERN COIL & ELECTRICAL	WC-10 RADIODYNE	1925	BAT	TABLE LP 2 DIAL	90
WESTERN COIL & ELECTRICAL	WC-11 RADIODYNE	1924	BAT	CONSOLE LB 6 TUBE 3 DIAL	
WESTERN COIL & ELECTRICAL	WC-11B	1924	BAT	TABLE HP 6 TUBE 3 DIAL	
WESTERN COIL & ELECTRICAL	WC-12 RADIODYNE	1924	BAT	CONSOLE LB 5 TUBE 3 DIAL	
WESTERN COIL & ELECTRICAL	WC-12B	1924	BAT	TABLE HP 5 TUBE 3 DIAL	
WESTERN COIL & ELECTRICAL	WC-14 RADIODYNE	1924	BAT	CONSOLE 5 TUBE 3 DIAL	
WESTERN COIL & ELECTRICAL	WC-14A	1924	BAT	TABLE LP 5 TUBE 3 DIAL	75
WESTERN COIL & ELECTRICAL	WC-14B	1924	BAT	TABLE HP 5 TUBE 3 DIAL	
WESTERN COIL & ELECTRICAL	WC-15 RADIODYNE	1924	BAT	TABLE SLOPE PANEL 5 T 2 DL	

MANUFACTURER	MODEL	YEAR	POWER	STYLE	PRICE
WESTERN COIL & ELECTRICAL	WC-17A	1924	BAT	TABLE LP 5 TUBE 2 DIAL	75
WESTERN COIL & ELECTRICAL	WC-18 RADIODYNE	1925	BAT	CONSOLE LB 5 TUBE 2 DIAL	
WESTERN COIL & ELECTRICAL	WC-18B	1925	BAT	TABLE LP 5 TUBE 2 DIAL	
WESTERN COIL & ELECTRICAL	WC-70G	1926	BAT	TABLE LP 9 TUBE 2 DIAL	160
WESTERN ELECTRIC	4B	1923	BAT	TABLE LP 2 DIAL	475
WESTFALL-KELLEY RADIO	AE	1925	BAT	TABLE LP 5 TUBE 3 DIAL	
WESTINGHOUSE	482	1940	AC	CONSOLE RADIO/PHONO	
WESTINGHOUSE	484	1940	AC	CONSOLE RADIO/PHONO	
WESTINGHOUSE	486	1940	AC	CONSOLE R/P W/RECORDER	
WESTINGHOUSE	WR-4	1930	AC	CONSOLE LB	
WESTINGHOUSE	WR-5	1930	AC	CONSOLE LB	
WESTINGHOUSE	WR-6	1930	AC	CONSOLE HB RC	
WESTINGHOUSE	WR-7	1930	AC	CONSOLE LB RADIO/PHONO RC	
WESTINGHOUSE	WR-8	1931	AC	CONSOLE COLUMAIRE (GFC)	200
WESTINGHOUSE	WR-10	1931	AC	TABLE COLUMETTE	70
WESTINGHOUSE	WR-12	1931	AC	CONSOLE GRANDFATHER CLOCK	170
WESTINGHOUSE	WR-12-X-3	1941	AC	CONSOLE RADIO/PHONO BC/SW	45
WESTINGHOUSE	WR-12-X-3	1941	AC	TABLE	30
WESTINGHOUSE	WR-12-X-4	1941	AC	TABLE PB	30
WESTINGHOUSE	WR-12-X-7	1941	AC	TABLE PB BC/SW	40
WESTINGHOUSE	WR-12-X-16	1941	AC	TABLE PB BC/SW	30
WESTINGHOUSE	WR-14	1931	AC	CATHEDRAL	150
WESTINGHOUSE	WR-15	1931	AC	CONSOLE GRANDFATHER CLOCK	190
WESTINGHOUSE	WR-20	1934	AC	TABLE	40
WESTINGHOUSE	WR-21	1934	AC	TABLE BC/SW	40
WESTINGHOUSE	WR-22	1934	AC	TABLE BC/SW	65
WESTINGHOUSE	WR-23	1934	AC	TOMBSTONE BC/SW	95
WESTINGHOUSE	WR-24	1934	AC	CONSOLE BC/SW	145
WESTINGHOUSE	WR-27	1934	AC	TABLE	45
WESTINGHOUSE	WR-28	1934	AC	TOMBSTONE BC/SW	100
WESTINGHOUSE	WR-29	1934	AC	CONSOLE BC/SW	145
WESTINGHOUSE	WR-30	1934	AC	CONSOLE BC/SW	150
WESTINGHOUSE	WR-42-X-4	1941	AC	CONSOLE RADIO/PHONO BC/SW	45
WESTINGHOUSE	WR-62-K-1	1941	AC	PORTABLE AIRPLANE LUGGAGE	15
WESTINGHOUSE	WR-100	1935	AC	TABLE BC/SW	45
WESTINGHOUSE	WR-101	1935	AC	TABLE BC/SW	35
WESTINGHOUSE	WR-102	1936	AC	TABLE BC/SW	
WESTINGHOUSE	WR-103	1936	AC	TABLE BC/SW	
WESTINGHOUSE	WR-116	1936	AC	TABLE BC/SW	
WESTINGHOUSE	WR-120	1937	AC	TABLE BC/SW	
WESTINGHOUSE	WR-140	1938	AC	TABLE BC/SW	
WESTINGHOUSE	WR-150	1938	AC	MANTLE	
WESTINGHOUSE	WR-152	1938	AC	TABLE	
WESTINGHOUSE	WR-154	1938	AC	TABLE	

MANUFACTURER	MODEL	YEAR	POWER	STYLE	PRICE
WESTINGHOUSE	WR-165 SERIES	1939	AC	MANTLE	40
WESTINGHOUSE	WR-166	1939	AC	MANTLE PLASTIC	25
WESTINGHOUSE	WR-168	1939	AC	TABLE BC/SWTIC	25
WESTINGHOUSE	WR-169	1939	AC	MANTLE PLASTIC PB	
WESTINGHOUSE	WR-170	1939	AC	MANTLE PLASTIC PB	
WESTINGHOUSE	WR-172	1939	AC	TABLE PLASTIC PB BC/SW	
WESTINGHOUSE	WR-173	1940	AC	MANTLE PLASTIC	20
WESTINGHOUSE	WR-175	1940	AC	MANTLE PLASTIC	25
WESTINGHOUSE	WR-182	1940	AC	TABLE BC/SW	35
WESTINGHOUSE	WR-201	1935	AC	TABLE BC/SW	45
WESTINGHOUSE	WR-203	1935	AC	TABLE BC/SW	45
WESTINGHOUSE	WR-204	1935	AC	TABLE BC/SW	45
WESTINGHOUSE	WR-205	1935	AC	TABLE BC/SW	45
WESTINGHOUSE	WR-209	1936	AC	TABLE BC/SW	50
WESTINGHOUSE	WR-210	1936	AC	TABLE BC/SW	
WESTINGHOUSE	WR-211	1936	AC	TABLE BC/SW	
WESTINGHOUSE	WR-212	1936	AC	TABLE BC/SW	
WESTINGHOUSE	WR-214	1936	AC	TABLE BC/SW/LW	
WESTINGHOUSE	WR-217	1937	AC	TABLE BC/SW	
WESTINGHOUSE	WR-222	1937	AC	TABLE BC/SW	
WESTINGHOUSE	WR-224	1937	AC	TABLE BC/SW	
WESTINGHOUSE	WR-226	1937	AC	TABLE BC/SW	
WESTINGHOUSE	WR-228	1937	AC	TABLE BC/SW	
WESTINGHOUSE	WR-256	1938	AC	TABLE	40
WESTINGHOUSE	WR-258	1938	AC	MANTLE	50
WESTINGHOUSE	WR-260	1938	AC	TABLE BC/SW	
WESTINGHOUSE	WR-262	1938	AC	TABLE BC/SW	
WESTINGHOUSE	WR-264	1938	AC	TABLE BC/SW	
WESTINGHOUSE	WR-270	1939	AC	TABLE PLASTIC PB	
WESTINGHOUSE	WR-272	1939	AC	TABLE PLASTIC PB BC/SW	45
WESTINGHOUSE	WR-274	1939	AC	TABLE PLASTIC PB BC/SW	40
WESTINGHOUSE	WR-288	1940	AC	TABLE PB BC/SW	45
WESTINGHOUSE	WR-303	1935	AC	CONSOLE BC/SW	95
WESTINGHOUSE	WR-304	1935	AC	CONSOLE BC/SW	85
WESTINGHOUSE	WR-305	1935	AC	CONSOLE BC/SW	85
WESTINGHOUSE	WR-306	1935	AC	CONSOLE BC/SW	95
WESTINGHOUSE	WR-308	1940	AC	CONSOLE PB BC/SW	135
WESTINGHOUSE	WR-310	1936	AC	CONSOLE BC/SW	
WESTINGHOUSE	WR-311	1936	AC	CONSOLE BC/SW	
WESTINGHOUSE	WR-312	1936	AC	CONSOLE BC/SW	
WESTINGHOUSE	WR-314	1936	AC	CONSOLE BC/SW/LW	
WESTINGHOUSE	WR-315	1936	AC	CONSOLE TUNING EYE BC/SW	
WESTINGHOUSE	WR-316	1936	AC	CONSOLE BC/SW	
WESTINGHOUSE	WR-326	1937	AC	CONSOLE BC/SW	
WESTINGHOUSE	WR-328	1937	AC	CONSOLE BC/SW	
WESTINGHOUSE	WR-330	1937	AC	CONSOLE BC/SW	
WESTINGHOUSE	WR-332	1937	AC	CONSOLE BC/SW	
WESTINGHOUSE	WR-334	1937	AC	CONSOLE BC/SW	
WESTINGHOUSE	WR-336	1937	AC	CONSOLE BC/SW	
WESTINGHOUSE	WR-338	1937	AC	CONSOLE ARMCHAIR	

MANUFACTURER	MODEL	YEAR	POWER	STYLE	PRICE
WESTINGHOUSE	WR-342	1938	AC	CONSOLE BC/SW	
WESTINGHOUSE	WR-366	1938	AC	CONSOLE BC/SW	
WESTINGHOUSE	WR-368	1938	AC	CONSOLE PB BC/SW	160
WESTINGHOUSE	WR-370	1938	AC	CONSOLE BC/SW	
WESTINGHOUSE	WR-372	1939	AC	CONSOLE PB BC/SW	
WESTINGHOUSE	WR-373	1939	AC	CONSOLE PB BC/SW	
WESTINGHOUSE	WR-374	1939	AC	CONSOLE PB BC/SW	95
WESTINGHOUSE	WR-375	1940	AC	CONSOLE PB BC/SW	95
WESTINGHOUSE	WR-468	1939	AC	TABLE RADIO/PHONO	
WESTINGHOUSE	WR-470	1939	AC	TABLE RADIO/PHONO PB	
WESTINGHOUSE	WR-472	1938	AC	TABLE RADIO/PHONO	
WESTINGHOUSE	WR-473	1939	AC	TABLE RADIO/PHONO PB BC/SW	
WESTINGHOUSE	WR-474	1939	AC	TABLE RADIO/PHONO PB BC/SW	45
WESTINGHOUSE	WR-475	1940	AC	TABLE RADIO/PHONO	15
WESTINGHOUSE	WR-476	1940	AC	CONSOLE R/P PB BC/SW	45
WESTINGHOUSE	WR-601	1935	2/DC	TABLE BC/SW	80
WESTINGHOUSE	WR-602	1935	6/DC	TABLE BC/SW	80
WESTINGHOUSE	WR-603	1936	6/DC	TABLE BC/SW	
WESTINGHOUSE	WR-604	1936	2/DC	TABLE BC/SW	
WESTINGHOUSE	WR-605	1936	2/DC	TABLE BC/SW	
WESTINGHOUSE	WR-606	1936	6/DC	CONSOLE BC/SW	
WESTINGHOUSE	WR-607	1936	2/DC	CONSOLE BC/SW	
WESTINGHOUSE	WR-608	1936	2/DC	CONSOLE BC/SW	
WESTINGHOUSE	WR-610	1937	6/DC	TABLE BC/SW	
WESTINGHOUSE	WR-675A	1939	BAT	PORTABLE	
WESTINGHOUSE	WR-676	1940	AC	PORTABLE	15
WILCOX-GAY	2T5-30	1932	AC	TABLE BC/SW	
WILCOX-GAY	2V7-31	1932	AC	TABLE	
WILCOX-GAY	2V7-510	1932	AC	CONSOLE	
WILCOX-GAY	2VB7-67	1933	AC	COMPACT	
WILCOX -GAY	2W10-515	1932	AC	CONSOLE	
WILCOX-GAY	3F7-67	1933	AC	TABLE BC/LW	
WILCOX-GAY	3J4-11	1935	2/DC	TABLE	
WILCOX-GAY	3J5-55	1933	AC	TABLE	
WILCOX-GAY	3JE5-93	1935	AC	TABLE BC/SW	
WILCOX-GAY	3KE5-26	1934	AC	TABLE	
WILCOX-GAY	3LB7-206	1933	AC	CONSOLE BC/SW RC	
WILCOX-GAY	3LB7-22-706	1933	AC	CONSOLE BC/SW RC	
WILCOX-GAY	3PA6-66	1933	AC	TABLE BC/LW	
WILCOX-GAY	3S5-66	1933	AC	TABLE BC/SW	
WILCOX-GAY	3T6-66	1933	AC	TABLE BC/SW	
WILCOX-GAY	3VB6-73	1934	AC	TABLE BC/SW	
WILCOX-GAY	3VB6-710	1934	AC	CONSOLE BC/SW	
WILCOX-GAY	4CD5-29	1934	AC	TABLE BC/SW	
WILCOX-GAY	4G8-840	1935	AC	CONSOLE	
WILCOX-GAY	4H13-840	1935	AC	CONSOLE BC/SW	
WILCOX-GAY	4JC6-84	1935	AC	TABLE BC/SW	
WILCOX-GAY	4JC6-780	1935	AC	CONSOLE BC/SW	

MANUFACTURER	MODEL	YEAR	POWER	STYLE	PRICE
WILCOX-GAY	5A6-89	1935	2/DC	TABLE	
WILCOX-GAY	5A6-810	1935	2/DC	CONSOLE	
WILCOX-GAY	5B5-89	1935	AC	TABLE BC/SW	
WILCOX-GAY	5B5-800	1935	AC	CONSOLE BC/SW	
WILCOX-GAY	5E7-91	1935	AC	TABLE BC/SW	
WILCOX-GAY	5E7-780	1935	AC	CONSOLE BC/SW	
WILCOX-GAY	5F8-90	1935	6/DC	TABLE BC/SW	
WILCOX-GAY	5F8-790	1935	6/DC	CONSOLE BC/SW	
WILCOX-GAY	35B-566	1934	AC	TABLE BC/SW	
WILCOX-GAY	255-30	1932	AC	TABLE	
WILCOX-GAY	A-11	1936	AC	TABLE BC/SW	
WILCOX-GAY	A-111	1941	AC	CONSOLE RADIO/PHONO BC/FM	50
WILCOX-GAY	A-114	1941	AC	CONSOLE R/P BC/FM	45
WILCOX-GAY	A-15	1936	AC	TABLE BC/SW	
WILCOX-GAY	A-16	1936	AC	TABLE BC/SW	
WILCOX-GAY	A-17	1936	AC	TABLE CIRCULAR EBONY BC/SW	
WILCOX-GAY	A-18	1936	AC	TABLE BC/SW	
WILCOX-GAY	A-19	1936	AC	CONSOLE BC/SW	
WILCOX-GAY	A-20	1936	AC	CONSOLE BC/SW	
WILCOX-GAY	A-21	1936	AC	CONSOLE BC/SW	
WILCOX-GAY	A-22	1936	2/DC	TABLE	
WILCOX-GAY	A-23	1937	2/DC	CONSOLE	
WILCOX-GAY	A-24	1937	6/DC	TABLE	
WILCOX-GAY	A-25	1937	6/DC	CONSOLE	
WILCOX-GAY	A-29	1937	AC	TABLE	
WILCOX-GAY	A-30	1937	AC	WALL	
WILCOX-GAY	A-31	1937	AC	TABLE	
WILCOX-GAY	A-32	1937	AC	TABLE	
WILCOX-GAY	A-33	1937	AC	TABLE	
WILCOX-GAY	A-34	1937	AC	TABLE	
WILCOX-GAY	A-35	1937	AC	TABLE	
WILCOX-GAY	A-36	1937	AC	CONSOLE ARMCHAIR	
WILCOX-GAY	A-37	1937	AC	CONSOLE	
WILCOX-GAY	A-38	1937	AC	CONSOLE	
WILCOX-GAY	A-43	1938	AC	MANTLE	
WILCOX-GAY	A-44	1938	AC	MANTLE	
WILCOX-GAY	A-48	1938	AC	MANTLE	
WILCOX-GAY	A-50	1938	AC	MANTLE	
WILCOX-GAY	A-51	1938	AC	MANTLE	20
WILCOX-GAY	A-52	1938	AC	MANTLE	30
WILCOX-GAY	A-53	1939	AC	MANTLE	25
WILCOX-GAY	A-54	1939	AC	TABLE PB BC/SW	40
WILCOX-GAY	A-55	1939	AC	CONSOLE PB BC/SW	55
WILCOX-GAY	A-68	1939	BAT	PORTABLE	
WILCOX-GAY	A-69	1939	AC	CONSOLE RADIO/PHONO	35
WILCOX-GAY	A-70	1939	AC	CONSOLE BC/SW	
WILCOX-GAY	A-71	1939	AC	PORTABLE	
WILCOX-GAY	A-76	1940	AC	TABLE	30

MANUFACTURER	MODEL	YEAR	POWER	STYLE	PRICE
WILCOX-GAY	A-93	1940	AC	CONSOLE	45
WILCOX-GAY	A-93	1940	AC	CONSOLE R/P W/REC. BC/SW	45
WILCOX-GAY	A-94	1940	AC	CONSOLE R/P W/RECORDER	60
WILCOX-GAY	CAMEO	1933	BAT	PORTABLE	
WILCOX-GAY	CANTATA	1933	AC	COMPACT	
WILCOX-GAY	CARILLON	1933	AC	MIDGET	
WILCOX-GAY	CORONA	1933	AC	CONSOLE	
WILCOX-GAY	CORONET	1933	AC	CONSOLE HB	
WILCOX-GAY	HILO	1933	AC	COMPACT	
WINDSOR FURNITURE	WINDSOR	1924	BAT	CONSOLE LB TABLE	250
WIRELESS SPECIALITY	IP-501-A	1922	BAT	TABLE HP 2 DIAL	600
WOLVERINE	25B	1923	BAT	TABLE LP 4 TUBE 3 DIAL	
WORK RITE	16	1926	BAT	TABLE LP 6 TUBE 2 DIAL	110
WORK RITE	17	1927	BAT	TABLE LP 6 TUBE 1 DIAL	95
WORK RITE	18	1928	AC	TABLE LP 7 TUBE 1 DIAL	
WORK RITE	26	1926	BAT	CONSOLE LB 6 TUBE 2 DIAL	135
WORK RITE	28	1928	AC	CONSOLE HB DOORS 7 T 1 DL	145
WORK RITE	33	1929	AC	CONSOLE LB 8 TUBE 1 DIAL	155
WORK RITE	35	1929	AC	CONSOLE LB 8 TUBE 1 DIAL	
WORK RITE	36	1926	BAT	CONSOLE DESK SPEAKER 6T 2 DIAL	200
WORK RITE	37	1927	BAT	TABLE LP 8 TUBE 1 DIAL	110
WORK RITE	37	1929	AC	CONSOLE HB 8 TUBE 1 DIAL	
WORK RITE	38	1928	AC	TABLE LP 9 TUBE 1 DIAL	
WORK RITE	39	1929	AC	COSOLE HB 8 TUBE 1 DIAL	
WORK RITE	40	1929	AC	CONSOLE RADIO/PHONO 8T 1 DIAL	
WORK RITE	48	1928	AC	CONSOLE 9 TUBE 1 DIAL	
WORK RITE	AIR MASTER	1924	BAT	TABLE HP SLOPE PAN 5T 3 DL	
WORK RITE	ARISTOCRAT	1924	BAT	CONSOLE HB 5 TUBE 3 DIAL	250
WORK RITE	CHUM	1924	BAT	TABLE HP EXPOSED 3 T 2 DL	240
WORK RITE	RADIO KING	1924	BAT	TABLE HP W/SPEAKER 5T 3 DL	225
WORLD WIRELESS	SUPER-FOUR	1923	BAT	TABLE LP 4 TUBE 2 DIAL	
WRIGHT RADIO	5-A	1925	BAT	TABLE LP 5 TUBE 3 DIAL	
WRIGHT RADIO	A	1923	BAT	TABLE LP 4 TUBE 2 DIAL	
WRIGHT RADIO	B	1923	BAT	TABLE LP 2 TUBE 2 DIAL	
WRIGHT RADIO	VII	1925	BAT	TABLE LP 7 TUBE 2 DIAL	

Z

MANUFACTURER	MODEL	YEAR	POWER	STYLE	PRICE
ZANEY-GILL	CLARION	1930	AC	CATHEDRAL	130
ZANEY-GILL	CLARIONETTE	1930	AC	CATHEDRAL	130
ZENITH	1-R	1922	BAT	TABLE LP 1 TUBE 3 DIAL	350
ZENITH	3-R	1923	BAT	TABLE LP 4 TUBE 1 DIAL	395
ZENITH	4-B-131	1936	6/DC	TABLE	55
ZENITH	4-B-132	1936	6/DC	TABLE	
ZENITH	4-B-231	1937	2/DC	TABLE	50
ZENITH	4-B-314	1939	BAT	TABLE PLASTIC PB	
ZENITH	4-F-227	1937	2/DC	TABLE	
ZENITH	4-K-310	1939	BAT	TABLE PLASTIC	
ZENITH	4-K-321	1939	BAT	TABLE	

MANUFACTURER	MODEL	YEAR	POWER	STYLE	PRICE
ZENITH	4-K-400D,M,Y	1939	BAT	PORTABLE CLOTH	35
ZENITH	4-K-400L,S	1939	BAT	PORTABLE	35
ZENITH	4-K-402D,M,Y	1939	BAT	PORTABLE CLOTH	35
ZENITH	4-K-402L	1939	BAT	PORTABLE LEATHERETTE	40
ZENITH	4-R	1923	BAT	TABLE LP 4 TUBE 1 DIAL	350
ZENITH	4-T-26	1935	AC	TABLE BC/SW	
ZENITH	4-T-51	1935	AC	CONSOLE BC/SW	
ZENITH	4-V-31	1935	6/DC	TABLE	
ZENITH	4-V-59	1935	6/DC	CONSOLE	
ZENITH	5-D-625	1941	AC	TABLE	40
ZENITH	5-F-233	1937	2/DC	TABLE BC/SW	
ZENITH	5-F-251	1937	2/DC	CONSOLE BC/SW	
ZENITH	5-G-401D,M	1939	AC	PORTABLE	25
ZENITH	5-G-401L,Y	1939	AC	PORTABLE	25
ZENITH	5-G-441	1939	AC	TABLE PB	
ZENITH	5-G-442	1939	AC	TABLE PB	
ZENITH	5-G-461	1939	AC	CONSOLE PB	
ZENITH	5-G-484M	1939	AC	PORTABLE	
ZENITH	5-J-217	1937	AC	TABLE BC/SW	125
ZENITH	5-J-247	1937	AC	CONSOLE ARMCHAIR BC/SW	
ZENITH	5-J-255	1937	AC	CONSOLE BC/SW	
ZENITH	5-R-135	1937	AC	TABLE	85
ZENITH	5-R-216	1937	AC	TABLE	
ZENITH	5-R-226	1937	AC	CONSOLE CHILDREN'S	
ZENITH	5-R-236	1937	AC	CONSOLE ARMCHAIR	
ZENITH	5-R-312	1938	AC	TABLE PLASTIC PB	45
ZENITH	5-R-316	1938	AC	TABLE PB	
ZENITH	5-R-317	1938	AC	TABLE PB	70
ZENITH	5-R-337	1938	AC	CONSOLE ARMCHAIR PB	
ZENITH	5-S-29	1935	AC	TABLE BC/SW	
ZENITH	5-S-56	1935	AC	CONSOLE BC/SW	
ZENITH	5-S-119	1936	AC	TABLE BC/SW	75
ZENITH	5-S-126	1936	AC	TABLE BC/SW	85
ZENITH	5-S-127	1936	AC	TABLE BC/SW	75
ZENITH	5-S-150	1936	AC	CONSOLE BC/SW	
ZENITH	5-S-151	1936	AC	CONSOLE BC/SW	
ZENITH	5-S-161	1936	AC	CONSOLE BC/SW	
ZENITH	5-S-218	1937	AC	TABLE BC/SW	
ZENITH	5-S-220	1937	AC	TABLE BC/SW	
ZENITH	5-S-228	1937	AC	TABLE BC/SW	
ZENITH	5-S-237	1937	AC	CONSOLE ARMCHAIR BC/SW	150
ZENITH	5-S-250	1937	AC	CONSOLE BC/SW	
ZENITH	5-S-252	1937	AC	CONSOLE BC/SW	
ZENITH	5-S-319	1937	AC	TABLE	50
ZENITH	5-S-327	1938	AC	TABLE BC/SW	
ZENITH	6-A-19	1935	AC	TABLE	40
ZENITH	6-B-129	1936	6/DC	TOMBSTONE BC/SW	65
ZENITH	6-B-164	1936	6/DC	CONSOLE BC/SW	
ZENITH	6-D-116	1936	AC	TABLE BC/SW	
ZENITH	6-D-117	1936	AC	TABLE BC/SW	140

MANUFACTURER	MODEL	YEAR	POWER	STYLE	PRICE
ZENITH	6-D-118	1936	AC	TABLE BC/SW	
ZENITH	6-D-219	1937	AC	TABLE BC/SW	
ZENITH	6-D-221	1937	AC	TABLE BC/SW	
ZENITH	6-D-311	1938	AC	TABLE PLASTIC	20
ZENITH	6-D-311W	1938	AC	TABLE PLASTIC	20
ZENITH	6-D-311Y	1938	AC	TABLE PLASTIC	20
ZENITH	6-D-312	1938	AC	TABLE PLASTIC	
ZENITH	6-D-312W	1938	AC	TABLE PLASTIC	
ZENITH	6-D-312Y	1938	AC	TABLE PLASTIC	
ZENITH	6-D-316	1938	AC	TABLE PB	
ZENITH	6-D-317	1938	AC	TABLE	40
ZENITH	6-D-326	1938	AC	CONSOLE CHILD'S	
ZENITH	6-D-336	1938	AC	CONSOLE ARMCHAIR	
ZENITH	6-D-337	1938	AC	CONSOLE ARMCHAIR PB	
ZENITH	6-D-360	1938	AC	CONSOLE	
ZENITH	6-D-410	1939	AC	MANTLE PLASTIC	
ZENITH	6-D-411	1939	AC	MANTLE PLASTIC	
ZENITH	6-D-412	1939	AC	MANTLE PLASTIC	
ZENITH	6-D-413	1939	AC	MANTLE PLASTIC PB	
ZENITH	6-D-414	1939	AC	MANTLE PLASTIC PB	
ZENITH	6-D-425	1939	AC	TABLE	
ZENITH	6-D-426	1939	AC	TABLE PB	
ZENITH	6-D-427	1939	AC	TABLE PB	
ZENITH	6-D-455	1939	AC	CONSOLE PERIOD PB	
ZENITH	6-D-456	1939	AC	CONSOLE PB	
ZENITH	6-D-466	1939	AC	CONSOLE CHAIRSIDE PB	
ZENITH	6-D-480	1939	AC	TABLE RADIO/PHONO	
ZENITH	6-D-481	1939	AC	TABLE RADIO/PHONO	
ZENITH	6-D-485	1939	AC	CONSOLE RADIO/PHONO PB	
ZENITH	6-D-510	1941	AC	TABLE PLASTIC	20
ZENITH	6-D-610	1939	AC	TABLE PLASTIC	30
ZENITH	6-D-614	1940	AC	TABLE	50
ZENITH	6-G-601M	1941	AC	PORTABLE	25
ZENITH	6-J-257	1937	AC	CONSOLE BC/SW	
ZENITH	6-P-416	1939	AC	MANTLE PLASTIC	
ZENITH	6-P-417	1939	AC	MANTLE PLASTIC	
ZENITH	6-P-428	1939	AC	TABLE	
ZENITH	6-P-429	1939	AC	TABLE PB	
ZENITH	6-P-430	1939	AC	TABLE PB	
ZENITH	6-P-447	1939	AC	CONSOLE CHAIRSIDE PB	
ZENITH	6-P-448	1939	AC	CONSOLE CHAIRSIDE PB	
ZENITH	6-P-457	1939	AC	CONSOLE PB	
ZENITH	6-S-27	1935	AC	TABLE BC/SW	75
ZENITH	6-S-52	1935	AC	CONSOLE BC/SW	85
ZENITH	6-S-128	1936	AC	TOMBSTONE BC/SW	135
ZENITH	6-S-137	1936	AC	CONSOLE END TABLE BC/SW	
ZENITH	6-S-147	1936	AC	CONSOLE END TABLE BC/SW	155
ZENITH	6-S-152	1936	AC	CONSOLE BC/SW	125
ZENITH	6-S-157	1936	AC	CONSOLE BC/SW	
ZENITH	6-S-203	1937	AC	CONSOLE RADIO/PHONO BC/SW	

MANUFACTURER	MODEL	YEAR	POWER	STYLE	PRICE
ZENITH	6-S-222	1937	AC	TABLE BC/SW	75
ZENITH	6-S-223	1937	AC	TABLE BC/SW	
ZENITH	6-S-229	1937	AC	TABLE BC/SW	70
ZENITH	6-S-230	1937	AC	TABLE BC/SW	
ZENITH	6-S-239	1937	AC	CONSOLE ARMCHAIR BC/SW	200
ZENITH	6-S-241	1937	AC	CONSOLE RADIO BAR BC/SW	
ZENITH	6-S-254	1937	AC	CONSOLE BC/SW	
ZENITH	6-S-256	1937	AC	CONSOLE BC/SW	
ZENITH	6-S-321	1939	AC	TABLE BC/SW	45
ZENITH	6-S-330	1938	AC	TABLE PB BC/SW	
ZENITH	6-S-341	1938	AC	CONSOLE ARM CHAIR PB BC/SW	
ZENITH	6-S-361	1938	AC	CONSOLE PB BC/SW	
ZENITH	6-S-362	1938	AC	CONSOLE PB BC/SW	
ZENITH	6-V-27	1935	6/DC	TOMBSTONE BC/SW	65
ZENITH	6-V-62	1935	6/DC	CONSOLE	
ZENITH	7-D-119	1936	AC	COMPACT BC/SW	
ZENITH	7-D-126	1936	AC	TABLE BC/SW	
ZENITH	7-D-127	1936	AC	TOMBSTONE BC/SW	85
ZENITH	7-D-138	1936	AC	TABLE ZEPHYR BC/SW	
ZENITH	7-D-148	1936	AC	CONSOLE END TABLE BC/SW	
ZENITH	7-D-151	1936	AC	CONSOLE BC/SW	
ZENITH	7-D-162	1936	AC	CONSOLE BC/SW	
ZENITH	7-D-168	1936	AC	CONSOLE BC/SW	
ZENITH	7-D-203	1937	AC	CONSOLE RADIO/PHONO BC/SW	
ZENITH	7-D-222	1937	AC	TABLE BC/SW	
ZENITH	7-D-223	1937	AC	TABLE BC/SW	
ZENITH	7-D-229	1937	AC	TABLE BC/SW	
ZENITH	7-D-239	1937	AC	CONSOLE ARMCHAIR BC/SW	
ZENITH	7-D-241	1937	AC	CONSOLE ARMCHAIR BC/SW	
ZENITH	7-D-243	1937	AC	CONSOLE ARMCHAIR BC/SW	
ZENITH	7-D-253	1937	AC	CONSOLE BC/SW	
ZENITH	7-G-605	1941	AC	PORTABLE TRANS-OCEANIC	135
ZENITH	7-J-232	1937	AC	TABLE BC/SW	
ZENITH	7-J-259	1937	AC	CONSOLE BC/SW	
ZENITH	7-S-28	1935	AC	TOMBSTONE BC/SW	145
ZENITH	7-S-53	1935	AC	CONSOLE BC/SW	
ZENITH	7-S-204	1937	AC	CONSOLE	230
ZENITH	7-S-232	1937	AC	TABLE BC/SW	
ZENITH	7-S-240	1937	AC	CONSOLE ARMCHAIR BC/SW	
ZENITH	7-S-242	1937	AC	CONSOLE ARMCHAIR BC/SW	
ZENITH	7-S-258	1937	AC	CONSOLE BC/SW	
ZENITH	7-S-260	1937	AC	CONSOLE BC/SW	
ZENITH	7-S-261	1937	AC	CONSOLE BC/SW	
ZENITH	7-S-432	1939	AC	TABLE PB	
ZENITH	7-S-433	1939	AC	TABLE PB	
ZENITH	7-S-434	1939	AC	CONSOLE PB	
ZENITH	7-S-449	1939	AC	CONSOLE CHAIRSIDE PB	
ZENITH	7-S-450	1939	AC	CONSOLE CHAIRSIDE R/P PB	

MANUFACTURER	MODEL	YEAR	POWER	STYLE	PRICE
ZENITH	7-S-458	1939	AC	CONSOLE PB	235
ZENITH	7-S-459	1939	AC	CONSOLE PERIOD PB	
ZENITH	7-S-460	1939	AC	CONSOLE PERIOD PB	
ZENITH	7-S-461	1939	AC	CONSOLE PB	
ZENITH	7-S-462	1939	AC	CONSOLE PERIOD PB	240
ZENITH	7-S-487	1939	AC	CONSOLE RADIO/PHONO PB	
ZENITH	7-S-488	1939	AC	CONSOLE RADIO/PHONO PB	
ZENITH	7-S-490	1939	AC	CONSOLE CHAIRSIDE R/P PB	
ZENITH	7-S-529	1941	AC	TABLE PB BC/SW	65
ZENITH	7-S-598	1941	AC	CONSOLE R/P PB BC/SW	80
ZENITH	8-H-61	1935	AC	CONSOLE	50
ZENITH	8-S-129	1936	AC	TABLE BC/SW	
ZENITH	8-S-154	1936	AC	CONSOLE BC/SW	230
ZENITH	9-S-30	1935	AC	TABLE BC/SW	
ZENITH	9-S-54	1935	AC	CONSOLE BC/SW	
ZENITH	9-S-55	1935	AC	CONSOLE BC/SW	
ZENITH	9-S-204	1937	AC	CONSOLE RADIO/PHONO BC/SW	90
ZENITH	9-S-232	1937	AC	TABLE BC/SW	
ZENITH	9-S-242	1937	AC	CONSOLE ARMCHAIR BC/SW	215
ZENITH	9-S-244	1937	AC	CONSOLE ARMCHAIR BC/SW	160
ZENITH	9-S-262	1937	AC	CONSOLE BC/SW	150
ZENITH	9-S-263	1937	AC	CONSOLE BC/SW	225
ZENITH	9-S-264	1937	AC	CONSOLE BC/SW	
ZENITH	9-S-307	1938	AC	CONSOLE	250
ZENITH	9-S-365	1938	AC	CONSOLE PB BC/SW	
ZENITH	9-S-367	1938	AC	CONSOLE	450
ZENITH	10-H-551	1941	AC	CONSOLE CHAIRSIDE BC/SW/FM	95
ZENITH	10-H-571	1941	AC	CONSOLE PB BC/SW/FM	190
ZENITH	10-H-573	1941	AC	CONSOLE PB BC/SW/FM	175
ZENITH	10-S-130	1936	AC	TABLE BC/SW	
ZENITH	10-S-147	1936	AC	CONSOLE END TABLE BC/SW	
ZENITH	10-S-153	1936	AC	CONSOLE BC/SW	230
ZENITH	10-S-155	1936	AC	CONSOLE BC/SW	
ZENITH	10-S-156	1936	AC	CONSOLE BC/SW	
ZENITH	10-S-157	1936	AC	CONSOLE BC/SW	
ZENITH	10-S-160	1936	AC	CONSOLE BC/SW	200
ZENITH	11	1927	BAT	TABLE LP 6 TUBE 1 DIAL	150
ZENITH	11-E	1927	AC	TABLE LP 7 TUBE 1 DIAL	125
ZENITH	12	1927	BAT	TABLE LP 6 TUBE 1 DIAL	140
ZENITH	12-A-57	1935	AC	CONSOLE BC/SW	
ZENITH	12-A-58	1935	AC	CONSOLE BC/SW	
ZENITH	12-H-670	1941	AC	CONSOLE BC/SW/FM	150
ZENITH	12-H-679	1941	AC	CONSOLE R/P PB BC/SW/FM	70
ZENITH	12-S-205	1937	AC	CONSOLE RADIO/PHONO BC/SW	
ZENITH	12-S-232	1937	AC	TABLE BC/SW	100
ZENITH	12-S-245	1937	AC	CONSOLE ARMCHAIR BC/SW	
ZENITH	12-S-265	1937	AC	CONSOLE BC/SW	150

MANUFACTURER	MODEL	YEAR	POWER	STYLE	PRICE
ZENITH	12-S-266	1937	AC	CONSOLE BC/SW	
ZENITH	12-S-267	1937	AC	CONSOLE BC/SW	
ZENITH	12-S-268	1937	AC	CONSOLE BC/SW	
ZENITH	12-S-569	1937	AC	CONSOLE	350
ZENITH	12-U-158	1936	AC	CONSOLE BC/SW	
ZENITH	12-U-159	1936	AC	CONSOLE BC/SW	
ZENITH	14	1927	BAT	CONSOLE 6 TUBE 1 DIAL	
ZENITH	14-E	1927	AC	CONSOLE 7 TUBE 1 DIAL	
ZENITH	15	1927	BAT	TABLE LP 8 TUBE 1 DIAL	
ZENITH	15-E	1927	AC	TABLE LP 9 TUBE 1 DIAL	
ZENITH	15-U-246	1937	AC	CONSOLE ARMCHAIR BC/SW	
ZENITH	15-U-269	1937	AC	CONSOLE BC/SW	
ZENITH	15-U-271	1937	AC	CONSOLE BC/SW	
ZENITH	15-U-272	1937	AC	CONSOLE BC/SW	
ZENITH	15-U-273	1937	AC	CONSOLE BC/SW	
ZENITH	16	1927	BAT	CONSOLE 8 TUBE 1 DIAL	
ZENITH	16-A-61	1935	AC	CONSOLE BC/SW	
ZENITH	16-A-63	1935	AC	CONSOLE BC/SW	
ZENITH	16-E	1927	AC	CONSOLE 9 TUBE 1 DIAL	
ZENITH	17	1927	BAT	TABLE LP 6 TUBE 1 DIAL	90
ZENITH	17-E	1927	AC	CONSOLE 17-E W/SPINET BASE	125
ZENITH	17-E	1927	AC	TABLE LP 10 TUBE 1 DIAL	
ZENITH	31	1928	BAT	TABLE LP 6 TUBE 1 DIAL	90
ZENITH	32	1928	BAT	CONSOLE LB 6 TUBE 1 DIAL	
ZENITH	33	1928	AC	TABLE LP 7 TUBE 1 DIAL	85
ZENITH	33-X	1928	AC	TABLE LP 7 TUBE 1 DIAL	
ZENITH	34	1928	AC	CONSOLE LB 7 TUBE 1 DIAL	
ZENITH	34-P	1928	AC	CONSOLE 8 TUBE 1 DIAL	95
ZENITH	35	1928	AC	CONSOLE HB 7 TUBE 1 DIAL	
ZENITH	35-A	1928	AC	CONSOLE 7 TUBE 1 DIAL	
ZENITH	35-AP	1928	AC	CONSOLE 8 TUBE 1 DIAL	
ZENITH	35-PX	1928	AC	CONSOLE 9 TUBE 1 DIAL	
ZENITH	37-A	1928	AC	CONSOLE RADIO/PHONO 9T 1 DIAL	
ZENITH	39	1928	AC	CONSOLE LB 10 TUBE 1 DIAL	
ZENITH	39-A	1928	AC	CONSOLE W/AUTOMATIC TUNING	
ZENITH	40-A	1929	AC	CONSOLE RADIO/PHONO 10T 1 DIAL	145
ZENITH	41	1929	AC	TABLE 7 TUBE 1 DIAL	
ZENITH	42	1929	AC	CONSOLE LB 7 TUBE 1 DIAL	
ZENITH	52	1929	AC	CONSOLE 9 TUBE 1 DIAL	
ZENITH	53	1929	AC	CONSOLE 9 TUBE 1 DIAL	350
ZENITH	54	1929	AC	CONSOLE 9 TUBE 1 DIAL	360
ZENITH	55	1929	AC	CONSOLE W/REMOTE TUNING	330
ZENITH	60	1930	AC	CONSOLE LB "SUPER-MIDGET"	
ZENITH	61	1930	AC	CONSOLE LB 9 TUBE 1 DIAL	275
ZENITH	62	1930	AC	CONSOLE LB 9 TUBE 1 DIAL	370
ZENITH	64	1930	AC	CONSOLE LB 9 TUBE 1 DIAL	

MANUFACTURER	MODEL	YEAR	POWER	STYLE	PRICE
ZENITH	67	1930	AC	CONSOLE LB 9 TUBE 1 DIAL	
ZENITH	71	1930	AC	CONSOLE LB 9 TUBE 1 DIAL	
ZENITH	72	1930	AC	CONSOLE LB 9 TUBE 1 DIAL	370
ZENITH	73	1930	AC	CONSOLE LB 9 TUBE 1 DIAL	
ZENITH	74	1930	AC	CONSOLE 9 TUBE 1 DIAL	
ZENITH	75	1930	AC	CONSOLE RADIO/PHONO 9T 1 DIAL	
ZENITH	90	1931	AC	CONSOLE HB	
ZENITH	91	1931	AC	CONSOLE HB	
ZENITH	92	1931	AC	CONSOLE HB	
ZENITH	103	1931	AC	CONSOLE HB	
ZENITH	125-U-270	1937	AC	CONSOLE BC/SW	
ZENITH	210	1932	AC	TABLE BC/SW	
ZENITH	210-5	1932	AC	TABLE BC/SW/LW	
ZENITH	220	1932	AC	CONSOLE LB BC/SW	
ZENITH	230	1932	AC	CATHEDRAL	
ZENITH	240	1932	AC	CONSOLE LB	
ZENITH	245	1932	AC	CONSOLE LB	250
ZENITH	250	1932	AC	TABLE BC/SW	
ZENITH	260	1932	AC	CONSOLE LB BC/SW	
ZENITH	270	1932	AC	CONSOLE LB RADIO/PHONO	
ZENITH	293	1934	AC	CONSOLE	
ZENITH	342	1928	AC	TABLE LP 7 TUBE 1 DIAL	
ZENITH	352	1928	AC	TABLE LP 7 TUBE 1 DIAL	
ZENITH	353-A	1928	AC	CONSOLE 7 TUBE 1 DIAL	
ZENITH	410	1932	AC	CONSOLE LB BC/SW	
ZENITH	411	1933	AC	CONSOLE	
ZENITH	411	1932	AC	CONSOLE LB BC/SW	
ZENITH	420	1932	AC	CONSOLE LB BC/SW	
ZENITH	430	1932	AC	CONSOLE SEMI-HIGH BC/SW	
ZENITH	440	1932	AC	CONSOLE SEMI-HIGH BC/SW	
ZENITH	441	1933	AC	CONSOLE ORNATE	270
ZENITH	473	1933	AC	CONSOLE RADIO/PHONO	
ZENITH	476-A	1933	AC	CONSOLE RADIO/PHONO	
ZENITH	563	1930	AC	CONSOLE	
ZENITH	701	1933	AC	TABLE	
ZENITH	705	1933	AC	TABLE	
ZENITH	706	1933	AC	TABLE	
ZENITH	707	1933	AC	TABLE	
ZENITH	711	1933	AC	TABLE	
ZENITH	715	1933	AC	TABLE	
ZENITH	725	1934	AC	CONSOLE	
ZENITH	730	1934	AC	MIDGET	
ZENITH	735	1934	AC	CONSOLE	
ZENITH	750	1933	AC	CONSOLE	
ZENITH	755	1933	AC	CONSOLE	220
ZENITH	760	1933	AC	CONSOLE	
ZENITH	770	1933	AC	CONSOLE	
ZENITH	775	1933	AC	CONSOLE	
ZENITH	801	1934	AC	TABLE	

MANUFACTURER	MODEL	YEAR	POWER	STYLE	PRICE
ZENITH	805	1934	AC	TABLE BC/SW	80
ZENITH	807	1934	AC	TOMBSTONE BC/SW	55
ZENITH	808	1934	AC	TOMBSTONE BC/SW	100
ZENITH	809	1934	AC	TABLE BC/SW	
ZENITH	811	1934	AC	TABLE BC/SW	
ZENITH	812	1934	AC	TABLE	50
ZENITH	825	1934	AC	TABLE BC/SW	
ZENITH	827	1934	AC	TABLE BC/SW	75
ZENITH	829	1934	AC	TABLE BC/SW	
ZENITH	835	1934	AC	TABLE BC/SW	25
ZENITH	845	1934	AC	CONSOLE BC/SW	
ZENITH	847	1934	AC	CONSOLE BC/SW	
ZENITH	850	1934	AC	CONSOLE BC/SW	
ZENITH	860	1934	AC	CONSOLE BC/SW	
ZENITH	861	1934	AC	CONSOLE BC/SW	
ZENITH	870	1934	AC	CONSOLE BC/SW	
ZENITH	871	1934	AC	CONSOLE BC/SW	
ZENITH	880	1934	AC	CONSOLE BC/SW	
ZENITH	945	1935	AC	MIDGET	
ZENITH	970	1934	AC	CONSOLE	
ZENITH	975	1934	AC	CONSOLE	
ZENITH	990	1935	AC	CONSOLE	
ZENITH	1000-2	1934	AC	CONSOLE	
ZENITH	1000-Z	1935	AC	CONSOLE STRATOSPHERE BC/SW	2000
ZENITH	1117	1935	AC	TOMBSTONE BC/SW	140
ZENITH	CHINESE DELUXE	1925	BAT	CONSOLE LB 6 TUBE 1 DIAL	195
ZENITH	CHINESE DELUXE	1926	AC	CONSOLE 10 TUBE 1 DIAL	
ZENITH	COLONIAL DLX	1926	AC	CONSOLE 10 TUBE 1 DIAL	
ZENITH	DELUXE	1934	AC	CONSOLE BC/SW	
ZENITH	ENGLISH DELUXE	1926	AC	CONSOLE 10 TUBE 1 DIAL	
ZENITH	ITALIAN DELUXE	1924	BAT	CONSOLE LB 6 TUBE 1 DIAL	250
ZENITH	ITALIAN DELUXE	1926	AC	CONSOLE 10 TUBE 1 DIAL	
ZENITH	IX	1924	BAT	CONSOLE BUREAU 6 T 2 DIAL	200
ZENITH	LH	1931	AC	TABLE	
ZENITH	MH	1931	AC	CONSOLE HB	
ZENITH	SPANISH DELUXE	1924	BAT	CONSOLE HB 6 TUBE 1 DIAL	250
ZENITH	SPANISH DELUXE	1926	AC	CONSOLE 10 TUBE 1 DIAL	
ZENITH	SUPER PORTABLE	1924	BAT	PORTABLE 6 TUBE 2 DIAL	440
ZENITH	VII SUPER	1924	BAT	TABLE LP 6 TUBE 2 DIAL	300
ZENITH	VIII SUPER	1924	BAT	CONSOLE-VII W/LEGS	195
ZENITH	WH	1931	AC	CONSOLE	
ZENITH	X	1924	BAT	CONSOLE BUREAU 6 T 2 DIAL	200

PART III

The TRADE DIRECTORY

Part III is divided into two major sections. Section 1 of Part III lists trade names alphabetically with reference to manufacturer. Section 2 of Part III lists manufacturers alphabetically with reference to their trade names.

The two sections provide for quickly cross-referencing trade names and manufacturers. There is no rule of thumb by which to determine whether the nameplate of a model carries the name of its manufacturer, the trade name of its manufacturer, or, as happens sometimes, both names. However, in a great many instances, only the trade name appears. Since it is easy to confuse the two primary sources of identification, readers who are in doubt about the actual identity of a manufacturer may resolve the dilemma by consulting the two comprehensive directories presented here.

SECTION 1

TRADE NAMES

TRADE NAME	MANUFACTURER
A	
A-C DAYTON	A-C ELECTRICAL MFG. CO. (DAYTON, OHIO)
A-P	ATLANTIC-PACIFIC RADIO SUPPLIES CO.
A. T. R.	AMERICAN TELEPHONE & RADIO CO.
ABBEY	SPLITDORF ELECTRICAL MANUFACTURING CO.
ABC	ABC RADIO LABORATORIES, FREEPORT RADIO CORP.
ACE	ACE RADIO CO., CROSLEY RADIO CORP., RADIO CIRCULAR CO.
ACME	ACME APPARATUS CO., ACME ELECTRIC & MANUFACTURING CO., SEARS MANUFACTURING CO., WRIGHT RADIO MANUFACTURING CO., WRIGHT-De COSTER RADIO, INC.
ACMEDYNE	AMSCO PRODUCTS CO.
ACMEFLEX	ACME APPARATUS CO.
ACMEPHONE	ACME APPARATUS CO.
ACOUSTIC	RCA VICTOR DIVISION OF RADIO CORP. OF AMERICA
ACRATEST	FEDERATED PURCHASER, INC.
ACRATONE	FEDERATED PURCHASER, INC.
ADLER-ROYALE	ADLER MANUFACTURING CO.
ADMINISTRATOR	CROSLEY RADIO CORP.
ADMIRAL	CONTINENTAL RADIO & TELEVISION CORP.
ADMIRALITY	NORDEN HAUCK RADIO CO.
ADROLA	ADROIT TOOL CO.
ADVANCE	ADVANCE ELECTRIC CO.
ADVENTURER	CROSLEY RADIO CORP.
AEOLIAN HALL	AEOLIAN CO.
AERIOLA	RCA VICTOR DIVISION OF RADIO CORP. OF AMERICA
AERIPHONE	NATIONAL RADIO CO.
AERO	HOODWIN, CHARLES, CO.
AERO PRODUCTS	HOODWIN, CHARLES, CO.
AERODYNE	AERODYNE CO.
AEROMASTER	STEWART-WARNER RADIO CORP.
AERONAUTIC	MISSION BELL RADIO MANUFACTURING CO.
AEROVOICE	AEROVOICE, INC.
AETNA	AETNA RADIO CORP.
AIR CHIEF	FIRESTONE TIRE & RUBBER CO.
AIR FLIGHT	SILVER-MARSHALL MANUFACTURING CO.
AIR HOUND	PILOT RADIO & TUBE CORP.
AIR KING	TWENTIETH CENTURY RADIO CO.
AIR MASTER	AUTOMATIC RADIO MANUFACTURING CO., INC., WORKRITE MANUFACTURING CO.

THE TRADE DIRECTORY

TRADE NAME	MANUFACTURER
AIR PILOT	SPIELMAN ELECTRIC CO.
AIR QUEEN	AIR-KING PRODUCTS CO., INC.
AIR ROAMER	KILBOURNE & CLARK MANUFACTURING CO.
AIR RULER	KILBOURNE & CLARK MANUFACTURING CO.
AIR SCOUT	PILOT RADIO & TUBE CORP.
AIR-KING	AIR-KING PRODUCTS CO., INC.
AIR-WAY	AIR-WAY ELECTRIC APPLIANCE CORP.
AIRCASTLE	SPIEGEL, MAY STERN
AIREX	AIREX CO.
AIRITE	INTERNATIONAL RADIO CORP.
AIRLINE	MONTGOMERY WARD & CO.
AIRO-MASTER	AIRO-MASTER CORP., ORTHODYNE RADIO MANUFACTURING CO.
AIRPHONE	NATIONAL AIRPHONE CORP.
AIRTROLA	DALCO RADIO CO.
AJAX	AJAX ELECTRIC SPECIALITY CO., PHILMORE MANUFACTURING CO.
AKRADYNE	SUNBEAM RADIO SALES CORP.
ALADDIN	MOE MANUFACTURING CO.
ALADYNE	ALADDIN MANUFACTURING CO.
ALDERMAN	CROSLEY RADIO CORP.
ALEMITE	STEWART-WARNER RADIO CORP.
ALHAMBRA	RCA VICTOR DIVISION OF RADIO CORP. OF AMERICA
ALL AMERICAN	ALL AMERICAN MOHAWK CORP.
ALL STAR	WARWICK MANUFACTURING CO.
ALL WORLD	SILVER-MARSHALL MANUFACTURING CO.
ALL-AMAX	ALL AMERICAN MOHAWK CORP.
ALLIED	ALLIED RADIO CORP.
ALONSON	COMMONWEALTH RADIO MANUFACTURING CO.
AMACO	AMERICAN APPARATUS CO.
AMBASSADOR	CROSLEY RADIO CORP.
AMBOROLA	AMERICAN BOSCH MAGNETO CORP.
AMBROLA	UNITED AMERICAN BOSCH CORP.
AMEREX	AMERICAN STANDARD ELECTRIC CO.
AMERICAN	GENERAL RADIO & TELEVISION CORP.
AMERICAN BEAUTY	AMERICAN AUTO AND RADIO MANUFACTURING CO., AMERICAN RADIO MANUFACTURING CO.
AMERICAN BOSCH	AMERICAN BOSCH MAGNETO CORP., UNITED AMERICAN BOSCH CORP.
AMERICAN WIRELESS	AMERICAN WIRELESS
AMERICAN WORLD	PACIFIC RADIO CORP.
AMERTRAN	AMERICAN TRANSFORMER CO.
AMES	PICARD RADIO CORP.
AMPERION	CAPEHART CORP., INC.
AMPIOLA	CADILLAC RADIO MANUFACTURING CO.
AMRAD	AMERICAN RADIO & RESEARCH CORP.
ANDREA	ANDREA RADIO CORP.
ANGELUS	DAVISON HAYNES CO., RADIO SERVICE CO.
ANNOUNCER	CROSLEY RADIO CORP.
ANSLEY	ANSLEY RADIO CORP.
APEX	UNITED STATES RADIO & TELEVISION CORP.

TRADE NAME	MANUFACTURER
APPEL-HENDERSON	APPEL-HENDERSON RADIO CO.
ARAGAIN	AUTOMETAL CORP.
ARBITER	CROSLEY RADIO CORP.
ARBORPHONE	CONSOLIDATED RADIO CORP., MACHINE SPECIALITY CO., PRECISION PRODUCTS CORP.
ARC	AUTOMATIC RADIO MANUFACTURING CO., INC.
ARC-LININGER	AMERICAN RADIO CORP.
ARCADIA	WELLS-GARDNER & CO.
ARGUS	ARGUS RADIO CORP.
ARGYLE	UNITED STATES RADIO & TELEVISION CORP.
ARIA	ALLIED STORES (NEW YORK)
ARISTOCRAT	RADIOCOMPAK CO.
ARKAY	R. K. RADIO LABORATORIES, INC.
ARLINGTON	WARWICK MANUFACTURING CO., WELLS MANUFACTURING CO. (Fond du Lac)
AROUND THE WORLD	SILVER-MARSHALL MANUFACTURING CO.
ARROW	ECOPHONE RADIO CORP.
ARVIA	NOBLITT-SPARKS INDUSTRIES, INC.
ARVIN	NOBLITT-SPARKS INDUSTRIES, INC.
ATCHINSON	ATCHINSON RADIO MANUFACTURING CO.
ATEC	ABLE TOOL & ENGINEERING CO.
ATLAS	AIR-KING PRODUCTS CO., INC.
ATWATER KENT	ATWATER KENT MANUFACTURING CO.
AUDIODYNE	AUDIOLA RADIO CORP.
AUDIOLA	AUDIOLA RADIO CORP.
AURORA	AURORA INC.
AUSTIN	AUSTIN RADIO MANUFACTURING CO.
AUTO TONE	WARWICK MANUFACTURING CO.
AUTO-ROLA	AUTOMATIC RADIO MANUFACTURING CO., INC.
AUTOCRAT	AUTOCRAT RADIO CO.
AUTOMASTER	AUTOMATIC RADIO MANUFACTURING CO., INC.
AUTOMATIC	AUTOMATIC RADIO MANUFACTURING CO., INC.
AUTRA	WARWICK MANUFACTURING CO.
AZTEC	STEIN, FRED W., CO.
B	
BABY GRAND	PHILCO RADIO & TELEVISION CORP.
BABYDYNE	KEN-MAR RADIO CORP.
BAIRD	SHORTWAVE & TELEVISION LABORATORIES, INC.
BALBOA	MONTGOMERY WARD & CO.
BALDER	BALDER RADIO CORP.
BALDWIN	BALDWIN, NATHANIEL, INC., PFANSTIEHL RADIO CO.
BALDWINETTE	BALDWIN, NATHANIEL, INC.
BALKEIT	BALKEIT RADIO CORP.
BALKEIT-NATIONAL	NATIONAL TRANSFORMER MANUFACTURING CO.
BANDBOX	CROSLEY RADIO CORP.
BANTAM	WARE RADIO CORP.
BAT-RY-LES	BATTERYLESS RADIO CORP.
BEAR CAT	WELLS RADIO MANUFACTURING CO. (Chicago)
BEARCAT	CHELSEA RADIO CO., SILVER-MARSHALL MANUFACTURING CO.
BEAVERTONE	BEAVER ELECTRIC CORP.

TRADE NAME	MANUFACTURER
BEL CANTO	AMERICAN RADIO & RESEARCH CORP.
BELL	BELL RADIO & TELEVISION CO.
BELL TONE	AIR-KING PRODUCTS CO., INC.
BELL-KNAPP	BELL-KNAPP HARDWARE & MANUFACTURING CO.
BELLE CLAIRE	R-B RADIO CO.
BELMONT	BELMONT RADIO CORP.
BENDIX	BENDIX RADIO MANUFACTURING CO.
BEST	PACIFIC RADIO CORP.
BESTONE	HYMAN, HENRY, & COMPANY
BEVERLY	WARWICK MANUFACTURING CO.
BIGFELLA	CROSLEY RADIO CORP.
BLACKBIRD	PHILMORE MANUFACTURING CO.
BLACKHAWK	WELLS-GARDNER & CO.
BLAIR	BLAIR RADIO LABORATORIES
BLAIRCO	BLAIR RADIO LABORATORIES
BLUE BIRD	GUTHRIE CO., PHILMORE MANUFACTURING CO.
BLUE RIBBON	CHICAGO NIPPLE MANUFACTURING CO.
BLUEBIRD	AUTOMATIC RADIO MANUFACTURING CO., INC., HANSEN STORAGE CO.
BOB-O-LINK	GUTHRIE CO.
BONNIBOY	CROSLEY RADIO CORP.
BOOK CASE	CROSLEY RADIO CORP.
BORGIA	RCA VICTOR DIVISION OF RADIO CORP. OF AMERICA
BOSCH	AMERICAN BOSCH MAGNETO CORP., UNITED AMERICAN BOSCH CORP.
BOSTONIAN	CLINTON RADIO MANUFACTURING CO.
BOSWORTH	BOSWORTH ELECTRIC MANUFACTURING CO.
BRANDES	KOLSTER RADIO CORP.
BRANDOLA	BRANDEIS, J. F., CORP.
BREMER TULLY	BREMER TULLY MANUFACTURING CO.
BRENTWOOD	GRIGSBY-GRUNOW CORP.
BROWNING	DEUTSCHMANN, TOBE, CORP.
BROWNING LABS	BROWNING LABORATORIES
BROWNING-DRAKE	BROWNING-DRAKE CORP., NATIONAL CO., INC.
BRUCEWOOD	GRIGSBY-GRUNOW CORP.
BRUNSWICK	BRUNSWICK-BALKE-COLLENDER CO., MERSHMAN BROS. CORP.
BUBBY-BOY	CROSLEY RADIO CORP.
BUCCANEER	CROSLEY RADIO CORP.
BUCKINGHAM	BUCKINGHAM RADIO CORP., CLINTON RADIO MANUFACTURING CO., PACIFIC RADIO CORP.
BUDDY	CROSLEY RADIO CORP.
BUELL	BUELL MANUFACTURING CO.
BULOVA	BULOVA WATCH CO.
BURAD	BUCKWALTER RADIO CORP.
BURLEY	SENTINEL RADIO CORP.
BUSH & LANE	BUSH & LANE PIANO CO.
C	
CABLE-NELSON	HOWARD RADIO MANUFACTURING CO.
CABRIOLET	CROSLEY RADIO CORP.
CADDY	MODERNOLA CO.

TRADE NAME	MANUFACTURER
CADILLAC	CADILLAC RADIO MANUFACTURING CO.
CAMDEN	RADIO PRODUCTS CORP.
CAMEO	GENERAL AMERICAN RADIO MANUFACTURING CO.
CANADIAN WESTINGHOUSE	CANADIAN WESTINGHOUSE CO., LTD.
CAPEHART	CAPEHART CORP., INC., FARNSWORTH TELEVISION & RADIO CORP.
CAPTAIN-KIDD	PILOT RADIO & TUBE CORP.
CARAVAN	WARWICK MANUFACTURING CO.
CARCO	CARTER MANUFACTURING CO.
CARDINAL	CARDINAL RADIO MANUFACTURING CO.
CARDINOLA	CARDINOLA RADIO CORP.
CARDWELL	CARDWELL, ALLEN D., MANUFACTURING CORP.
CARIDAL	LONG RADIO CO.
CARLILE TABLE	WARWICK MANUFACTURING CO.
CARLTON	UNITED STATES RADIO & TELEVISION CORP.
CAROL-DYNE	SOUTHEASTERN RADIO CORP.
CAROLER	CROSLEY RADIO CORP.
CARTERET	CARTERET RADIO LABS
CARUSO	LAUREHK RADIO MANUFACTURING CO.
CASA	CROSLEY RADIO CORP.
CASCADE	KELLOGG SWITCHBOARD & SUPPLY CO., NUNN-LANDON CO.
CASE	CASE ELECTRIC CORP., INDIANA ELECTRIC MANUFACTURING CO., MAJESTIC RADIO & TELEVISION CORP., UNITED STATES RADIO & TELEVISION CORP.
CASSCO	CASS, PHILIP, & CO.
CATHAY	WARWICK MANUFACTURING CO.
CATHEDRAL TONE	FEDERATED PURCHASER, INC.
CAVALCADE	CAVALCADE RADIO CO. (Chicago), CAVALCADE RADIO CORP. (San Francisco)
CAVALIER	MONTGOMERY WARD & CO.
CELESTE	LAKESIDE SUPPLY CO.
CENTURY	CENTURY RADIO PRODUCTS CO.
CHALLENGER	MONTGOMERY WARD & CO., TILMAN RADIO CORP.
CHAMPION	CHAMPION RADIO LABORATORIES
CHAMPION, JR.	GULBRANSEN RADIO CO.
CHELTENHAM	FRANKLIN RADIO CORP.
CHELTENWOOD	GRIGSBY-GRUNOW CORP.
CHERIO	CROSLEY RADIO CORP.
CHIEF	CROSLEY RADIO CORP.
CHIEFTAIN	ALL AMERICAN MOHAWK CORP.
CHUM	CROSLEY RADIO CORP.
CLAGO	CLAGO RADIO CORP.
CLARADYNE	CROUCH RADIO CO.
CLARATONE	EQUITABLE RADIO CORP.
CLARION	CLARION RADIO CORP., TRANSFORMER CORPORATION OF AMERICA, WARWICK MANUFACTURING CO., ZANEY-GILL CORP.
CLARIONETTE	ZANEY-GILL CORP.
CLASSIC	INSULINE CORPORATION OF AMERICA
CLASSIQUE	BELMONT RADIO CORP.
CLASSMATE	CROSLEY RADIO CORP.

THE TRADE DIRECTORY

TRADE NAME	MANUFACTURER
CLEARBROOK	HEIGES BROTHERS
CLEARFIELD	SHERMAN RADIO MANUFACTURING CORP.
CLEARODYNE	CLEARTONE RADIO CORP.
CLEARTONE	CLEARTONE RADIO CORP.
CLEARVIEW	KAESS MANUFACTURING CORP.
CLEVELAND	CLEVELAND PRODUCTS CO.
CLIMAX	CLIMAX RADIO & TELEVISION CO.
CLINTON	CLINTON RADIO MANUFACTURING CO.
CLIPPER	CROSLEY RADIO CORP., RADIO PRODUCTS CORP.
CO-OP	RADIO PRODUCTS CORP.
COAST-TO-COAST	SILVER-MARSHALL MANUFACTURING CO.
COLLEGIAN	MONTGOMERY WARD & CO.
COLONIAL	ATLAS COLONIAL CORP., COLONIAL RADIO CORP.
COLUMBIA	COLUMBIA PHONOGRAPH CO., COLUMBIA RADIO CO., MAJESTIC RADIO & TELEVISION CORP., RCA VICTOR DIVISION OF RADIO CORP. OF AMERICA
COM-RAD	COMMONWEALTH RADIO MANUFACTURING CO.
COMBIDYNE	WOLVERINE RADIO CO.
COMET	KENNEDY, COLIN B., CO.
COMMANDER	MONTGOMERY WARD & CO.
COMMISSIONER	CROSLEY RADIO CORP.
COMMODORE	MONTGOMERY WARD & CO.
COMPANION	CROSLEY RADIO CORP.
CON-RAD	CONSUMERS RADIO CO.
CONCERDYNE	JONES ELECTRIC & RADIO MANUFACTURING CO.
CONCERT GRAND	CONCERT RADIOPHONE CO.
CONGRESSMAN	CROSLEY RADIO CORP.
CONQUEROR	INSULINE CORPORATION OF AMERICA
CONSOMELLO	MAZDA RADIO MANUFACTURING CO.
CONSTITUTION	CROSLEY RADIO CORP.
CONTINENTAL	CONTINENTAL RADIO & TELEVISION CORP., RADIO PRODUCTS CORP.
CORDOVA	BRUNSWICK-BALKE-COLLENDER CO.
CORONA	CLINTON RADIO MANUFACTURING CO., CORONA RADIO & TELEVISION CORP., SEEBURG, J. P., CORP.
CORONADO	CORONADO MANUFACTURING CO., GAMBLE STORES, INC., KINGSTON RADIO CO., INC., MONTGOMERY WARD & CO.
CORONET	KENNEDY, COLIN B., CO.
CORSAIR	CROSLEY RADIO CORP.
CORTEZ	MONTGOMERY WARD & CO.
COSMOPOLITAN	COSMOPOLITAN PHUSIFORMER
COSMOTONE	COSMOTONE, INC.
COTO SYMPHONIC	COTO-COIL CO.
COUNTERPHASE	BREMER TULLY MANUFACTURING CO.
COURIER	GRAY ELECTRIC CO., UNITED REPRODUCERS CORP.
CRESCENTYNE	CRESCENT RADIO MANUFACTURING CO.
CROMWELL	RCA VICTOR DIVISION OF RADIO CORP. OF AMERICA
CRONY	CROSLEY RADIO CORP.
CROSLEY	CROSLEY RADIO CORP.
CROWN	CUSTOM BUILT RADIO MANUFACTURING CO.
CROYDEN	WARWICK MANUFACTURING CO.

TRADE NAME	MANUFACTURER
CRUISER	CROSLEY RADIO CORP., UNITED AMERICAN BOSCH CORP.
CRUSADER	BELMONT RADIO CORP., CLINTON RADIO MANUFACTURING CO., PACIFIC RADIO CORP., RADIO PRODUCTS CORP.
CUB	CLINTON RADIO MANUFACTURING CO.
CUNNINGHAM	CLIMAX RADIO & TELEVISION CO.
D	
DALCOPHONE	DALCO RADIO CO.
DAY RADIA	GENERAL MOTORS RADIO CORP.
DAYCRAFT	DAYFAN ELECTRIC CO., DAYTON FAN & MOTOR CO. (Dayton, Ohio), GENERAL MOTORS RADIO CORP.
DAYFAN	DAYFAN ELECTRIC CO., DELCO RADIO CORP., GENERAL MOTORS RADIO CORP.
DAYMAR	GENERAL MOTORS RADIO CORP.
DAYOLA	DAYTON FAN & MOTOR CO. (Dayton, Ohio), GENERAL MOTORS RADIO CORP.
DAYRAD	DAYTON RADIO CORP., RADIO PRODUCTS CORP.
DAYROYAL	DAYFAN ELECTRIC CO.
DAYTONIA	DAYTON RADIO CO., GENERAL MOTORS RADIO CORP.
De WALD	De WALD RADIO MANUFACTURING CO.
DeFOREST	RCA VICTOR DIVISION OF RADIO CORP. OF AMERICA
DELANO	MODERNOLA CO.
DELCO	UNITED MOTOR SERVICE
DELTA DE LUXE	WARWICK MANUFACTURING CO.
DELUXE	WARWICK MANUFACTURING CO.
DERESNADYNE	ANDREWS RADIO CO.
DETROLA	DETROLA RADIO & TELEVISION CORP.
DeWALD	PIERCE AIRO CO.
DEXAPHONE	MOGUL ELECTRIC CO.
DEXTER	DEXTER METAL MANUFACTURING CO.
DIAMADYNE	NORBERT RADIO SERVICE
DIAMOND-T	DIAMOND-T RADIO MANUFACTURERS
DICTATOR	MONTGOMERY WARD & CO.
DICTOGRAPH	DICTOGRAPH PRODUCTS CO.,INC.
DINAH	WARWICK MANUFACTURING CO.
DIRECTOR	CROSLEY RADIO CORP.
DISCOVERER	CROSLEY RADIO CORP.
DISTANTONE	DISTANTONE RADIOS, INC.
DOLORES	GOERDES, FRED, W., CO.
DOMINAIRE	MACLAREN MANUFACTURING CO.
DORAPHONE	SETCHELL CARLSON, INC.
DORON SUPER-EQUIDYNE	SILVA MANUFACTURING CO.
DRAGON	PILOT RADIO & TUBE CORP.
DREXOLA	DREXEL RADIO MANUFACTURING CO.
DRUID	WARWICK MANUFACTURING CO.
DUAL TONE	WARWICK MANUFACTURING CO.
DUBILIER	DUBILIER CLOCK CORP., DUMONT ELECTRIC CO., INC.
DUET	AMERICAN RADIO & RESEARCH CORP.
DUETTE	CORONA RADIO & TELEVISION CORP.
DUNN-HALL	DUNN-HALL CO.
DUO SYMPHONIC	WHOLESALE RADIO SERVICE CO., INC.

TRADE NAME	MANUFACTURER
DUODYNE	GLOBE ELECTRIC CO.
DUOTONE	THOMPSON, R. E., MANUFACTURING CO.
DYMAC	ELECTRICAL PRODUCTS MANUFACTURING CO.
DYNAPHONE	ANSLEY RADIO CORP.
DYNERGY	DYNAMOTIVE RADIO CORP.
E	
EAGLE	ALL AMERICAN MOHAWK CORP., EAGLE RADIO CO., WURLITZER, RUDOLPH, CO.
EAGLET	EAGLE RADIO CO.
EARL	EARL RADIO CORP.
ECHO	WAUGH ELECTRIC MANUFACTURING CO.
ECHOETTE	ECOHPHONE RADIO MANUFACTURING CO.
ECHOPHONE	ECHOPHONE RADIO MANUFACTURING CO., RADIO SHOP
ECKHARMONIC	ECKHARDT CORP.
ECODYNE	ECODYNE RADIO CO.
ECOPHONE	ECOPHONE RADIO CORP.
EDISON	EDISON, THOMAS A., INC.
EDISON-BELL	EDISON-BELL RADIO & PHONOGRAPH CO.
EL REY	EL REY RADIO MANUFACTURING CO.
ELECTRAD	ELECTRAD CO., INC.
ELECTRADYNE	EAGLE CHARGER CORP., FOREIGH & DOMESTIC ELECTRICAL COMMODITIES
ELECTROLA	AMERICAN SPECIALITY MANUFACTURING CO., RCA VICTOR DIVISION OF RADIO CORP. OF AMERICA
ELECTROLA RADIOLA	RCA VICTOR DIVISION OF RADIO CORP. OF AMERICA
ELECTROMATIC	ELECTROMATIC DISTRIBUTORS, INC.
ELECTROPHONE	HARRIS MANUFACTURING CO.
ELECTROSTATIC	UNITED REPRODUCERS CORP.
ELECTROTONE	HARRIS MANUFACTURING CO.
ELF	CROSLEY RADIO CORP.
ELGIN	ELGIN RADIO SUPPLY CO.
ELKAY	ELKAY MANUFACTURING CO. , LANGBEIN-KAUFMANN CO.
ELLSWOOD	GRIGSBY-GRUNOW CORP.
ELMCO	ELECTRIC MACHINE CORP.
EMERSON	EMERSON RADIO & PHONOGRAPH CORP.
EMERSONETTE	EMERSON RADIO & PHONOGRAPH CORP.
EMPIRE	EMPIRE ELECTRICAL PRODUCTS CO.
ENDURANCE	WESTERN TIRE & AUTO STORES, INC.
ENSEMBLE	PREMIER ELECTRIC CORP.
ENSIGN	ESPEY MANUFACTURING CO., INC.
ENVOY	INSULINE CORPORATION OF AMERICA
ENVOYETTE	INSULINE CORPORATION OF AMERICA
EQUAPHASE	FRESHMAN, CHARLES, CO.
ERAC	FRESHMAN, CHARLES, CO.
ERLA	ELECTRICAL RESEARCH LABORATORIES, INC., SENTINEL RADIO CORP.
ESPEY	ESPEY MANUFACTURING CO., INC.
ESSEX	UNITED AMERICAN BOSCH CORP.
ETHERPHONE	AMERICAN ETHERPHONE CORP., RADIO APPARATUS CO.
ETHOVOX	BURNDEPT WIRELESS CORP. OF AMERICA

TRADE NAME	MANUFACTURER
EVEREADY	NATIONAL CARBON CO.
EVERYMAN	DeFOREST RADIO CO.
EXCELLO	EXCELLO PRODUCTS CORP.
F	
FADA	ANDREA, F. A. D., INC., FADA RADIO & ELECTRIC CO.
FADA NEUTROLA	ANDREA, F. A. D., INC.
FADALETTE	FADA RADIO & ELECTRIC CO.
FAIRBANKS-MORSE	FAIRBANKS-MORSE & CO.
FALCK	ADVANCE ELECTRIC CO.
FANSTEEL	BALKEIT RADIO CORP.
FANTASY	MONTGOMERY WARD & CO.
FARNSWORTH	FARNSWORTH TELEVISION & RADIO CORP.
FARRAND	FARRAND MANUFACTURING CO.
FEARNOLA	CONTINENTAL RADIO & TELEVISION CORP., WARWICK MANUFACTURING CO., WELLS-GARDNER & CO.
FEDERAL	CLINTON RADIO MANUFACTURING CO., FEDERAL RADIO CORP.
FENWAY	FERGUSON RADIO CORP.
FERGUSON	FERGUSON RADIO CORP.
FERRANTI	FERRANTI, INC.
FERRODYNE	STEWART-WARNER RADIO CORP.
FIDEL-O-MATIC	LEHMAN RADIO SALON, INC., PORT-O-MATIC CORP.
FIDELITONE	WARWICK MANUFACTURING CO., WELLS-GARDNER & CO.
FINE-ARTS	PLAZA MUSIC CO.
FIRESIDE	FREELY CORP.
FIRESTONE	FIRESTONE TIRE & RUBBER CO.
FIRST NATIONAL	BALKEIT RADIO CORP.
FIVER	CROSLEY RADIO CORP.
FLEETWOOD	FLEETWOOD RADIO CORP.
FLEXODYNE	HALL, CHARLES, CO.
FLIVER	RADIO PANEL SHOP
FLORENZA	RCA VICTOR DIVISION OF RADIO CORP. OF AMERICA
FORACO	FORACO CORP.
FORD	WELLSTON RADIO CORP.
FORD-MAJESTIC	GRIGSBY-GRUNOW CORP.
FORDSON	FORDSON RADIO MANUFACTURING CO.
FRANKLIN	FRANKLIN RADIO CORP.
FREED	FREED RADIO & TELEVISION CORP.
FREED-EISEMANN	FREED RADIO & TELEVISION CORP.
FRESHMAN	BELMONT RADIO CORP., FRESHMAN, CHARLES. CO.
FROST-MINTON	FROST-MINTON CORP.
G	
GALLEON	CROSLEY RADIO CORP.
GAMBLE	KINGSTON RADIO CO., INC.
GAMBLE-SKOGMO	KINGSTON RADIO CO., INC.
GAROD	GAROD RADIO CORP.
GAYLORD	GAYLORD MANUFACTURING CO.
GEISHA	SPLITDORF ELECTRICAL MANUFACTURING CO.
GEMBOX	CROSLEY RADIO CORP.
GEMCHEST	CROSLEY RADIO CORP.

TRADE NAME	MANUFACTURER
GENERAL	GENERAL RADIO & TELEVISION CORP., RADOLEK CO., WARWICK MANUFACTURING CO., WESTERN AUTO SUPPLY CO.
GENERAL ELECTRIC	GENERAL ELECTRIC CO.
GENEVA	CORDONIC MANUFACTURING CO.
GIBLIN	STANDARD RADIO & ELECTRIC CO.
GILBERT	GILBERT, R. W., CO.
GILFILLAN	GIFILLAN BROTHERS, INC.
GLOBE	GLOBE ELECTRIC CO.
GLOBE TROTTER	CONTINENTAL RADIO & TELEVISION CORP., GLOBE TROTTER RADIO CO., KENNEDY, COLIN B., CO.
GLORITONE	UNITED STATES RADIO & TELEVISION CORP.
GOLD MEDAL	GOLD MEDAL RADIO CORP., WARWICK MANUFACTURING CO.
GOLD SEAL	GOLD SEAL ELECTRIC CO.
GOLD STAR	KODEL RADIO CORP.
GOLD TEST	WARWICK MANUFACTURING CO.
GOLDCREST	CLEARTONE RADIO CORP.
GOLDENTONE	FORDSON RADIO MANUFACTURING CO., WARWICK MANUFACTURING CO.
GOLDENVOICE	GALVIN MANFACTURING CORP.
GOLDFINCH	GUTHRIE CO.
GOODYEAR	GOODYEAR TIRE & RUBBER CO.
GOODYEAR WINGS	BELMONT RADIO CORP.
GOVERNOR	CROSLEY RADIO CORP.
GRANADA	WELLS-GARDNER & CO.
GRANDETTE	THOMPSON, R. E., MANUFACTURING CO.
GRAYBAR	GRAYBAR ELECTRIC CO.
GREAT NORTHERN	WELLS-GARDNER & CO.
GREBE	GREBE MANUFACTURING CO., INC.
GREEN DIAMOND	HOWARD RADIO MANUFACTURING CO.
GRUNOW	GENERAL HOUSEHOLD UTILITIES CO.
GULBRANSEN	GULBRANSEN RADIO CO., WELLS-GARDNER & CO.
GYPSY	GRAND RAPIDS RADIO MANUFACTURING CO., KINGSTON RADIO CO., INC.

H

TRADE NAME	MANUFACTURER
HALLERIO	HALLER, W. B., CO.
HALOWAT	HALLOCK & WATSON RADIO CORP.
HALSON	HALSON RADIO & TELEVISION CORP.
HAMILTON	HAMILTON RADIO CORP.
HAMILTON-LLOYD	PLAZA MUSIC CO.
HARMO-SONIC	AMPLEX INSTRUMENT LABORATORIES
HARMONIC	JONES RADIO CO.
HARMONSON	HARMON & SONS CO.
HARMONY	CLINTON RADIO MANUFACTURING CO., HARMONY MANUFACTURING CO.
HARTMAN	HARTMAN ELECTRICAL MANUFACTURING CO.
HATRY	HATRY & YOUNG
HAVENWOOD	GRIGSBY-GRUNOW CORP.
HEMCO	RICHARDS, GEORGE, & CO.
HERALD	INTERNATIONAL TRADING CORP.

TRADE NAME	MANUFACTURER
HERALDYNE	BRENARD MANUFACTURING CO.
HETRO	HETRO ELECTRICAL INDUSTRIES, INC.
HETROLA	BRANSTON, CHARLES A., CO.
HEXADYNE	HUNT, W. D., CO.
HI-Q	HAMMARLUND-ROBERTS, INC.
HILLTOP SIX	HITCHCOCK-HILL CO.
HOBBS	MISSION BELL RADIO MANUFACTURING CO.
HOLLYWOOD	CONSOLIDATED RADIO & TELEVISION CORP.
HOLLYWOOD MASTERPIECE	TURNER, K. M., RADIO CORP.
HOMER	HOMER PRODUCTS, INC.
HOODWIN	HOODWIN, CHARLES, CO.
HORN	HORN, HERBERT H., MANUFACTURING CO.
HOWARD	HOWARD RADIO MANUFACTURING CO.
HUBCO	BALTIMORE HUB-WHEEL & MFG. CO.
HUDSON	WELLS-GARDNER & CO.
HUNTINGTON	BUCKINGHAM RADIO CORP.
HY-PEX	CHATHAM RADIO CORP.
HYATT	ELECTROPHONE CORP., HYATT ELECTRIC CORP.
HYDAWAY	BALDWIN, NATHANIEL, INC.
HYPER-MERTON	ZENITH RADIO CORP.
HYPERDYNE	INDIANA ELECTRIC MANUFACTURING CO.
HYPERION	RCA VICTOR DIVISION OF RADIO CORP. OF AMERICA
I	
ICA	INSULINE CORPORATION OF AMERICA
IMPERIAL	BELMONT RADIO CORP., FINK INDUSTRIES, IMPERIAL FURNITURE CO., KING MANUFACTURING CORP., TYRMAN RADIO CO.
INDUCTROLE	AMERICAN RADIO & RESEARCH CORP.
INSULLETTE	INSULINE CORPORATION OF AMERICA
INTER-OCEAN	ZENITH RADIO CORP.
INTERNATIONAL	INTERNATIONAL RADIO CORP., KEN-MAR RADIO CORP., KOLSTER RADIO CORP.
INTERPANEL	DeFOREST RADIO CO.
INTRADYNE	NATIONAL RADIO CO.
INVERSE DUPLEX	GRIMES, DAVID, INC.
IRC	INTERNATIONAL RADIO CORP.
ISOFARAD	WALBERT MANUFACTURING CO.
ISOTONE	HIGH FREQUENCY LABORATORIES
J	
J. R. H.	FEDDERS MANUFACTURING CO.
JACKSON	JACKSON RESEARCH LABORATORIES
JACKSON BELL	JACKSON BELL CO., LTD.
JESSE FRENCH	FRENCH, JESSE & SONS, PIANO CO.
JEWEL	EMERSON RADIO & PHONOGRAPH CORP., GENERAL AMERICAN RADIO MANUFACTURING CO., INTERNATIONAL RADIO CORP., JEWEL MANUFACTURING CO., SPARKS-WITHINGTON CO.
JEWELBOX	CROSLEY RADIO CORP.
JEWELCASE	CROSLEY RADIO CORP.

TRADE NAME	MANUFACTURER
JEWETT	JEWETT RADIO & PHONOGRAPH CO.
JONES	JONES, JOSEPH W., RADIO MANUFACTURING CO.
JUMBO	PILOT RADIO & TUBE CORP.
JUST-RITE	AIR-OLA RADIO CO.
K	
KADETTE	EMERSON RADIO & PHONOGRAPH CORP., INTERNATIONAL RADIO CORP., KADETTE RADIO CORP.
KARRYADIO	ARMLEY RADIO CORP.
KELLER-FULLER	KELLER-FULLER MANUFACTURING CO.
KELLOGG	KELLOGG SWITCHBOARD & SUPPLY CO.
KEMPER	KEMPER RADIO LABORATORIES
KEMPER KOMPAK	KEMPER RADIO LABORATORIES
KEN-MAR	KEN-MAR RADIO CORP.
KENMAN	KENMAN ELECTRIC CO.
KENNEDY	GOLDBLATT STORES, KENNEDY, COLIN B., CO.
KENOLA	KENOLA RADIO CORP.
KENT	KLITZEN RADIO MANUFACTURING CO.
KEYSTONE	RADOLEK CO.
KING	KING MANUFACTURING CORP., RADOLEK CO., WORKRITE MANUFACTURING CO.
KING COLE	ANYLITE ELECTRIC CO.
KING-HINNERS	KING QUALITY PRODUCTS CO.
KINGSBURY	WARWICK MANUFACTURING CO.
KINGSTON	KINGSTON RADIO CO., INC.
KNICKERBOCKER	BALDER RADIO CORP.
KNIGHT	ALLIED RADIO CORP.
KODEL	KODEL RADIO CORP.
KOLSTER	FEDERAL TELEGRAPH OF CALIFORNIA, FEDERAL-BRANDES, INC., KOLSTER RADIO CORP.
KOMPAKT	PHILMORE MANUFACTURING CO.
KOMPENTROL	KARDON PRODUCTS CO., INC.
KYLECTRON	UNITED REPRODUCERS CORP.
L	
L'TATRO	ECKSTEIN RADIO & TELEVISION CO., L'TATRO MANUFACTURING CO.
LA FRANCE	FAIR STORE, THE
LA SALLE	BUCKINGHAM RADIO CORP., CHAMPION RADIO LABORATORIES
LACAULT	LACAULT, R. E., CO.
LAFAYETTE	LAFAYETTE RADIO CORP., MISSISSIPPI VALLEY RADIO CO., RADIO WIRE & TELEVISION, INC., WHOLESALE RADIO SERVICE CO., INC.
LANCO	LANGHORNE & CO.
LANG	LANG RADIO CORP.
LanSing	UNITED ENGINE CO.
LARK	SEELEY ELECTRIC CO.
LASHER CAPACIDYNE	FAIRVIEW ELECTRIC SHOP
LAUREHK	LAUREHK RADIO MANUFACTURING CO.
LAZY X	PHILCO RADIO & TELEVISION CORP.
LEADER	CROSLEY RADIO CORP.

TRADE NAME	MANUFACTURER
LEGISLATOR	CROSLEY RADIO CORP.
LEMCO	LEE ELECTRIC & MANUFACTURING CO.
LEROY	RADIO SHOP
LEUTZ	LEUTZ, C. R., INC.
LEWOL	LEWOL MANUFACTURING CO.
LEXINGTON	WELLS-GARDNER & CO.
LIBERTY	LIBERTY MUSIC SHOP, LIBERTY TRANSFORMER CO.
LIGHTWEIGHT CHAMPION	SENTINEL RADIO CORP.
LINCOLN	LINCOLN RADIO CORP.
LINCOLN SUPER	OZARKA RADIO CORP.
LINCOLN SYMPHONIC	LINCOLN RADIO CORP.
LINCOLN-PHILCO	PHILCO RADIO & TELEVISION CORP.
LITTLE GIANT	Q. T. RADIO PRODUCTS
LITTLE WONDER	PHILMORE MANUFACTURING CO.
LOGODYNE	KODEL RADIO CORP.
LULLABY	MITCHELL MANUFACTURING CO.
LUMITONE	MITCHELL MANUFACTURING CO.
LUXOR	LUXOR RADIO MANUFACTURING CO.
LYRIC	ALL AMERICAN MOHAWK CORP., WURLITZER, RUDOLPH, CO.
M	
M. R. C.	MICHIGAN RADIO CORP.
MACK SINCROFLEX	MACK CO.
MACY	MACY, R. H., CO.
MAGADYNE	NASSAU RADIO CO.
MAGICOLOR	I. C. A. EXPORT CO., INC.
MAGNAFORMER	RADIART LABORATORIES
MAGNAVOX	MAGNAVOX CO., INC.
MAJESCO	R. K. RADIO LABORATORIES, INC.
MAJESTIC	GRIGSBY-GRUNOW CORP., MAJESTIC RADIO & TELEVISION CORP.
MALONE-LEMMON	CARYLOYD ELECTRIC & RADIO CO.
MANTOLA	GOODRICH RUBBER & TIRE CO., UNITED STATES RADIO & TELEVISION CORP.
MARCONIPHONE	MARCONI BROS., INC.
MARLODYNE	MARLOW, ERNEST, CO.
MARQUETTE	MARQUETTE RADIO CO.
MARSHALL	CLINTON RADIO MANUFACTURING CO., CONTINENTAL RADIO & TELEVISION CORP.
MARTI	MARTI ELECTRIC RADIO, CO., INC.
MARTINOLA	VIBROPLEX CO., INC.
MARV-O-DYNE	AMBER MANUFACTURING CO.
MARVEL	INSULINE CORPORATION OF AMERICA
MARVELO	RADOLEK CO.
MARVELTONE	AJAX ELECTRIC SPECIALITY CO.
MARWOL	MARWOL RADIO CORP.
MARWOOD	MARWOOD RADIO CORP.
MASCOT	INSULINE CORPORATION OF AMERICA
MASTER	MASTER RADIO MANUFACTURING CO., INC., UNION CONSTRUCTION CO.
MASTER MIDGET	MASTER RADIO MANUFACTURING CO., INC.

TRADE NAME	MANUFACTURER
MASTERPIECE	FRESHMAN, CHARLES, CO., McMURDO SILVER CORP.
MASTERTONE	BOSTON RADIO MANUFACTURING CO., HIGH FREQUENCY LABORATORIES
MATE	CROSLEY RADIO CORP.
MAY	MAY RADIO & REFRIGERATION CO.
MAYFAIR	CLAGO RADIO CORP.
MAYFLOWER	CLAGO RADIO CORP.
MAYOR	CROSLEY RADIO CORP.
McCULLOUGH	PITTSBURGH RADIO SUPPLY CO.
McMILLAN	McMILLAN RADIO CO.
McMURDO SILVER	McMURDO SILVER CORP.
MECO	METROPOLITAN ELECTRIC CO.
MEISSNER	MEISSNER MANUFACTURING CO.
MELBURN	MELBURN RADIO MANUFACTURING CO.
MELCO SUPREME	AMSCO PRODUCTS CO.
MELL-O-TONE	TYMESON RADIO CO.
MELODEE	NORBERT RADIO SERVICE, ROTH RADIO, INC.
MELODY	BETTER RADIO PRODUCTS
MELODYNE	EATON-EDWARDS RADIO CO.
MELROSE	UNITED STATES RADIO & TELEVISION CORP.
MERRIMAC	CROSLEY RADIO CORP.
MERRYMAKER	CROSLEY RADIO CORP.
METRODYNE	METRO ELECTRIC CO.
MICHIGAN	MICHIGAN RADIO CORP.
MICKEY-MOUSE	EMERSON RADIO & PHONOGRAPH CORP.
MICRODYNE	APEX ELECTRIC MANUFACTURING CO., LAKESIDE SUPPLY CO.
MIDWEST	MIDWEST RADIO CORP.
MIGHTY ATOM	WHOLESALE RADIO SERVICE CO., INC.
MIKADO	SPLITDORF ELECTRICAL MANUFACTURING CO.
MINERVA	MINERVA RADIO CO.
MINSTREL	CROSLEY RADIO CORP., MONTGOMERY WARD & CO.
MINUET	THOMPSON, R. E., MANUFACTURING CO.
MINUETTE	AMERICAN RADIO & RESEARCH CORP.
MINUTE MAN	PATHE RADIO & TELEVISION CORP.
MIRACO	MIDWEST RADIO CORP.
MISSION	CONSOLIDATED RADIO & TELEVISION CORP.
MISSION BELL	MISSION BELL RADIO MANUFACTURING CO.
MOCKING BIRD	GUTHRIE CO.
MODYNE	MORRDYNE RADIO MANUFACTURING CO.
MOHAWK	ALL AMERICAN MOHAWK CORP., MOHAWK ELECTRIC CO.
MONARCH	CHAMPION RADIO LABORATORIES, KING MANUFACTURING CORP.
MONODYNE	NATIONAL AIRPHONE CORP.
MONOTRAD	CROSLEY RADIO CORP.
MONOTROL	SLEEPER RADIO CORP.
MONROE	BRAUN, W. C., CO.
MONROE-MARVELO	RADOLEK CO.
MONTGOMERY WARD	MONTGOMERY WARD & CO.
MONTROSET	MONTROSE RADIO LABORATORIES
MOON	MOON RADIO CORP.
MORRADYNE	MORRISON RADIOPHONE CO., INC.

TRADE NAME	MANUFACTURER
MOTO-MASTER	WARWICK MANUFACTURING CO.
MOTO-MIDGET	ACME ELECTRIC & MANUFACTURING CO.
MOTO-VOX	MOTO-METER GAUGE & EQUIPMENT CO.
MOTOMASTER	MOTOMASTER RADIO CO.
MOTOROLA	GALVIN MANFACTURING CORP.
MOTOSET	FADA RADIO & ELECTRIC CO.
MU-RAD	MU-RAD RADIO CORP.
MULTIFLEX	RADIO RECEPTOR CO.
MURDOCK	MURDOCK, WILLIAM J., CO.
MUSAPHONIC	GENERAL ELECTRIC CO.
MUSIC BOX	ZANEY-GILL CORP.
MUSIC MASTER	MUSIC MASTER CORP., SLEEPER RADIO CORP., WARE RADIO CORP.
MUSICAIRE	SENTINEL RADIO CORP.
MUSIQUE	LAUREHK RADIO MANUFACTURING CO.
MY OWN	MY OWN RADIO CO.
N	
NAMCO	NAMCO MANUFACTURING CO.
NASSAU	NASSAU RADIO CO.
NATIONAL UNION	NATIONAL UNION RADIO CORP.
NATURELLE	MONTROSE-HAST RADIO LABORATORIES
NAVAJO	ALL AMERICAN MOHAWK CORP.
NAVIGATOR	A-C ELECTRICAL MFG. CO. (DAYTON, OHIO)
NEO-DYNE	STEINWAY RADIO CORP.
NEUTROFLEX	TANNER, C. D., CO.
NEUTROWOUND	ADVANCE AUTOMOBILE ACCESSORIES CORP., NEUTROWOUND RADIO MANUFACTURING CO.
NEW YORKER	GRIMES, DAVID, INC.
NEWLANDS	INCANDESCENT SUPPLY CO.
NIGHTENGALE	CLINTON RADIO MANUFACTURING CO.
NIGHTINGALE	GUTHRIE CO., HANSEN STORAGE CO.
NO-BAT-RY	BATTERYLESS RADIO CORP.
NO-DIAL	OHIO STAMPING & ENGINEERING CO.
NOCTURNE	SPLITDORF ELECTRICAL MANUFACTURING CO.
NORBERT	NORBERT RADIO SERVICE
NORCO	NORTHWESTERN RADIO MANUFACTURING CO., REMLER CO., LTD.
NORDEN-HAUCK	NORDEN-HAUCK RADIO CO.
NYACCO	NEW YORK ALBUM & CARD CO.
NYACCOFLEX	NEW YORK ALBUM & CARD CO.
O	
OLYMPIA	CROSLEY RADIO CORP.
OPERADIO	OPERADIO MANUFACTURING CO.
ORACLE	CROSLEY RADIO CORP.
ORIOLE	CONTINENTAL RADIO & TELEVISION CORP., ROBERTS, HENRY CO., RADIO CORP., W-K ELECTRIC CO.
ORPHEUS	McMURDO SILVER CORP., ROTH-DOWNS MANUFACTURING CO.
ORTHONOLA	LYON RADIO CO.
ORTHOPHONIC	RCA VICTOR DIVISION OF RADIO CORP. OF AMERICA

TRADE NAME	MANUFACTURER
ORTHOSONIC	FEDERAL RADIO CORP.
OVERTONE	PFANSTIEHL RADIO CO.
OZARKA	OZARKA RADIO CORP., RADOLEK CO.
P	
PACENT	PACENT ENGINEERING CORP.
PACIFIC	PACIFIC RADIO CORP.
PACKARD	PACKARD RADIO CORP.
PACKARD-BELL	PACKARD-BELL CO.
PACKARD-PHILCO	PHILCO RADIO & TELEVISION CORP.
PAGE-FIVE	R-B RADIO CO.
PAL	CROSLEY RADIO CORP.
PAM	SAMPSON PAM CO.
PANAMUSE	CAPEHART CORP., INC.
PANATROPE	BRUNSWICK-BALKE-COLLENDER CO.
PARADYNE	ADAMS-MORGAN CO.
PARAGON	ADAMS-MORGAN CO., PARAGON ELECTRIC CO.
PARAMOUNT	RADIO SUPPLY HOUSE
PARTNER	CROSLEY RADIO CORP.
PATHE	PATHE RADIO & TELEVISION CORP.
PATTERSON	PATTERSON RADIO CORP., PIERSON-DELANE, INC.
PEARSON	ELECTRICAL RESEARCH LABORATORIES, INC., PEARSON DIVISION ELECTRICAL RESEARCH LABS
PEE WEE	DETROLA RADIO & TELEVISION CORP., WESTERN TIRE & AUTO STORES, INC.
PEERLESS	PACIFIC RADIO CORP., UNITED REPRODUCERS CORP.
PENN C	PENNSYLVANIA RADIO MANUFACTURING CO.
PENN-C	P & M RADIO CO.
PEP BOYS	GALVIN MANFACTURING CORP.
PERASCO	PERRY RADIO SUPPLY CO.
PERFECTONE	PERFECTONE, INC.
PETER PAN	JACKSON BELL CO., LTD.
PETITE	BEN RADIO CORP.
PFANSTIEHL	PFANSTIEHL RADIO CO.
PHILCO	PHILADELPHIA STORAGE BATTERY CO., PHILCO RADIO & TELEVISION CORP.
PHILHARMONIC	PHILHARMONIC RADIO CO.
PHILMORE	PHILMORE MANUFACTURING CO.
PHOTOPHONE	RCA VICTOR DIVISION OF RADIO CORP. OF AMERICA
PIED-PIPER	INSULINE CORPORATION OF AMERICA
PIERSON	PIERSON, THE, CO.
PILGRIM	PILGRIM ELECTRIC CORP.
PILOT	PILOT RADIO & TUBE CORP.
PILOTONE	PILOT RADIO & TUBE CORP.
PIONEER	PIONEER RADIO CORP., RADOLEK CO.
PLAYBOY	CROSLEY RADIO CORP.
PLAYTIME	CROSLEY RADIO CORP.
PLAZA	PLAZA MUSIC CO.
PLIO-6	GOLDEN-LEUTZ CORP., LEUTZ, C. R., INC.
PLIODYNE	GOLDEN-LEUTZ CORP., LEUTZ, C. R., INC.
PLYMOUTH	PLYMOUTH RADIO CORP.

TRADE NAME	MANUFACTURER
POCKET RADIO	AUTO INDICATOR CO.
POLONAISE	SPLITDORF ELECTRICAL MANUFACTURING CO.
POLY	ROYAL RADIO CORP.
POLYDYNE	A-C ELECTRICAL MANUFACTURING CO., ALLIED RADIO CORP.
PORT-O-MATIC	LEHMAN RADIO SALON, INC.
PORTO-O-RADIO	AMERICAN EUROPEAN RADIO CORP.
PORTOLA	PORTOLA RADIO CO.
POWELL	POWELL MANUFACTURING CO.
POWEROLA	POWEROLA RADIO CORP.
PREMIER	PREMIER ELECTRIC CORP.
PRIESS	PRIESS RADIO CORP.
PRINCESS	MONTGOMERY WARD & CO.
PRIVATEER	CROSLEY RADIO CORP.
PRMCO	PORTS MANUFACTURING CO.
PUP	CROSLEY RADIO CORP.
Q	
QUAD	RADIO DOC
QUALIPHONE	WARNER ENGINEERING CORP.
R	
R F L	RADIO FREQUENCY LABORATORIES
R. E. L.	RADIO ENGINEERING LABORATORY, INC.
RADA-CHRON	RADIO-CHRON CORP.
RADAK	CLAPP-EASTHAM CO.
RADIART	RADIART CORP.
RADIETTE	KELLER-FULLER MANUFACTURING CO.
RADIO ELECTROLA	RCA VICTOR DIVISION OF RADIO CORP. OF AMERICA
RADIO KNIGHT	CHISHOLM-BARFIELD CORP.
RADIO MASTER	RADIO MASTER CORP.
RADIO PAL	STANRAD CORP.
RADIO-ALARM-O-GRAPH	PINTO SPEAKING CLOCK CO.
RADIO-CRAFT	DeFOREST RADIO CO.
RADIO-KEG	R. K. RADIO LABORATORIES, INC.
RADIOBAR	RADIOBAR COMPANY OF AMERICA
RADIOCHRON	CLAGO RADIO CORP.
RADIODYNE	WESTERN COIL & ELECTRICAL CO.
RADIOEAR	STANDARD RADIO & ELECTRIC CO.
RADIOGRAPH	RADIOGRAPH CORP.
RADIOLA	BRUNSWICK-BALKE-COLLENDER CO., RCA VICTOR DIVISION OF RADIO CORP. OF AMERICA
RADIOLETTE	RCA VICTOR DIVISION OF RADIO CORP. OF AMERICA
RADIOMASTER	CONTINENTAL RADIO & TELEVISION CORP.
RADIOPHONE	DeFOREST RADIO CO.
RADIOTROPE	UNITED STATES RADIO & TELEVISION CORP.
RADOLEK	RADOLEK CO.
RAINBOW	PILOT RADIO & TUBE CORP.
RAMWAY	ELECTRICAL RESEARCH LABORATORIES, INC.
RANDOLPH	RANDOLPH RADIO CORP.
RANGER	BROWN & MANHART CO., WESTFALL-KELLEY RADIO SALES
RANGER FIVE	AINSWORTH-GATES RADIO CO.

TRADE NAME	MANUFACTURER
RAULAND	RAULAND CORP.
RAYCRAFT	MALONE-LEMMON PRODUCTS INC.
RCA-VICTOR	RCA VICTOR DIVISION OF RADIO CORP. OF AMERICA
RECEPTOR	RCA VICTOR DIVISION OF RADIO CORP. OF AMERICA
RECEPTRAD	RADIO RECEPTOR CO.
RECITAL	RADIO RECEPTOR CO.
RECORDIO	WILCOX-GAY CORP.
REGAL	AMERICAN SPECIALITY MANUFACTURING CO., KINGSTON RADIO CO., INC.
REGENODYNE	CHELSEA RADIO CO.
REGENOFLEX	BRUNSWICK-BALKE-COLLENDER CO.
REGENT	HIGH FREQUENCY LABORATORIES, REGENT RADIO CORP.
REMINGTON	BUCKINGHAM RADIO CORP., DUNN-HALL CO.
REMLER	GRAY-DANIELSON MANUFACTURING CO., REMLER CO., LTD.
RENAISSANCE	DeFOREST RADIO CO.
REPOSE	CROSLEY RADIO CORP.
REPUBLIC	AMERICAN SPECIALITY MANUFACTURING CO., REPUBLIC INDUSTRIES
REVELATION	HOLLAND RADIO CORP.
REVELER	CROSLEY RADIO CORP.
REVERE	REVERE RADIO CORP.
REXCO MASTERDYNE	REX RADIO SPECIALITY
RHAPSODY	SPLITDORF ELECTRICAL MANUFACTURING CO.
RICO-DYNE	RADIO INDUSTRIES CORP.
RITA De LUX	GRANT, H. R., CO.
RIVERSIDE	WARWICK MANUFACTURING CO.
RIVOLI	RADIO INDUSTRIES CORP.
ROAD CHIEF	DETROLA RADIO & TELEVISION CORP.
ROAD MASTER	DETROLA RADIO & TELEVISION CORP.
ROAMER	ALLIED RADIO CORP., FIRESTONE TIRE & RUBBER CO.
ROAMIO	CROSLEY RADIO CORP.
ROCHESTER	CHICAGO NIPPLE MANUFACTURING CO.
RONDEAU	CROSLEY RADIO CORP.
ROTOFOR	INTERNATIONAL RADIO CORP.
ROYAL	CONSOLIDATED RADIO & TELEVISION CORP., KENNEDY, COLIN B., CO., KING MANUFACTURING CORP.
ROYALE	GRIFFIN SMITH MANUFACTURING CO., ROYALE RADIO MANUFACTURING CO.
RPM ODYNE	RADIO PRODUCTS CORP.
RUTIC	RUCKELSHAUS, JOHN G., INC.
S	
SAMPSON	SAMPSON ELECTRIC CO.
SARA	SARA, J. & L., CO., INC.
SARGENT	SARGENT, E. M., CO.
SARGENT-RAYMOND	SILVER-MARSHALL MANUFACTURING CO.
SATTERLEE	MOON RADIO CORP.
SCOTT	SCOTT TRANSFORMER CO., SCOTT, E. H., RADIO LABORATORIES, INC.
SCOUT	INSULINE CORPORATION OF AMERICA, SLEEPER RADIO CORP.
SEARS, ROEBUCK	SEARS, ROEBUCK & COMPANY

TRADE NAME	MANUFACTURER
SECRETARY	CROSLEY RADIO CORP.
SECTRON	SECTRON RADIO CO.
SENATOR	CROSLEY RADIO CORP.
SENTINEL	COLUMBIA RADIO CO., ELECTRICAL RESEARCH LABORATORIES, INC., SENTINEL RADIO CORP., UNITED AIR CLEANER CO.
SEPTET	CROSLEY RADIO CORP.
SERENADER	HEART OF AMERICA RADIO CO., MONTGOMERY WARD & CO., PARAMOUNT MANUFACTURING CO., SLEEPER RADIO CORP.
SERENATA	AMERICAN RADIO & RESEARCH CORP.
SEVEN SEAS	LEUTZ, C. R., INC.
SEVETTE	CROSLEY RADIO CORP.
SEXTET	CROSLEY RADIO CORP.
SEXTETTE	ALL AMERICAN MOHAWK CORP.
SHAMROCK	SHAMROCK MANUFACTURING CO.
SHELDON	SHELDON RADIO CO.
SHEPCO	SHEPPARD POTTER CO., INC.
SHERATON	CROSLEY RADIO CORP., MODERNOLA CO.
SHERIDAN	WARWICK MANUFACTURING CO.
SHERWOOD	GRIGSBY-GRUNOW CORP.
SHOWBOX	CROSLEY RADIO CORP.
SHOWBOY	CROSLEY RADIO CORP.
SHOWCHEST	CROSLEY RADIO CORP.
SILVER	McMURDO SILVER CORP.
SILVER GHOST	LEUTZ, C. R., INC.
SILVER KING	APPEL-HENDERSON RADIO CO.
SILVER-MARSHALL	SILVER-MARSHALL MANUFACTURING CO.
SILVERTONE	SEARS, ROEBUCK & COMPANY
SIMPLEX	SIMPLEX RADIO CO.
SIMPLIFORM	RADIO MASTER CORP.
SINCLAIR	LAUREL MOTORS CORP.
SIX	CROSLEY RADIO CORP.
SIXTY	CROSLEY RADIO CORP.
SKELCO	WELLS-GARDNER & CO.
SKY HAWK	REPUBLIC INDUSTRIES
SKY ROVER	ELECTRICAL RESEARCH LABORATORIES, INC., WARWICK MANUFACTURING CO., WELLS-GARDNER & CO.
SLAGLE	CONTINENTAL RADIO & TELEVISION CORP.
SLEEPER	SLEEPER RADIO CORP.
SOLAR	WARWICK MANUFACTURING CO., WELLS-GARDNER & CO.
SOLO	MONTGOMERY WARD & CO.
SOLODYNE	CUMMINGS RADIO CORP.
SONDO	CROSLEY RADIO CORP.
SONGSTER	CROSLEY RADIO CORP.
SONNESTER	CROSLEY RADIO CORP.
SONORA	SONORA PHONOGRAPH CO., SONORA RADIO & TELEVISION CORP.
SOVEREIGN	E & N PRODUCTS CO., MONTGOMERY WARD & CO., SOVEREIGN ELECTRIC MANUFACTURING CO.
SPARTON	SPARKS-WITHINGTON CO.
SPLITDORF	SPLITDORF ELECTRICAL MANUFACTURING CO.
SPORTER	GALVIN MANFACTURING CORP.

TRADE NAME	MANUFACTURER
ST. REGIS	INTERNATIONAL RADIO CORP.
STANDARD	PILOT RADIO & TUBE CORP., PLAZA MUSIC CO., STANDARD RADIO CO. (Kansas City, Mo.), STANRAD CORP., WARWICK MANUFACTURING CO.
STANDARDYNE	STANDARD RADIO CORP. (Worcester, Ma.)
STAR	WELLS-GARDNER & CO.
STAR RAIDER	COMMONWEALTH RADIO MANUFACTURING CO., CONTINENTAL RADIO & TELEVISION CORP.
STARCK CLASSIQUE	BELMONT RADIO CORP.
STATES RADIO	STATES RADIO CORP.
STEINITE	STEINITE RADIO CO., WARWICK MANUFACTURING CO.
STERLING	GIBSON-SEARS RADIO CORP., NAYLOR RADIO CORP., STERLING MANUFACTURING CO.
STEWART-WARNER	STEWART-WARNER RADIO CORP.
STORY & CLARK	STORY & CLARK RADIO CORP.
STRATOSPHERE	ZENITH RADIO CORP.
STROBODYNE	LACAULT, R. E., CO.
STROMBERG-CARLSON	STROMBERG-CARLSON TELEPHONE MANUFACTURING CO.
SUN	SUN RADIO MANUFACTURING CO.
SUN SET	SUN RADIO MANUFACTURING CO.
SUNBEAM	CONTINENTAL RADIO & TELEVISION CORP.
SUNBEAM RADIOMASTER	CORONA RADIO & TELEVISION CORP.
SUPER	SUPER-ANTENNA CO.
SUPER DRAGON	PILOT RADIO & TUBE CORP.
SUPER SINGLE	PLYMOUTH RADIO & PHONOGRAPH CO.
SUPER WASP	PILOT RADIO & TUBE CORP.
SUPER-EQUIDYNE	DORON BROTHERS ELECTRIC CO.
SUPER-FLEX	SYPHER MANUFACTURING CO.
SUPER-MERTON	ZENITH RADIO CORP.
SUPER-REINARTZ	ELGIN RADIO SUPPLY CO.
SUPER-ZENITH	ZENITH RADIO CORP.
SUPERADIO	DeWITT-LaFRANCE CO.
SUPERDYNE	TUSKA, C. D., CO.
SUPERETTE	RCA VICTOR DIVISION OF RADIO CORP. OF AMERICA
SUPERFIVE	PRECEL RADIO MANUFACTURING CO.
SUPERFLEX	BENSEN ENGINEERING CO., RADIO PRODUCTS CORP.
SUPERIOR	PORTER, FRANK R. CO., SOUTH HILLS WIRELESS SHOP, SUPERIOR CABINET CORP.
SUPERIORDYNE	ELECTRICAL RESEARCH & MANUFACTURING CO.
SUPERIORFLEX	ELECTRICAL RESEARCH & MANUFACTURING CO.
SUPERLATONE	BAUMAN, WARREN J., CO.
SUPERPHONIC	RADIO EQUIPMENT CO.
SUPERPOWER	RADIO RECEPTOR CO.
SUPERSONIC	SPIEGEL, MAY STERN
SUPERTONE	BANFER PRODUCTS CORP., BUCKWALTER RADIO CORP., CLEVELAND AUTOMOBILE ACCESORIES CORP., PHILMORE MANUFACTURING CO., RADIO PRODUCTS CORP., WELLS-GARDNER & CO.
SUPERUNIT	HANSCOM RADIO DEVICES
SUPREMACY	MACY, R. H., CO.
SUPREME	PATTERSON RADIO CORP.

TRADE NAME	MANUFACTURER
SYLFAN	BAKER-SMITH CO.
SYMPHONIC	CORONA RADIO & TELEVISION CORP.
SYMPHONY	CROSLEY RADIO CORP., JONES RADIO CO., KINGSTON RADIO CO., INC.
SYNCHROPHASE	GREBE MANUFACTURING CO., INC.
SYNCRATONE	GAYLORD MANUFACTURING CO.
T	
TANNERFLEX	TANNER, C. D., CO.
TATRO	L'TATRO MANUFACTURING CO.
TCA	TRANSFORMER CORPORATION OF AMERICA
TE-LEK-TOR-ET	STROMBERG-CARLSON TELEPHONE MANUFACTURING CO.
TEA TABLE	WARWICK MANUFACTURING CO.
TECHNIDYNE	APEX ELECTRIC MANUFACTURING CO.
TELEDYNE	CUTTING & WASHINGTON RADIO CORP.
TELEMACO	TELEPHONE MAINTENANCE CO.
TELETONE	TELETONE CORP. OF AMERICA
TELEX	TELEX RADIO & TELEVISION CO.
TELOFLEX	PRIESS RADIO CORP.
TEMPLE	TEMPLE RADIO CORP., TRANSFORMER CORPORATION OF AMERICA
TEMPLETON	WARWICK MANUFACTURING CO.
TEMPLETONE	TRANSFORMER CORPORATION OF AMERICA
TEMPOTONE	BARKER BROS.
TEN	CROSLEY RADIO CORP.
TENACE	CROSLEY RADIO CORP.
TENSTRIKE	CROSLEY RADIO CORP.
TERAFONE	MOON RADIO CORP.
THERMIODYNE	ALGONQUIN ELECTRIC CO., SHEPPARD POTTER CO., INC., THERMIODYNE RADIO CORP.
THOMPSON	THOMPSON, R. E., MANUFACTURING CO.
THOROLA	REICHMANN CO.
THOROLA ISLODYNE	REICHMANN CO.
THRILL BOX	NATIONAL CO., INC.
TIFFANY TONE	HORN, HERBERT H., MANUFACTURING CO.
TINY TIM	DUNN-HALL CO.
TOBE TUNER	DEUTSCHMANN, TOBE, CORP.
TODD	TODD, INC.
TOM THUMB	AUTOMATIC RADIO MANUFACTURING CO., INC.
TONE MASTER	ELECTRICAL RESEARCH LABORATORIES, INC., WARWICK MANUFACTURING CO.
TONE-A-DYNE	RESAS, INC.
TORODYNE	AINSWORTH RADIO CO., SEARS MANUFACTURING CO.
TOTEM	CROSLEY RADIO CORP.
TRANSATLANTIC	I. C. A. EXPORT CO., INC.
TRANSCONTINENTAL	BETTS & BETTS CORP.
TRANSITONE	PHILCO RADIO & TELEVISION CORP.
TRANSOCEANIC	GOLDEN-LEUTZ CORP., LEUTZ, C. R., INC.
TRANSPACIFIC	I. C. A. EXPORT CO., INC.
TRAV-LER	TRAV-LER RADIO & TELEVISION CORP.
TRAVEL-AIR	TRAV-LER RADIO & TELEVISION CORP.

TRADE NAME	MANUFACTURER
TRAVEL-AIRE	TRAV-LER RADIO & TELEVISION CORP.
TRAVETTE	CROSLEY RADIO CORP.
TRAVO	CROSLEY RADIO CORP.
TREASURE CHEST	GRIGSBY-GRUNOW CORP.
TREBOR	TREBOR RADIO CO.
TREGODYNE	TREGO RADIO MANUFACTURING CO.
TRF	FERGUSON RADIO CORP.
TRIANGLE	WESTERN AUTO SUPPLY CO.
TRIANON	BLAIR RADIO LABORATORIES
TRINITY	BEACON RADIO MANUFACTURING CO.
TRIPLEX	MU-RAD RADIO CORP.
TRIRDYN	CROSLEY RADIO CORP.
TROJAN	TROJAN RADIO CORP.
TROUBADOR	CROSLEY RADIO CORP., SLEEPER RADIO CORP., ULTRAMAR MANUFACTURING CORP., WARWICK MANUFACTURING CO.
TROY	TROY RADIO & TELEVISION CO.
TRUE VALUE	WELLS-GARDNER & CO.
TRUE-DIAL	MISSION BELL RADIO MANUFACTURING CO.
TRUETEST	WARWICK MANUFACTURING CO.
TRUETONE	BELMONT RADIO CORP., WARWICK MANUFACTURING CO., WELLS-GARDNER & CO., WESTERN AUTO SUPPLY CO.
TRUPHONIC	CHELSEA RADIO CO.
TRUTONE	BOSTON RADIO MANUFACTURING CO.
TUSCANY	RCA VICTOR DIVISION OF RADIO CORP. OF AMERICA
TWELVE	CROSLEY RADIO CORP.
TWENTIETH CENTURY	PILOT RADIO & TUBE CORP.
TWIN 6	WARWICK MANUFACTURING CO.
TYRMAN	TYRMAN RADIO CO.
U	
ULTRA-MARVEL	INDUSTRIAL RADIO SERVICE
ULTRADYNE	LACAULT, R. E., CO., PHENIX RADIO CORP., RADIO RESEARCH LABORATORIES
ULTRAMAR	ULTRAMAR MANUFACTURING CORP.
ULTROLA	INDUSTRIAL RADIO SERVICE
UNIDYNE	UNITED MANUFACTURING & DISTRIBUTION CO.
UNIPAK	SILVER-MARSHALL MANUFACTURING CO.
UNITROL	KODEL RADIO CORP.
UNITROLA	Du BOIS RADIO CO.
UNITUNE	NASSAU RADIO CO.
UNIVERSAL	CLINTON RADIO MANUFACTURING CO., UNIVERSAL BATTERY CO., WELLS-GARDNER & CO.
UNIVERSAL COMPANION	I. C. A. EXPORT CO., INC.
USL	USL RADIO INC.
V	
VAGABOND	CROSLEY RADIO CORP.
VALLEY	VALLEY ELECTRIC CO.
VALLEYTONE	VALLEY ELECTRIC CO.
VERIPHONE	METROPOLITAN ENGINEERING SERVICE CO.

TRADE NAME	MANUFACTURER
VERNON	VERNON RADIO CO.
VICTOR	RCA VICTOR DIVISION OF RADIO CORP. OF AMERICA
VICTOREEN	VICTOREEN RADIO CO., WALKER, G. W., CO.
VICTORY	CHAMPION RADIO LABORATORIES, UNITED RADIO MANUFACTURING CORP.
VICTROLA RADIOLA	RCA VICTOR DIVISION OF RADIO CORP. OF AMERICA
VIKING	CROSLEY RADIO CORP., OZARKA RADIO CORP., WARWICK MANUFACTURING CO.
VITA-TONE	GALVIN MANFACTURING CORP.
VITATONE	ZANEY-GILL CORP.
VOCELESTE	GENERAL AMERICAN RADIO MANUFACTURING CO.
VOGUE	DAVIS MANUFACTURING CO.
W	
W. O. R.	WARWICK MANUFACTURING CO.
WALTHAM	WALTHAM RADIO CORP.
WALTON	NORTHWEST SPECIALITY SERVICE, LTD.
WARE	WARE RADIO CORP.
WARWICK	DETROLA RADIO & TELEVISION CORP., FORDSON RADIO MANUFACTURING CO., WARWICK MANUFACTURING CO.
WASHINGTON	CROSLEY RADIO CORP.
WASP	PILOT RADIO & TUBE CORP.
WATTERSON	WARWICK MANUFACTURING CO.
WAVE MASTER	KELLOGG SWITCHBOARD & SUPPLY CO.
WEBSTER	WEBSTER ELECTRICAL CORP.
WELLS-GARDNER	WELLS-GARDNER & CO.
WESTERN CABLE	WESTERN CABLE & LIGHT CO.
WESTGALE	WESTGALE ELECTRIC CO.
WESTINGHOUSE	CANADIAN WESTINGHOUSE CO., LTD., WESTINGHOUSE ELECTRIC SUPPLY CO.
WESTMINSTER	CLINTON RADIO MANUFACTURING CO.
WEXTARK	ALLIED RADIO CORP.
WHELCO	WELLS-GARDNER & CO.
WHITE	WHITE WIND ELECTRIC CO.
WHITESTONE	BOISSONAULT, G., CO.
WIDE WORLD	BIG THREE RADIO CORP.
WIGIT	CROSLEY RADIO CORP.
WILCOX-GAY	WILCOX-GAY CORP.
WINDHAM	WARWICK MANUFACTURING CO.
WINDSOR	WARWICK MANUFACTURING CO., WINDSOR FURNITURE CO.
WINGS	BELMONT RADIO CORP. , SENTINEL RADIO CORP.
WIRELESS SPECIALITY	WIRELESS SPECIALITY APPARATUS CO.
WOLVERINE	KLITZEN RADIO MANUFACTURING CO.
WONDERTONE	CONTINENTAL RADIO & TELEVISION CORP., KEYSTONE RADIO CO., WONDERTONE RADIO MANUFACTURING CO.
WORKRITE	UNITED STATES ELECTRIC CORP., WORKRITE MANUFACTURING CO.
WORLD SEVEN	HOWARD RADIO MANUFACTURING CO.
WORLD WIDE	PACIFIC RADIO CORP.
WORLD WIDE NINE	SILVER-MARSHALL MANUFACTURING CO.

TRADE NAME	MANUFACTURER
WORNER-HETEROPLEX	ELVIN RADIO MANUFACTURING CO.
WQXR RECEIVER	ESPEY MANUFACTURING CO., INC.
WRIGHT	WRIGHT RADIO MANUFACTURING CO.
WURLITZER	WURLITZER, RUDOLPH, CO.
Y	
YORKER	ESPEY MANUFACTURING CO., INC.
Z	
Z-NITH	CHICAGO RADIO LABORATORIES
ZENETTE	ZENITH RADIO CORP.
ZENITH	CHICAGO RADIO LABORATORIES, ZENITH RADIO CORP.
ZEPHYR	ZEPHYR RADIO CO.

SECTION 2

RADIO MANUFACTURERS

MANUFACTURER	TRADE NAME
A	
A-C ELECTRICAL MANUFACTURING CO.	POLYDYNE
A-C ELECTRICAL MFG. CO. (DAYTON, OHIO)	A-C DAYTON, NAVIGATOR
ABC RADIO LABORATORIES	ABC
ABLE TOOL & ENGINEERING CO.	ATEC
ACE RADIO CO.	ACE
ACME APPARATUS CO.	ACME, ACMEFLEX, ACMEPHONE
ACME ELECTRIC & MANUFACTURING CO.	ACME, MOTO-MIDGET
ADAMS-MORGAN CO.	PARADYNE, PARAGON
ADLER MANUFACTURING CO.	ADLER-ROYALE
ADROIT TOOL CO.	ADROLA
ADVANCE AUTOMOBILE ACCESSORIES CORP.	NEUTROWOUND
ADVANCE ELECTRIC CO.	ADVANCE, FALCK
AEOLIAN CO.	AEOLIAN HALL
AERODYNE CO.	AERODYNE
AEROVOICE, INC.	AEROVOICE
AETNA RADIO CORP.	AETNA
AINSWORTH RADIO CO.	TORODYNE
AINSWORTH-GATES RADIO CO.	RANGER FIVE
AIR-KING PRODUCTS CO., INC.	AIR QUEEN, AIR-KING, ATLAS, BELL TONE
AIR-OLA RADIO CO.	JUST-RITE
AIR-WAY ELECTRIC APPLIANCE CORP.	AIR-WAY
AIREX CO.	AIREX
AIRO-MASTER CORP.	AIRO-MASTER
AJAX ELECTRIC SPECIALITY CO.	AJAX, MARVELTONE
ALADDIN MANUFACTURING CO.	ALADYNE
ALGONQUIN ELECTRIC CO.	THERMIODYNE
ALL AMERICAN MOHAWK CORP.	ALL AMERICAN, ALL-AMAX, CHIEFTAIN, EAGLE, LYRIC, MOHAWK, NAVAJO, SEXTETTE
ALLIED RADIO CORP.	ALLIED, KNIGHT, POLYDYNE, ROAMER, WEXTARK
ALLIED STORES (NEW YORK)	ARIA
AMBER MANUFACTURING CO.	MARV-O-DYNE
AMERICAN APPARATUS CO.	AMACO
AMERICAN AUTO AND RADIO MANUFACTURING CO.	AMERICAN BEAUTY
AMERICAN BOSCH MAGNETO CORP.	AMBOROLA, AMERICAN BOSCH, BOSCH
AMERICAN ETHERPHONE CORP.	ETHERPHONE

MANUFACTURER	TRADE NAME
AMERICAN EUROPEAN RADIO CORP.	PORTO-O-RADIO
AMERICAN RADIO & RESEARCH CORP.	AMRAD, BEL CANTO, DUET, INDUCTROLE, MINUETTE, SERENATA
AMERICAN RADIO CORP.	ARC-LININGER
AMERICAN RADIO MANUFACTURING CO.	AMERICAN BEAUTY
AMERICAN SPECIALITY MANUFACTURING CO.	ELECTROLA, REGAL, REPUBLIC
AMERICAN STANDARD ELECTRIC CO.	AMEREX
AMERICAN TELEPHONE & RADIO CO.	A. T. R.
AMERICAN TRANSFORMER CO.	AMERTRAN
AMERICAN WIRELESS	AMERICAN WIRELESS
AMPLEX INSTRUMENT LABORATORIES	HARMO-SONIC
AMSCO PRODUCTS CO.	ACMEDYNE, MELCO SUPREME
ANDREA RADIO CORP.	ANDREA
ANDREA, F. A. D., INC.	FADA, FADA NEUTROLA
ANDREWS RADIO CO.	DERESNADYNE
ANSLEY RADIO CORP.	ANSLEY, DYNAPHONE
ANYLITE ELECTRIC CO.	KING COLE
APEX ELECTRIC MANUFACTURING CO.	MICRODYNE, TECHNIDYNE
APPEL-HENDERSON RADIO CO.	APPEL-HENDERSON, SILVER KING
ARGUS RADIO CORP.	ARGUS
ARMLEY RADIO CORP.	KARRYADIO
ATCHINSON RADIO MANUFACTURING CO.	ATCHINSON
ATLANTIC-PACIFIC RADIO SUPPLIES CO.	A-P
ATLAS COLONIAL CORP.	COLONIAL
ATWATER KENT MANUFACTURING CO.	ATWATER KENT
AUDIOLA RADIO CORP.	AUDIODYNE, AUDIOLA
AURORA INC.	AURORA
AUSTIN RADIO MANUFACTURING CO.	AUSTIN
AUTO INDICATOR CO.	POCKET RADIO
AUTOCRAT RADIO CO.	AUTOCRAT
AUTOMATIC RADIO MANUFACTURING CO., INC.	AIR MASTER, ARC, AUTO-ROLA, AUTOMASTER, AUTOMATIC, BLUEBIRD, TOM THUMB
AUTOMETAL CORP.	ARAGAIN

B

MANUFACTURER	TRADE NAME
BAKER-SMITH CO.	SYLFAN
BALDER RADIO CORP.	BALDER, KNICKERBOCKER
BALDWIN, NATHANIEL, INC.	BALDWIN, BALDWINETTE, HYDAWAY
BALKEIT RADIO CORP.	BALKEIT, FANSTEEL, FIRST NATIONAL
BALTIMORE HUB-WHEEL & MFG. CO.	HUBCO
BANFER PRODUCTS CORP.	SUPERTONE
BARKER BROS.	TEMPOTONE
BATTERYLESS RADIO CORP.	BAT-RY-LES, NO-BAT-RY
BAUMAN, WARREN J., CO.	SUPERLATONE
BEACON RADIO MANUFACTURING CO.	TRINITY
BEAVER ELECTRIC CORP.	BEAVERTONE
BELL RADIO & TELEVISION CO.	BELL
BELL-KNAPP HARDWARE & MANUFACTURING CO.	BELL-KNAPP

MANUFACTURER	TRADE NAME
BELMONT RADIO CORP.	BELMONT, CLASSIQUE, CRUSADER, FRESHMAN, GOODYEAR WINGS, IMPERIAL, STARCK CLASSIQUE, TRUETONE, WINGS
BEN RADIO CORP.	PETITE
BENDIX RADIO MANUFACTURING CO.	BENDIX
BENSEN ENGINEERING CO.	SUPERFLEX
BETTER RADIO PRODUCTS	MELODY
BETTS & BETTS CORP.	TRANSCONTINENTAL
BIG THREE RADIO CORP.	WIDE WORLD
BLAIR RADIO LABORATORIES	BLAIR, BLAIRCO, TRIANON
BOISSONAULT, G., CO.	WHITESTONE
BOSTON RADIO MANUFACTURING CO.	MASTERTONE, TRUTONE
BOSWORTH ELECTRIC MANUFACTURING CO.	BOSWORTH
BRANDEIS, J. F., CORP.	BRANDOLA
BRANSTON, CHARLES A., CO.	HETROLA
BRAUN, W. C., CO.	MONROE
BREMER TULLY MANUFACTURING CO.	BREMER TULLY, COUNTERPHASE
BRENARD MANUFACTURING CO.	HERALDYNE
BROWN & MANHART CO.	RANGER
BROWNING LABORATORIES	BROWNING LABS
BROWNING-DRAKE CORP.	BROWNING-DRAKE
BRUNSWICK-BALKE-COLLENDER CO.	BRUNSWICK, CORDOVA, PANATROPE, RADIOLA, REGENOFLEX
BUCKINGHAM RADIO CORP.	BUCKINGHAM, HUNTINGTON, LA SALLE, REMINGTON
BUCKWALTER RADIO CORP.	BURAD, SUPERTONE
BUELL MANUFACTURING CO.	BUELL
BULOVA WATCH CO.	BULOVA
BURNDEPT WIRELESS CORP. OF AMERICA	ETHOVOX
BUSH & LANE PIANO CO.	BUSH & LANE
C	
CADILLAC RADIO MANUFACTURING CO.	AMPIOLA, CADILLAC
CANADIAN WESTINGHOUSE CO., LTD.	CANADIAN WESTINGHOUSE, WESTINGHOUSE
CAPEHART CORP., INC.	AMPERION, CAPEHART, PANAMUSE
CARDINAL RADIO MANUFACTURING CO.	CARDINAL
CARDINOLA RADIO CORP.	CARDINOLA
CARDWELL, ALLEN D., MANUFACTURING CORP.	CARDWELL
CARTER MANUFACTURING CO.	CARCO
CARTERET RADIO LABS	CARTERET
CARYLOYD ELECTRIC & RADIO CO.	MALONE-LEMMON
CASE ELECTRIC CORP.	CASE
CASS, PHILIP, & CO.	CASSCO
CAVALCADE RADIO CO. (Chicago)	CAVALCADE
CAVALCADE RADIO CORP. (San Francisco)	CAVALCADE
CENTURY RADIO PRODUCTS CO.	CENTURY
CHAMPION RADIO LABORATORIES	CHAMPION, LA SALLE, MONARCH, VICTORY
CHATHAM RADIO CORP.	HY-PEX

MANUFACTURER	TRADE NAME
CHELSEA RADIO CO.	BEARCAT, REGENODYNE, TRUPHONIC
CHICAGO NIPPLE MANUFACTURING CO.	BLUE RIBBON, ROCHESTER
CHICAGO RADIO LABORATORIES	Z-NITH, ZENITH
CHISHOLM-BARFIELD CORP.	RADIO KNIGHT
CLAGO RADIO CORP.	CLAGO, MAYFAIR, MAYFLOWER, RADIOCHRON
CLAPP-EASTHAM CO.	RADAK
CLARION RADIO CORP.	CLARION
CLEARTONE RADIO CORP.	CLEARODYNE, CLEARTONE, GOLDCREST
CLEVELAND AUTOMOBILE ACCESORIES CORP.	SUPERTONE
CLEVELAND PRODUCTS CO.	CLEVELAND
CLIMAX RADIO & TELEVISION CO.	CLIMAX, CUNNINGHAM
CLINTON RADIO MANUFACTURING CO.	BOSTONIAN, BUCKINGHAM, CLINTON, CORONA, CRUSADER, CUB, FEDERAL, HARMONY, MARSHALL, NIGHTENGALE, UNIVERSAL, WESTMINSTER
COLONIAL RADIO CORP.	COLONIAL
COLUMBIA PHONOGRAPH CO.	COLUMBIA
COLUMBIA RADIO CO.	COLUMBIA, SENTINEL
COMMONWEALTH RADIO MANUFACTURING CO.	ALONSON, COM-RAD, STAR RAIDER
CONCERT RADIOPHONE CO.	CONCERT GRAND
CONSOLIDATED RADIO & TELEVISION CORP.	HOLLYWOOD, MISSION, ROYAL
CONSOLIDATED RADIO CORP.	ARBORPHONE
CONSUMERS RADIO CO.	CON-RAD
CONTINENTAL RADIO & TELEVISION CORP.	ADMIRAL, CONTINENTAL, FEARNOLA, GLOBE TROTTER, MARSHALL, ORIOLE, RADIOMASTER, SLAGLE, STAR RAIDER, SUNBEAM, WONDERTONE
CORDONIC MANUFACTURING CO.	GENEVA
CORONA RADIO & TELEVISION CORP.	CORONA, DUETTE, SUNBEAM RADIOMASTER, SYMPHONIC
CORONADO MANUFACTURING CO.	CORONADO
COSMOPOLITAN PHUSIFORMER	COSMOPOLITAN
COSMOTONE, INC.	COSMOTONE
COTO-COIL CO.	COTO SYMPHONIC
CRESCENT RADIO MANUFACTURING CO.	CRESCENTYNE
CROSLEY RADIO CORP.	ACE, ADMINISTRATOR, ADVENTURER, ALDERMAN, AMBASSADOR, ANNOUNCER, ARBITER, BANDBOX, BIGFELLA, BONNIBOY, BOOK CASE, BUBBY-BOY, BUCCANEER, BUDDY, CABRIOLET, CAROLER, CASA, CHERIO, CHIEF, CHUM, CLASSMATE, CLIPPER, COMMISSIONER, COMPANION, CONGRESSMAN, CONSTITUTION, CORSAIR, CRONY, CROSLEY, CRUISER, DIRECTOR, DISCOVERER, ELF, FIVER, GALLEON,

MANUFACTURER	TRADE NAME
CROSLEY RADIO CORP. (Continued)	GEMBOX, GEMCHEST, GOVERNOR, JEWELBOX, JEWELCASE, LEADER, LEGISLATOR, MATE, MAYOR, MERRIMAC, MERRYMAKER, MINSTREL, MONOTRAD, OLYMPIA, ORACLE, PAL, PARTNER, PLAYBOY, PLAYTIME, PRIVATEER, PUP, REPOSE, REVELER, ROAMIO, RONDEAU, SECRETARY, SENATOR, SEPTET, SEVETTE, SEXTET, SHERATON, SHOWBOX, SHOWBOY, SHOWCHEST, SIX, SIXTY, SONDO, SONGSTER, SONNESTER, SYMPHONY, TEN, TENACE, TENSTRIKE, TOTEM, TRAVETTE, TRAVO, TRIRDYN, TROUBADOR, TWELVE, VAGABOND, VIKING, WASHINGTON, WIGIT
CROUCH RADIO CO.	CLARADYNE
CUMMINGS RADIO CORP.	SOLODYNE
CUSTOM BUILT RADIO MANUFACTURING CO.	CROWN
CUTTING & WASHINGTON RADIO CORP.	TELEDYNE

D

MANUFACTURER	TRADE NAME
DALCO RADIO CO.	AIRTROLA, DALCOPHONE
DAVIS MANUFACTURING CO.	VOGUE
DAVISON HAYNES CO.	ANGELUS
DAYFAN ELECTRIC CO.	DAYCRAFT, DAYFAN, DAYROYAL
DAYTON FAN & MOTOR CO. (Dayton, Ohio)	DAYCRAFT, DAYOLA
DAYTON RADIO CO.	DAYTONIA
DAYTON RADIO CORP.	DAYRAD
De WALD RADIO MANUFACTURING CO.	De WALD
DeFOREST RADIO CO.	EVERYMAN, INTERPANEL, RADIO-CRAFT, RADIOPHONE, RENAISSANCE
DELCO RADIO CORP.	DAYFAN
DETROLA RADIO & TELEVISION CORP.	DETROLA, PEE WEE, ROAD CHIEF, ROAD MASTER, WARWICK
DEUTSCHMANN, TOBE, CORP.	BROWNING, TOBE TUNER
DeWITT-LaFRANCE CO.	SUPERADIO
DEXTER METAL MANUFACTURING CO.	DEXTER
DIAMOND-T RADIO MANUFACTURERS	DIAMOND-T
DICTOGRAPH PRODUCTS CO.,INC.	DICTOGRAPH
DISTANTONE RADIOS, INC.	DISTANTONE
DORON BROTHERS ELECTRIC CO.	SUPER-EQUIDYNE
DREXEL RADIO MANUFACTURING CO.	DREXOLA
Du BOIS RADIO CO.	UNITROLA
DUBILIER CLOCK CORP.	DUBILIER
DUMONT ELECTRIC CO., INC.	DUBILIER
DUNN-HALL CO.	DUNN-HALL, REMINGTON, TINY TIM
DYNAMOTIVE RADIO CORP.	DYNERGY

MANUFACTURER	TRADE NAME
E	
E & N PRODUCTS CO.	SOVEREIGN
EAGLE CHARGER CORP.	ELECTRADYNE
EAGLE RADIO CO.	EAGLE, EAGLET
EARL RADIO CORP.	EARL
EATON-EDWARDS RADIO CO.	MELODYNE
ECHOPHONE RADIO MANUFACTURING CO.	ECHOPHONE
ECKHARDT CORP.	ECKHARMONIC
ECKSTEIN RADIO & TELEVISION CO.	L'TATRO
ECODYNE RADIO CO.	ECODYNE
ECOHPHONE RADIO MANUFACTURING CO.	ECHOETTE
ECOPHONE RADIO CORP.	ARROW, ECOPHONE
EDISON, THOMAS A., INC.	EDISON
EDISON-BELL RADIO & PHONOGRAPH CO.	EDISON-BELL
EL REY RADIO MANUFACTURING CO.	EL REY
ELECTRAD CO., INC.	ELECTRAD
ELECTRIC MACHINE CORP.	ELMCO
ELECTRICAL PRODUCTS MANUFACTURING CO.	DYMAC
ELECTRICAL RESEARCH & MANUFACTURING CO.	SUPERIORDYNE, SUPERIORFLEX
ELECTRICAL RESEARCH LABORATORIES, INC.	ERLA, PEARSON, RAMWAY, SENTINEL, SKY ROVER, TONE MASTER
ELECTROMATIC DISTRIBUTORS, INC.	ELECTROMATIC
ELECTROPHONE CORP.	HYATT
ELGIN RADIO SUPPLY CO.	ELGIN, SUPER-REINARTZ
ELKAY MANUFACTURING CO.	ELKAY
ELVIN RADIO MANUFACTURING CO.	WORNER-HETEROPLEX
EMERSON RADIO & PHONOGRAPH CORP.	EMERSON, EMERSONETTE, JEWEL, KADETTE, MICKEY-MOUSE
EMPIRE ELECTRICAL PRODUCTS CO.	EMPIRE
EQUITABLE RADIO CORP.	CLARATONE
ESPEY MANUFACTURING CO., INC.	ENSIGN, ESPEY, WQXR RECEIVER, YORKER
EXCELLO PRODUCTS CORP.	EXCELLO
F	
FADA RADIO & ELECTRIC CO.	FADA, FADALETTE, MOTOSET
FAIR STORE, THE	LA FRANCE
FAIRBANKS-MORSE & CO.	FAIRBANKS-MORSE
FAIRVIEW ELECTRIC SHOP	LASHER CAPACIDYNE
FARNSWORTH TELEVISION & RADIO CORP.	CAPEHART, FARNSWORTH
FARRAND MANUFACTURING CO.	FARRAND
FEDDERS MANUFACTURING CO.	J. R. H.
FEDERAL RADIO CORP.	FEDERAL, ORTHOSONIC
FEDERAL TELEGRAPH OF CALIFORNIA	KOLSTER
FEDERAL-BRANDES, INC.	KOLSTER
FEDERATED PURCHASER, INC.	ACRATEST, ACRATONE, CATHEDRAL TONE
FERGUSON RADIO CORP.	FENWAY, FERGUSON, TRF
FERRANTI, INC.	FERRANTI
FINK INDUSTRIES	IMPERIAL
FIRESTONE TIRE & RUBBER CO.	AIR CHIEF, FIRESTONE, ROAMER

MANUFACTURER	TRADE NAME
FLEETWOOD RADIO CORP.	FLEETWOOD
FORACO CORP.	FORACO
FORDSON RADIO MANUFACTURING CO.	FORDSON, GOLDENTONE, WARWICK
FOREIGH & DOMESTIC ELECTRICAL COMMODITIES	ELECTRADYNE
FRANKLIN RADIO CORP.	CHELTENHAM, FRANKLIN
FREED RADIO & TELEVISION CORP.	FREED, FREED-EISEMANN
FREELY CORP.	FIRESIDE
FREEPORT RADIO CORP.	ABC
FRENCH, JESSE & SONS, PIANO CO.	JESSE FRENCH
FRESHMAN, CHARLES, CO.	EQUAPHASE, ERAC, MASTERPIECE
FRESHMAN, CHARLES. CO.	FRESHMAN
FROST-MINTON CORP.	FROST-MINTON
G	
GALVIN MANFACTURING CORP.	GOLDENVOICE, MOTOROLA, PEP BOYS, SPORTER, VITA-TONE
GAMBLE STORES, INC.	CORONADO
GAROD RADIO CORP.	GAROD
GAYLORD MANUFACTURING CO.	GAYLORD, SYNCRATONE
GENERAL AMERICAN RADIO MANUFACTURING CO.	CAMEO, JEWEL, VOCELESTE
GENERAL ELECTRIC CO.	GENERAL ELECTRIC, MUSAPHONIC
GENERAL HOUSEHOLD UTILITIES CO.	GRUNOW
GENERAL MOTORS RADIO CORP.	DAY RADIA, DAYCRAFT, DAYFAN, DAYMAR, DAYOLA, DAYTONIA
GENERAL RADIO & TELEVISION CORP.	AMERICAN, GENERAL
GIBSON-SEARS RADIO CORP.	STERLING
GIFILLAN BROTHERS, INC.	GILFILLAN
GILBERT, R. W., CO.	GILBERT
GLOBE ELECTRIC CO.	DUODYNE, GLOBE
GLOBE TROTTER RADIO CO.	GLOBE TROTTER
GOERDES, FRED, W., CO.	DOLORES
GOLD MEDAL RADIO CORP.	GOLD MEDAL
GOLD SEAL ELECTRIC CO.	GOLD SEAL
GOLDBLATT STORES	KENNEDY
GOLDEN-LEUTZ CORP.	PLIO-6, PLIODYNE, TRANSOCEANIC
GOODRICH RUBBER & TIRE CO.	MANTOLA
GOODYEAR TIRE & RUBBER CO.	GOODYEAR
GRAND RAPIDS RADIO MANUFACTURING CO.	GYPSY
GRANT, H. R., CO.	RITA De LUX
GRAY ELECTRIC CO.	COURIER
GRAY-DANIELSON MANUFACTURING CO.	REMLER
GRAYBAR ELECTRIC CO.	GRAYBAR
GREBE MANUFACTURING CO., INC.	GREBE, SYNCHROPHASE
GRIFFIN SMITH MANUFACTURING CO.	ROYALE
GRIGSBY-GRUNOW CORP.	BRENTWOOD, BRUCEWOOD, CHELTENWOOD, ELLSWOOD, FORD-MAJESTIC, HAVENWOOD, MAJESTIC, SHERWOOD, TREASURE CHEST
GRIMES, DAVID, INC.	INVERSE DUPLEX, NEW YORKER

MANUFACTURER	TRADE NAME
GULBRANSEN RADIO CO.	CHAMPION, JR., GULBRANSEN
GUTHRIE CO.	BLUE BIRD, BOB-O-LINK, GOLDFINCH, MOCKING BIRD, NIGHTINGALE
H	
HALL, CHARLES, CO.	FLEXODYNE
HALLER, W. B., CO.	HALLERIO
HALLOCK & WATSON RADIO CORP.	HALOWAT
HALSON RADIO & TELEVISION CORP.	HALSON
HAMILTON RADIO CORP.	HAMILTON
HAMMARLUND-ROBERTS, INC.	HI-Q
HANSCOM RADIO DEVICES	SUPERUNIT
HANSEN STORAGE CO.	BLUEBIRD, NIGHTINGALE
HARMON & SONS CO.	HARMONSON
HARMONY MANUFACTURING CO.	HARMONY
HARRIS MANUFACTURING CO.	ELECTROPHONE, ELECTROTONE
HARTMAN ELECTRICAL MANUFACTURING CO.	HARTMAN
HATRY & YOUNG	HATRY
HEART OF AMERICA RADIO CO.	SERENADER
HEIGES BROTHERS	CLEARBROOK
HETRO ELECTRICAL INDUSTRIES, INC.	HETRO
HIGH FREQUENCY LABORATORIES	ISOTONE, MASTERTONE, REGENT
HITCHCOCK-HILL CO.	HILLTOP SIX
HOLLAND RADIO CORP.	REVELATION
HOMER PRODUCTS, INC.	HOMER
HOODWIN, CHARLES, CO.	AERO, AERO PRODUCTS, HOODWIN
HORN, HERBERT H., MANUFACTURING CO.	HORN, TIFFANY TONE
HOWARD RADIO MANUFACTURING CO.	CABLE-NELSON, GREEN DIAMOND, HOWARD, WORLD SEVEN
HUNT, W. D., CO.	HEXADYNE
HYATT ELECTRIC CORP.	HYATT
HYMAN, HENRY, & COMPANY	BESTONE
I	
I. C. A. EXPORT CO., INC.	MAGICOLOR, TRANSATLANTIC, TRANSPACIFIC, UNIVERSAL COMPANION
IMPERIAL FURNITURE CO.	IMPERIAL
INCANDESCENT SUPPLY CO.	NEWLANDS
INDIANA ELECTRIC MANUFACTURING CO.	CASE, HYPERDYNE
INDUSTRIAL RADIO SERVICE	ULTRA-MARVEL, ULTROLA
INSULINE CORPORATION OF AMERICA	CLASSIC, CONQUEROR, ENVOY, ENVOYETTE, ICA, INSULLETTE, MARVEL, MASCOT, PIED-PIPER, SCOUT
INTERNATIONAL RADIO CORP.	AIRITE, INTERNATIONAL, IRC, JEWEL, KADETTE, ROTOFOR, ST. REGIS
INTERNATIONAL TRADING CORP.	HERALD
J	
JACKSON BELL CO., LTD.	JACKSON BELL, PETER PAN

MANUFACTURER	TRADE NAME
JACKSON RESEARCH LABORATORIES	JACKSON
JEWEL MANUFACTURING CO.	JEWEL
JEWETT RADIO & PHONOGRAPH CO.	JEWETT
JONES ELECTRIC & RADIO MANUFACTURING CO.	CONCERDYNE
JONES RADIO CO.	HARMONIC, SYMPHONY
JONES, JOSEPH W., RADIO MANUFACTURING CO.	JONES
K	
KADETTE RADIO CORP.	KADETTE
KAESS MANUFACTURING CORP.	CLEARVIEW
KARDON PRODUCTS CO., INC.	KOMPENTROL
KELLER-FULLER MANUFACTURING CO.	KELLER-FULLER, RADIETTE
KELLOGG SWITCHBOARD & SUPPLY CO.	CASCADE, KELLOGG, WAVE MASTER
KEMPER RADIO LABORATORIES	KEMPER, KEMPER KOMPAK
KEN-MAR RADIO CORP.	BABYDYNE, INTERNATIONAL, KEN-MAR
KENMAN ELECTRIC CO.	KENMAN
KENNEDY, COLIN B., CO.	COMET, CORONET, GLOBE TROTTER, KENNEDY, ROYAL
KENOLA RADIO CORP.	KENOLA
KEYSTONE RADIO CO.	WONDERTONE
KILBOURNE & CLARK MANUFACTURING CO.	AIR ROAMER, AIR RULER
KING MANUFACTURING CORP.	IMPERIAL, KING, MONARCH, ROYAL
KING QUALITY PRODUCTS CO.	KING-HINNERS
KINGSTON RADIO CO., INC.	CORONADO, GAMBLE, GAMBLE-SKOGMO, GYPSY, KINGSTON, REGAL, SYMPHONY
KLITZEN RADIO MANUFACTURING CO.	KENT, WOLVERINE
KODEL RADIO CORP.	GOLD STAR, KODEL, LOGODYNE, UNITROL
KOLSTER RADIO CORP.	BRANDES, INTERNATIONAL, KOLSTER
L	
L'TATRO MANUFACTURING CO.	L'TATRO, TATRO
LACAULT, R. E., CO.	LACAULT, STROBODYNE, ULTRADYNE
LAFAYETTE RADIO CORP.	LAFAYETTE
LAKESIDE SUPPLY CO.	CELESTE, MICRODYNE
LANG RADIO CORP.	LANG
LANGBEIN-KAUFMANN CO.	ELKAY
LANGHORNE & CO.	LANCO
LAUREHK RADIO MANUFACTURING CO.	CARUSO, LAUREHK, MUSIQUE
LAUREL MOTORS CORP.	SINCLAIR
LEE ELECTRIC & MANUFACTURING CO.	LEMCO
LEHMAN RADIO SALON, INC.	FIDEL-O-MATIC, PORT-O-MATIC
LEUTZ, C. R., INC.	LEUTZ, PLIO-6, PLIODYNE, SEVEN SEAS, SILVER GHOST, TRANSOCEANIC
LEWOL MANUFACTURING CO.	LEWOL
LIBERTY MUSIC SHOP	LIBERTY
LIBERTY TRANSFORMER CO.	LIBERTY
LINCOLN RADIO CORP.	LINCOLN, LINCOLN SYMPHONIC
LONG RADIO CO.	CARIDAL
LUXOR RADIO MANUFACTURING CO.	LUXOR

MANUFACTURER	TRADE NAME
LYON RADIO CO.	ORTHONOLA
M	
MACHINE SPECIALITY CO.	ARBORPHONE
MACK CO.	MACK SINCROFLEX
MACLAREN MANUFACTURING CO.	DOMINAIRE
MACY, R. H., CO.	MACY, SUPREMACY
MAGNAVOX CO., INC.	MAGNAVOX
MAJESTIC RADIO & TELEVISION CORP.	CASE, COLUMBIA, MAJESTIC
MALONE-LEMMON PRODUCTS INC.	RAYCRAFT
MARCONI BROS., INC.	MARCONIPHONE
MARLOW, ERNEST, CO.	MARLODYNE
MARQUETTE RADIO CO.	MARQUETTE
MARTI ELECTRIC RADIO, CO., INC.	MARTI
MARWOL RADIO CORP.	MARWOL
MARWOOD RADIO CORP.	MARWOOD
MASTER RADIO MANUFACTURING CO., INC.	MASTER, MASTER MIDGET
MAY RADIO & REFRIGERATION CO.	MAY
MAZDA RADIO MANUFACTURING CO.	CONSOMELLO
McMILLAN RADIO CO.	McMILLAN
McMURDO SILVER CORP.	MASTERPIECE, McMURDO SILVER, ORPHEUS, SILVER
MEISSNER MANUFACTURING CO.	MEISSNER
MELBURN RADIO MANUFACTURING CO.	MELBURN
MERSHMAN BROS. CORP.	BRUNSWICK
METRO ELECTRIC CO.	METRODYNE
METROPOLITAN ELECTRIC CO.	MECO
METROPOLITAN ENGINEERING SERVICE CO.	VERIPHONE
MICHIGAN RADIO CORP.	M. R. C., MICHIGAN
MIDWEST RADIO CORP.	MIDWEST, MIRACO
MINERVA RADIO CO.	MINERVA
MISSION BELL RADIO MANUFACTURING CO.	AERONAUTIC, HOBBS, MISSION BELL, TRUE-DIAL
MISSISSIPPI VALLEY RADIO CO.	LAFAYETTE
MITCHELL MANUFACTURING CO.	LULLABY, LUMITONE
MODERNOLA CO.	CADDY, DELANO, SHERATON
MOE MANUFACTURING CO.	ALADDIN
MOGUL ELECTRIC CO.	DEXAPHONE
MOHAWK ELECTRIC CO.	MOHAWK
MONTGOMERY WARD & CO.	AIRLINE, BALBOA, CAVALIER, CHALLENGER, COLLEGIAN, COMMANDER, COMMODORE, CORONADO, CORTEZ, DICTATOR, FANTASY, MINSTREL, MONTGOMERY WARD, PRINCESS, SERENADER, SOLO, SOVEREIGN
MONTROSE RADIO LABORATORIES	MONTROSET
MONTROSE-HAST RADIO LABORATORIES	NATURELLE
MOON RADIO CORP.	MOON, SATTERLEE, TERAFONE
MORRDYNE RADIO MANUFACTURING CO.	MODYNE
MORRISON RADIOPHONE CO., INC.	MORRADYNE

MANUFACTURER	TRADE NAME
MOTO-METER GAUGE & EQUIPMENT CO.	MOTO-VOX
MOTOMASTER RADIO CO.	MOTOMASTER
MU-RAD RADIO CORP.	MU-RAD, TRIPLEX
MURDOCK, WILLIAM J., CO.	MURDOCK
MUSIC MASTER CORP.	MUSIC MASTER
MY OWN RADIO CO.	MY OWN
N	
NAMCO MANUFACTURING CO.	NAMCO
NASSAU RADIO CO.	MAGADYNE, NASSAU, UNITUNE
NATIONAL AIRPHONE CORP.	AIRPHONE, MONODYNE
NATIONAL CARBON CO.	EVEREADY
NATIONAL CO., INC.	BROWNING-DRAKE, THRILL BOX
NATIONAL RADIO CO.	AERIPHONE, INTRADYNE
NATIONAL TRANSFORMER MANUFACTURING CO.	BALKEIT-NATIONAL
NATIONAL UNION RADIO CORP.	NATIONAL UNION
NAYLOR RADIO CORP.	STERLING
NEUTROWOUND RADIO MANUFACTURING CO.	NEUTROWOUND
NEW YORK ALBUM & CARD CO.	NYACCOFLEX, NYACCO
NOBLITT-SPARKS INDUSTRIES, INC.	ARVIA, ARVIN
NORBERT RADIO SERVICE	DIAMADYNE, MELODEE, NORBERT
NORDEN HAUCK RADIO CO.	ADMIRALITY
NORDEN-HAUCK RADIO CO.	NORDEN-HAUCK
NORTHWEST SPECIALITY SERVICE, LTD.	WALTON
NORTHWESTERN RADIO MANUFACTURING CO.	NORCO
NUNN-LANDON CO.	CASCADE
O	
OHIO STAMPING & ENGINEERING CO.	NO-DIAL
OPERADIO MANUFACTURING CO.	OPERADIO
ORTHODYNE RADIO MANUFACTURING CO.	AIRO-MASTER
OZARKA RADIO CORP.	LINCOLN SUPER, OZARKA, VIKING
P	
P & M RADIO CO.	PENN-C
PACENT ENGINEERING CORP.	PACENT
PACIFIC RADIO CORP.	AMERICAN WORLD, BEST, BUCKINGHAM, CRUSADER, PACIFIC, PEERLESS, WORLD WIDE
PACKARD RADIO CORP.	PACKARD
PACKARD-BELL CO.	PACKARD-BELL
PARAGON ELECTRIC CO.	PARAGON
PARAMOUNT MANUFACTURING CO.	SERENADER
PATHE RADIO & TELEVISION CORP.	MINUTE MAN, PATHE
PATTERSON RADIO CORP.	PATTERSON, SUPREME
PEARSON DIVISION ELECTRICAL RESEARCH LABS	PEARSON
PENNSYLVANIA RADIO MANUFACTURING CO.	PENN C
PERFECTONE, INC.	PERFECTONE
PERRY RADIO SUPPLY CO.	PERASCO
PFANSTIEHL RADIO CO.	BALDWIN, OVERTONE, PFANSTIEHL

MANUFACTURER	TRADE NAME
PHENIX RADIO CORP.	ULTRADYNE
PHILADELPHIA STORAGE BATTERY CO.	PHILCO
PHILCO RADIO & TELEVISION CORP.	BABY GRAND, LAZY X, LINCOLN-PHILCO, PACKARD-PHILCO, PHILCO, TRANSITONE
PHILHARMONIC RADIO CO.	PHILHARMONIC
PHILMORE MANUFACTURING CO.	AJAX, BLACKBIRD, BLUE BIRD, KOMPAKT, LITTLE WONDER, PHILMORE, SUPERTONE
PICARD RADIO CORP.	AMES
PIERCE AIRO CO.	DeWALD
PIERSON, THE, CO.	PIERSON
PIERSON-DELANE, INC.	PATTERSON
PILGRIM ELECTRIC CORP.	PILGRIM
PILOT RADIO & TUBE CORP.	AIR HOUND, AIR SCOUT, CAPTAIN-KIDD, DRAGON, JUMBO, PILOT, PILOTONE, RAINBOW, STANDARD, SUPER DRAGON, SUPER WASP, TWENTIETH CENTURY, WASP
PINTO SPEAKING CLOCK CO.	RADIO-ALARM-O-GRAPH
PIONEER RADIO CORP.	PIONEER
PITTSBURGH RADIO SUPPLY CO.	McCULLOUGH
PLAZA MUSIC CO.	FINE-ARTS, HAMILTON-LLOYD, PLAZA, STANDARD
PLYMOUTH RADIO & PHONOGRAPH CO.	SUPER SINGLE
PLYMOUTH RADIO CORP.	PLYMOUTH
PORT-O-MATIC CORP.	FIDEL-O-MATIC
PORTER, FRANK R. CO.	SUPERIOR
PORTOLA RADIO CO.	PORTOLA
PORTS MANUFACTURING CO.	PRMCO
POWELL MANUFACTURING CO.	POWELL
POWEROLA RADIO CORP.	POWEROLA
PRECEL RADIO MANUFACTURING CO.	SUPERFIVE
PRECISION PRODUCTS CORP.	ARBORPHONE
PREMIER ELECTRIC CORP.	ENSEMBLE, PREMIER
PRIESS RADIO CORP.	PRIESS, TELOFLEX

Q

Q. T. RADIO PRODUCTS	LITTLE GIANT

R

R-B RADIO CO.	BELLE CLAIRE, PAGE-FIVE
R. K. RADIO LABORATORIES, INC.	ARKAY, MAJESCO, RADIO-KEG
RADIART CORP.	RADIART
RADIART LABORATORIES	MAGNAFORMER
RADIO APPARATUS CO.	ETHERPHONE
RADIO CIRCULAR CO.	ACE
RADIO DOC	QUAD
RADIO ENGINEERING LABORATORY, INC.	R. E. L.
RADIO EQUIPMENT CO.	SUPERPHONIC
RADIO FREQUENCY LABORATORIES	R F L

RADIO MANUFACTURERS

MANUFACTURER	TRADE NAME
RADIO INDUSTRIES CORP.	RICO-DYNE, RIVOLI
RADIO MASTER CORP.	RADIO MASTER, SIMPLIFORM
RADIO PANEL SHOP	FLIVER
RADIO PRODUCTS CORP.	CAMDEN, CLIPPER, CO-OP, CONTINENTAL, CRUSADER, DAYRAD, RPM ODYNE, SUPERFLEX, SUPERTONE
RADIO RECEPTOR CO.	MULTIFLEX, RECEPTRAD, RECITAL, SUPERPOWER
RADIO RESEARCH LABORATORIES	ULTRADYNE
RADIO SERVICE CO.	ANGELUS
RADIO SHOP	ECHOPHONE, LEROY
RADIO SUPPLY HOUSE	PARAMOUNT
RADIO WIRE & TELEVISION, INC.	LAFAYETTE
RADIO-CHRON CORP.	RADA-CHRON
RADIOBAR COMPANY OF AMERICA	RADIOBAR
RADIOCOMPAK CO.	ARISTOCRAT
RADIOGRAPH CORP.	RADIOGRAPH
RADOLEK CO.	GENERAL, KEYSTONE, KING, MARVELO, MONROE-MARVELO, OZARKA, PIONEER, RADOLEK
RANDOLPH RADIO CORP.	RANDOLPH
RAULAND CORP.	RAULAND
RCA VICTOR DIVISION OF RADIO CORP. OF AMERICA	ACOUSTIC, AERIOLA, ALHAMBRA, BORGIA, COLUMBIA, CROMWELL, DeFOREST, ELECTROLA, ELECTROLA RADIOLA, FLORENZA, HYPERION, ORTHOPHONIC, PHOTOPHONE, RADIO ELECTROLA, RADIOLA, RADIOLETTE, RCA-VICTOR, RECEPTOR, SUPERETTE, TUSCANY, VICTOR, VICTROLA RADIOLA
REGENT RADIO CORP.	REGENT
REICHMANN CO.	THOROLA, THOROLA ISLODYNE
REMLER CO., LTD.	NORCO, REMLER
REPUBLIC INDUSTRIES	REPUBLIC, SKY HAWK
RESAS, INC.	TONE-A-DYNE
REVERE RADIO CORP.	REVERE
REX RADIO SPECIALITY	REXCO MASTERDYNE
RICHARDS, GEORGE, & CO.	HEMCO
ROBERTS, HENRY CO., RADIO CORP.	ORIOLE
ROTH RADIO, INC.	MELODEE
ROTH-DOWNS MANUFACTURING CO.	ORPHEUS
ROYAL RADIO CORP.	POLY
ROYALE RADIO MANUFACTURING CO.	ROYALE
RUCKELSHAUS, JOHN G., INC.	RUTIC
S	
SAMPSON ELECTRIC CO.	SAMPSON
SAMPSON PAM CO.	PAM
SARA, J. & L., CO., INC.	SARA

MANUFACTURER	TRADE NAME
SARGENT, E. M., CO.	SARGENT
SCOTT TRANSFORMER CO.	SCOTT
SCOTT, E. H., RADIO LABORATORIES, INC.	SCOTT
SEARS MANUFACTURING CO.	ACME, TORODYNE
SEARS, ROEBUCK & COMPANY	SEARS, ROEBUCK, SILVERTONE
SECTRON RADIO CO.	SECTRON
SEEBURG, J. P., CORP.	CORONA
SEELEY ELECTRIC CO.	LARK
SENTINEL RADIO CORP.	BURLEY, ERLA, LIGHTWEIGHT CHAMPION, MUSICAIRE, SENTINEL, WINGS
SETCHELL CARLSON, INC.	DORAPHONE
SHAMROCK MANUFACTURING CO.	SHAMROCK
SHELDON RADIO CO.	SHELDON
SHEPPARD POTTER CO., INC.	SHEPCO, THERMIODYNE
SHERMAN RADIO MANUFACTURING CORP.	CLEARFIELD
SHORTWAVE & TELEVISION LABORATORIES, INC.	BAIRD
SILVA MANUFACTURING CO.	DORON SUPER-EQUIDYNE
SILVER-MARSHALL MANUFACTURING CO.	AIR FLIGHT, ALL WORLD, AROUND THE WORLD, BEARCAT, COAST-TO-COAST, SARGENT-RAYMOND, SILVER-MARSHALL, UNIPAK, WORLD WIDE NINE
SIMPLEX RADIO CO.	SIMPLEX
SLEEPER RADIO CORP.	MONOTROL, MUSIC MASTER, SCOUT, SERENADER, SLEEPER, TROUBADOR
SONORA PHONOGRAPH CO.	SONORA
SONORA RADIO & TELEVISION CORP.	SONORA
SOUTH HILLS WIRELESS SHOP	SUPERIOR
SOUTHEASTERN RADIO CORP.	CAROL-DYNE
SOVEREIGN ELECTRIC MANUFACTURING CO.	SOVEREIGN
SPARKS-WITHINGTON CO.	JEWEL, SPARTON
SPIEGEL, MAY STERN	AIRCASTLE, SUPERSONIC
SPIELMAN ELECTRIC CO.	AIR PILOT
SPLITDORF ELECTRICAL MANUFACTURING CO.	ABBEY, GEISHA, MIKADO, NOCTURNE, POLONAISE, RHAPSODY, SPLITDORF
STANDARD RADIO & ELECTRIC CO.	GIBLIN, RADIOEAR
STANDARD RADIO CO. (Kansas City, Mo.)	STANDARD
STANDARD RADIO CORP. (Worcester, Ma.)	STANDARDYNE
STANRAD CORP.	RADIO PAL, STANDARD
STATES RADIO CORP.	STATES RADIO
STEIN, FRED W., CO.	AZTEC
STEINITE RADIO CO.	STEINITE
STEINWAY RADIO CORP.	NEO-DYNE
STERLING MANUFACTURING CO.	STERLING
STEWART-WARNER RADIO CORP.	AEROMASTER, ALEMITE, FERRODYNE, STEWART-WARNER
STORY & CLARK RADIO CORP.	STORY & CLARK
STROMBERG-CARLSON TELEPHONE MANUFACTURING CO.	STROMBERG-CARLSON, TE-LEK-TOR-ET
SUN RADIO MANUFACTURING CO.	SUN, SUN SET
SUNBEAM RADIO SALES CORP.	AKRADYNE

RADIO MANUFACTURERS

MANUFACTURER	TRADE NAME
SUPER-ANTENNA CO.	SUPER
SUPERIOR CABINET CORP.	SUPERIOR
SYPHER MANUFACTURING CO.	SUPER-FLEX
T	
TANNER, C. D., CO.	NEUTROFLEX, TANNERFLEX
TELEPHONE MAINTENANCE CO.	TELEMACO
TELETONE CORP. OF AMERICA	TELETONE
TELEX RADIO & TELEVISION CO.	TELEX
TEMPLE RADIO CORP.	TEMPLE
THERMIODYNE RADIO CORP.	THERMIODYNE
THOMPSON, R. E., MANUFACTURING CO.	DUOTONE, GRANDETTE, MINUET, THOMPSON
TILMAN RADIO CORP.	CHALLENGER
TODD, INC.	TODD
TRANSFORMER CORPORATION OF AMERICA	CLARION, TCA, TEMPLE, TEMPLETONE
TRAV-LER RADIO & TELEVISION CORP.	TRAV-LER, TRAVEL-AIR, TRAVEL-AIRE
TREBOR RADIO CO.	TREBOR
TREGO RADIO MANUFACTURING CO.	TREGODYNE
TROJAN RADIO CORP.	TROJAN
TROY RADIO & TELEVISION CO.	TROY
TURNER, K. M., RADIO CORP.	HOLLYWOOD MASTERPIECE
TUSKA, C. D., CO.	SUPERDYNE
TWENTIETH CENTURY RADIO CO.	AIR KING
TYMESON RADIO CO.	MELL-O-TONE
TYRMAN RADIO CO.	IMPERIAL, TYRMAN
U	
ULTRAMAR MANUFACTURING CORP.	TROUBADOR, ULTRAMAR
UNION CONSTRUCTION CO.	MASTER
UNITED AIR CLEANER CO.	SENTINEL
UNITED AMERICAN BOSCH CORP.	AMBROLA, AMERICAN BOSCH, BOSCH, CRUISER, ESSEX
UNITED ENGINE CO.	LANSING
UNITED MANUFACTURING & DISTRIBUTION CO.	UNIDYNE
UNITED MOTOR SERVICE	DELCO
UNITED RADIO MANUFACTURING CORP.	VICTORY
UNITED REPRODUCERS CORP.	COURIER, ELECTROSTATIC, KYLECTRON, PEERLESS
UNITED STATES ELECTRIC CORP.	WORKRITE
UNITED STATES RADIO & TELEVISION CORP.	APEX, ARGYLE, CARLTON, CASE, GLORITONE, MANTOLA, MELROSE, RADIOTROPE
UNIVERSAL BATTERY CO.	UNIVERSAL
USL RADIO INC.	USL
V	
VALLEY ELECTRIC CO.	VALLEY, VALLEYTONE
VERNON RADIO CO.	VERNON
VIBROPLEX CO., INC.	MARTINOLA
VICTOREEN RADIO CO.	VICTOREEN

MANUFACTURER	TRADE NAME
W	
W-K ELECTRIC CO.	ORIOLE
WALBERT MANUFACTURING CO.	ISOFARAD
WALKER, G. W., CO.	VICTOREEN
WALTHAM RADIO CORP.	WALTHAM
WARE RADIO CORP.	BANTAM, MUSIC MASTER, WARE
WARNER ENGINEERING CORP.	QUALIPHONE
WARWICK MANUFACTURING CO.	ALL STAR, ARLINGTON, AUTO TONE, AUTRA, BEVERLY, CARAVAN, CARLILE TABLE, CATHAY, CLARION, CROYDEN, DELTA DE LUXE, DELUXE, DINAH, DRUID, DUAL TONE, FEARNOLA, FIDELITONE, GENERAL, GOLD MEDAL, GOLD TEST, GOLDENTONE, KINGSBURY, MOTO-MASTER, RIVERSIDE, SHERIDAN, SKY ROVER, SOLAR, STANDARD, STEINITE, TEA TABLE, TEMPLETON, TONE MASTER, TROUBADOR, TRUETEST, TRUETONE, TWIN 6, VIKING, W. O. R., WARWICK, WATTERSON, WINDHAM, WINDSOR
WAUGH ELECTRIC MANUFACTURING CO.	ECHO
WEBSTER ELECTRICAL CORP.	WEBSTER
WELLS MANUFACTURING CO. (Fond du Lac)	ARLINGTON
WELLS RADIO MANUFACTURING CO. (Chicago)	BEAR CAT
WELLS-GARDNER & CO.	ARCADIA, BLACKHAWK, FEARNOLA, FIDELITONE, GRANADA, GREAT NORTHERN, GULBRANSEN, HUDSON, LEXINGTON, SKELCO, SKY ROVER, SOLAR, STAR, SUPERTONE, TRUE VALUE, TRUETONE, UNIVERSAL, WELLS-GARDNER, WHELCO
WELLSTON RADIO CORP.	FORD
WESTERN AUTO SUPPLY CO.	GENERAL, TRIANGLE, TRUETONE
WESTERN CABLE & LIGHT CO.	WESTERN CABLE
WESTERN COIL & ELECTRICAL CO.	RADIODYNE
WESTERN TIRE & AUTO STORES, INC.	ENDURANCE, PEE WEE
WESTFALL-KELLEY RADIO SALES	RANGER
WESTGALE ELECTRIC CO.	WESTGALE
WESTINGHOUSE ELECTRIC SUPPLY CO.	WESTINGHOUSE
WHITE WIND ELECTRIC CO.	WHITE
WHOLESALE RADIO SERVICE CO., INC.	DUO SYMPHONIC, LAFAYETTE, MIGHTY ATOM
WILCOX-GAY CORP.	RECORDIO, WILCOX-GAY
WINDSOR FURNITURE CO.	WINDSOR
WIRELESS SPECIALITY APPARATUS CO.	WIRELESS SPECIALITY
WOLVERINE RADIO CO.	COMBIDYNE

RADIO MANUFACTURERS

MANUFACTURER	TRADE NAME
WONDERTONE RADIO MANUFACTURING CO.	WONDERTONE
WORKRITE MANUFACTURING CO.	AIR MASTER, KING, WORKRITE
WRIGHT RADIO MANUFACTURING CO.	ACME, WRIGHT
WRIGHT-De COSTER RADIO, INC.	ACME
WURLITZER, RUDOLPH, CO.	EAGLE, LYRIC, WURLITZER
Z	
ZANEY-GILL CORP.	CLARION, CLARIONETTE, MUSIC BOX, VITATONE
ZENITH RADIO CORP.	HYPER-MERTON, INTER-OCEAN, STRATOSPHERE, SUPER-MERTON, SUPER-ZENITH, ZENETTE, ZENITH
ZEPHYR RADIO CO.	ZEPHYR